T0342317

Human Motion Capture and Identification for Assistive Systems Design in Rehabilitation

Pubudu N. Pathirana
Saiyi Li
Yee Siong Lee
Trieu Pham

This edition first published 2021

Registered Offices
John Wiley & Sons, Inc., 111 River Street, Hoboken, NJ 07030, USA
John Wiley & Sons Ltd, The Atrium, Southern Gate, Chichester, West Sussex, PO19 8SQ, UK

Editorial Office
The Atrium, Southern Gate, Chichester, West Sussex, PO19 8SQ, UK

For details of our global editorial offices, customer services, and more information about Wiley products visit us at www.wiley.com.

Wiley also publishes its books in a variety of electronic formats and by print-on-demand. Some content that appears in standard print versions of this book may not be available in other formats.

Library of Congress Cataloging-in-Publication Data

Names: Pathirana, Pubudu N., author. | Li, Saiyi, author. | Lee, Yee Siong, author. | Pham, Trieu, author.
Title: Human motion capture and identification for assistive systems design in rehabilitation / Pubudu N. Pathirana, Saiyi Li, Yee Siong Lee, Trieu Pham.
Description: Hoboken, NJ : Wiley, 2021. | Includes bibliographical references and index.
Identifiers: LCCN 2020043708 (print) | LCCN 2020043709 (ebook) | ISBN 9781119515074 (cloth) | ISBN 9781119515234 (adobe pdf) | ISBN 9781119515210 (epub)
Subjects: LCSH: Rehabilitation technology. | Musculoskeletal system–Wounds and injuries–Treatment. | Neuromuscular diseases–Treatment. | Medical rehabilitation. | Telecommunication in medicine. | Human locomotion–Measurement. | Human locomotion–Computer simulation.
Classification: LCC RM950 .P38 2021 (print) | LCC RM950 (ebook) | DDC 617/.03–dc23
LC record available at https://lccn.loc.gov/2020043708
LC ebook record available at https://lccn.loc.gov/2020043709

Cover Design: Wiley
Cover Image: © Westend61/Getty Images

Set in 9.5/12.5pt STIXTwoText by SPi Global, Chennai, India
Printed and bound by CPI Group (UK) Ltd, Croydon, CR0 4YY

C9781119515074_260421

Contents

1

Introduction

1.1 Human Body – Kinematic Perspective

The human musculoskeletal system (also known as the locomotor system) is the combination of two main components, namely a passive skeletal system and an active muscular system to facilitate human movements with support and stability. The skeletal system initiates with 230 bones at birth and reduces to 206 due to fusion of some bones as the person reaches adulthood and these bones form the axial skeleton and the appendicular skeleton via 360 joints. There are four types of bones in the human body, namely short bones, flat bones, irregular bones and sesamoid bones, while there are three types of joints, immovable joints, slightly movable joints and freely movable joints. The muscular system includes muscles, tendons, ligaments, and other connecting tissues [108]. Figure 1.1 is a depiction of these two components.

The muscular system consists of over 700 muscles, which can be classified mainly into skeletal, cardiac and smooth muscles. Facial and tongue muscles are special cases with the tongue having the largest concentration of muscles. Based on the functionality, the muscles can also be classified as involuntary (cardiac, smooth) and voluntary (skeletal).

Human body kinematic movements can be classified into two main types – voluntary and involuntary. Execution of daily activities and specialised activities with complete cognitive control causes voluntary movements and such movement is the expression of a thought through action. Almost all areas of the central nervous system are involved in the execution of voluntary movements and the main flow of information may begin in cognitive cortical areas in the frontal lobe or in sensory cortical areas in the occipital, parietal and temporal lobes. Ultimately, information flows from motor areas in the frontal lobe through the brain stem and spinal cord to the motor neurones [110].

Muscles are contracted due to the signals received throughout the central nervous system. The peripheral nervous system that is associated with the skeletal muscle voluntary control of the body is referred to as the somatic nervous system. There are 43 segments of nerves in the human body and associated with each segment is a pair of sensory and motor nerves. The spinal cord has 31 segments of nerves while the remaining 12 are in the brain stem. The electrical signals instigated from the brain are transmitted through the nerves and prompt the release of the chemical acetylcholine from the presynaptic terminals. This chemical

Human Motion Capture and Identification for Assistive Systems Design in Rehabilitation, First Edition.
Pubudu N. Pathirana, Saiyi Li, Yee Siong Lee and Trieu Pham.

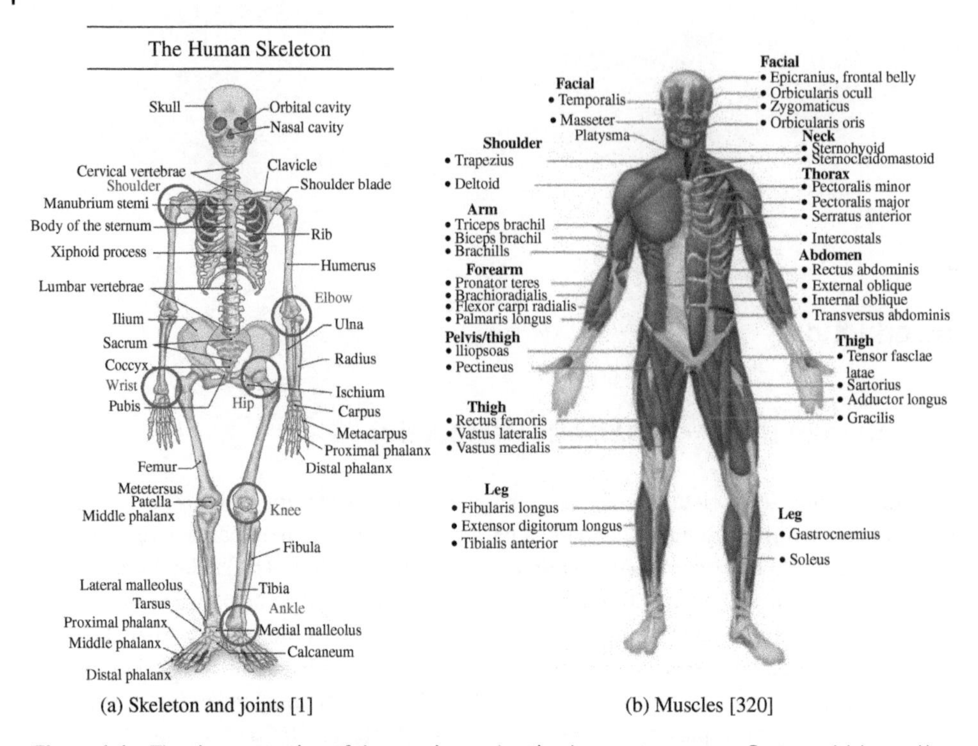

(a) Skeleton and joints [1]

(b) Muscles [320]

Figure 1.1 The demonstration of the passive and active locomotor system. Sources: (a) https://commons.wikimedia.org/wiki/File:Human_skeleton_front_en.svg; (b) https://commons.wikimedia.org/wiki/File:1105_Anterior_and_Posterior_Views_of_Muscles.jpg.

is picked up by special sensors (receptors) in the muscle tissue. If enough receptors are stimulated by acetylcholine, the muscles will contract and force the specific part of the body to move. Almost all voluntary motor functions are controlled by the area referred to as the motor cortex in the brain. The primary motor cortex generates neural impulses to activate the muscle contractions and the nerves cross the body midline so that the left hemisphere of the brain controls the right side of the body and vice versa. Other areas of the motor cortex include the posterior parietal cortex, the premotor cortex and the supplementary motor cortex. The posterior parietal cortex is involved in transforming visual information into motor commands and transmits the information to the premotor cortex and the supplementary motor cortex. The supplementary motor cortex is involved in planning complex motions and also coordinating the two hands, whereas the premotor cortex is invoked in sensory guidance of movement and controls the more proximal muscles and trunk muscles of the body. According to Taga [345], the neural system indeed formulates structured sequences of signals or can even be considered as programs for the human motion via activation of muscles. Therefore, certain physiological conditions or injuries that affect the primary motor cortex (Figure 1.2) can adversely affect the functionality of the locomotor system.

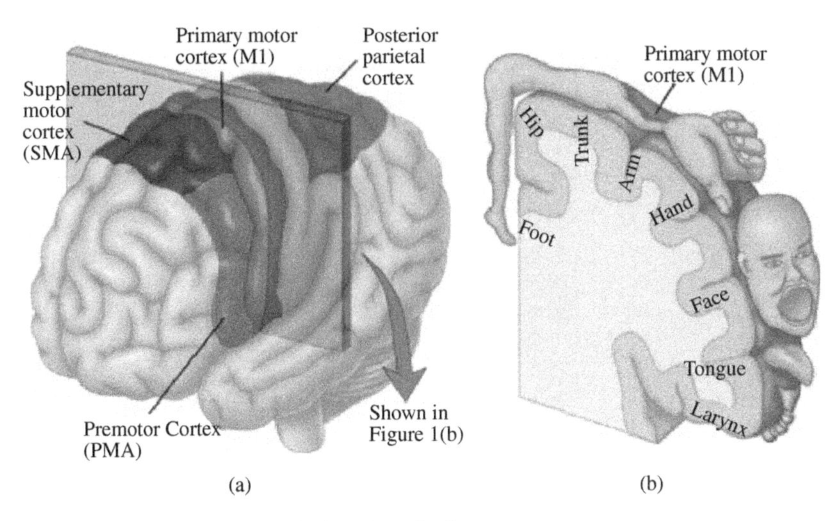

Figure 1.2 Functional description of the brain motor cortex.

1.2 Musculoskeletal Injuries and Neurological Movement Disorders

Movement difficulties can be due to a number of causes and generally are classified as neuromuscular disorders. The causes of these abnormal movements can be classified into two main categories: musculoskeletal injuries and neurological movement disorders.

1.2.1 Musculoskeletal injuries

These injuries are normally observed in joints associated with certain degrees of movement, such as shoulder, elbow and wrists in the upper extremities and hips, knees and ankles in lower ones, which may eventually lead to abnormal movements or event disabilities. Some examples of disorders are shown in Tables 1.1 and 1.2.

1.2.2 Neuromuscular disorders

The disorders that can be associated with the nervous and muscular systems affect the movements and can sometimes exhibit characteristic movement patterns associated with certain conditions. Neuromuscular disorders affect the nerves that control the voluntary muscles – muscles that can normally be controlled by the individual. Such disorders include motor neurone diseases, neuropathies, muscular dystrophies and neurodegenerative disorders. These disorders can be classified according to the area of the neuromuscular system that is affected.

Table 1.1 Examples of musculoskeletal injuries in joints of the upper extremities. Inj is for injury and ST is for studies about using physical rehabilitation to treat the conditions or stimulate recovery (the same as that in Table 1.2).

	Shoulder	Elbow	Wrist
Movements	Flexion, extension, abduction, adduction, internal and external rotation [84]	"Flexion and extension at the ulnohumeral and radiocapitellar articulations, while pronation and supination at the proximal radioulnar joint" [56]	Flexion, extension, radial deviation and ulnar deviation [271]
Inj1	Shoulder impingement	Tennis elbow	Carpal tunnel syndrome
Description	It "occurs against the anterior edge and undersurface of the anterior third of the acromion, the coracoacromial ligament, and, at times, the acromioclavicular joint" [255] and deemed as one of the factors that lead to shoulder disability [254].	Although it is not perfectly understood, it negatively influences "the attachment of the extensors of the forearm at the lateral side of the elbow", thereby leading to pain [365].	It usually is caused by the pressure on the median nerve on a wrist and leads to various conditions, such as pain, paraesthesiae, hypoaesthesia and so on [287].
ST	[238]	[366]	[251]
Inj2	Adhesive capsulitis		Scaphoid
Description	The general cause leading to this condition is described as "progressive fibrosis and ultimate contracture of the glenohumeral joint capsule" [258].		It is usually caused by a hyperextended and radially deviated wrist and seen in patients aged between 15 and 40 [167].
ST	[343]		[299]

Upper motor neurone disorders–

Conditions such as a cerebrovascular accident (stroke), Parkinson's disease, multiple sclerosis, Huntington's disease (Huntington's chorea) and Creutzfeldt-Jakob disease are examples of upper motor neurone diseases.

Lower motor neurone disorders – spinal muscular atrophies

Lower motor neurones are located in either the anterior grey column, anterior nerve roots (spinal lower motor neruones) or the cranial nerve nuclei of the brain stem and cranial nerves with a motor function (cranial nerve lower motor neurones) [1]. All voluntary movement relies on spinal lower motor neurones, which innervate skeletal muscle fibres and act as a link between the upper motor neurons and muscles [2, 3]. Cranial nerve lower motor neurones control movements of the eyes and tongue, and contribute to chewing, swallowing and vocalisation [4]. Damage to the lower motor neurone can lead to flaccid paralysis.

Table 1.2 Examples of musculoskeletal injuries in joints of the lower extremities.

	Hip	Knee	Ankle
Movements	Flexion, extension, abduction, adduction, internal and external rotation [149]	There are two ways to describe the degree of freedom (DOF) in a knee. One is with two DOFs (flexion-extension and axial rotation) [234] and the other is with six DOFs (flexion-extension, varus-valgus, internal-external rotation and mediolateral, anteroposterior and superoinferior translation around mediolateral, anteroposterior and superoinferior axis) [122].	Extension, flexion, valgus and varus [301]
Inj1	Hamstring strain	Patellofemoral pain syndrome	Achilles tendonitis
Description	It usually associated with lower extremity activities, like football, soccer, dancing and so on, while this condition occurs in different phases of motions in various types of activity [149].	It is an anterior knee pain and mainly resulted from "aberrant motion of the patella in the trochlear groove" [123].	The physical findings of this condition include soft tissue swelling, local tenderness and sometimes crepitus [256].
ST	[326]	[214]	[222]
Inj2	Groin pain	Anterior cruciate ligament (ACL) injury	Lateral sprain
Description	It contributes 2–5% of all sport injuries [243]. Vincent *et al.* [243] also mentioned that the diagnosis of this pain is hard because of its complex anatomy in the affected region, as well as the coexistence of multiple injuries.	The causes are in two major categories, including non-contact (usually resulted from sudden deceleration before changing direction or a landing motion) and contact (valgus collapse) [45].	It can be deemed as the most common injury in ankles [114], which is usually cause by inversion of the foot [114].
ST	[302]	[74]	[103]

Neuropathies

Neuropathies involve dysfunction of the peripheral nerves, which consist of: motor neurones, that carry the electrical signals directly from the spinal cord and brain stem to activate muscle movement; the sensory neurones, which convey sensory information such as pain, temperature, light touch, vibration and position to the brain; and the autonomic neurons, which go to the internal organs and control blood vessel reflexes.

Neuromuscular junction disorders

Myasthenia gravis and Lambert-Eaton syndrome are examples of neuromuscular junction disorders. Muscular dystrophies and inflammatory myopathies such as polymyositis are examples of primary muscular (myopathic) disorders.

Neurodegenarative classification encapsulates the progressive loss of structure or function of the neurone and a number of conditions exhibit this form of progression. Hence, amyotrophic lateral sclerosis (ALS), Parkinson's and Huntington's are classified as neurodegenerative diseases that affect movement. These conditions exhibit characteristically slower movement compared to healthy people – hypokinesis – or excessive and abnormal involuntary movements – hyperkinesis [156]. Some common examples of hypokinesis include bradykinesia, freezing, rigidity and stiff muscles, while those belonging to hyperkinesis are chorea, dyskinesia, myoclonus, tics and tremor [242].

The most common neurological [215] and adult movement disorder, essential tremor (ET), is about 20 times more prevalent than Parkinson's disease itself. Patients with ET are likely to have tremors with 4–12 Hz and the risk factors associated with this are age, ethnicity and family history [215]. The condition affects the performance of work-related tasks and activities of daily living (ADLs) and a number of medical and physical rehabilitation approaches are in use as treatments for the condition [39, 295].

An estimated 7 to 10 million people worldwide are living with the second-most common neurodegenerative disorder, Parkinson's disease (PD) [90]. Approximately 60 000 Americans are diagnosed with PD each year while in western Europe this figure is 160 for every 100 000 over the age of 80 [89]. In China, approximately 1.7 million above the age of 55 [396] are suffering from the condition. The movement disorders experienced by a PD patient can be classified into three stages [244]. In the initial stage, the patient may exhibit a forward stooped posture, festinating gait, rigidity, etc. During the first 10 years of PD, characteristic movements such as resting tremor, hypokinesia and micrographic handwriting are common. During the later phase, patients may exhibit dyskinesia, akinesia, postural instability, etc. In terms of treatment, various kinds of medical therapies, such as levodopa, as well as surgical approaches and deep brain stimulation are utilised to control symptoms in addition to physical rehabilitation therapies [171].

Although these two conditions are common and have a significant impact on the quality of life, they are not fatal diseases. In contrast, stroke is one of the most fatal conditions in developed countries [374]. However, a majority of stroke suffers may be alive after the initial injury, albeit losing some motor functions lifelong or for a prolonged period of time [193]. In 2005, there were 5.7 million deaths in low- and middle-income countries due to stroke, which has increased significantly to 6.5 million and 7.8 million, respectively, in 2015 [338]. Age, gender, race, ethnicity and heredity are considered as important markers of risk factors [273], while hypertension, cardiac disease, diabetes, glucose metabolism, lipids, cigarette smoking, alcohol, illicit drug use, lifestyle, etc., are considered to have an adverse influence on the likelihood of stroke [273]. Similar to other disorders, physiotherapy is widely used as a rehabilitation therapy to assist stroke patients to regain physical functionality [291].

Such a complex locomotor and neural system are frequently influenced by various injuries and disorders. Therefore, a number of approaches have been explored to treat people suffering from conditions leading to abnormal movements and even disabilities.

Among these methods, physical rehabilitation is commonly utilised to assist patients in recovery and reacquire ADLs.

Defined as "the treatment of disease, injury, or deformity by physical methods such as massage, heat treatment, and exercise rather than by drugs or surgery" [279], physiotherapy (also known as physical therapy) has been applied in clinics for thousands of years. A number of therapies are included in physical therapy, such as mechanotherapy, hydrotherapy, balneotherapy and so on [349], among which mechanotherapy was documented as early as the 1840s. In recent decades, physical therapy has been applied extensively for various musculoskeletal injuries and neurological movement disorders [107, 142]. The detailed examples can be found in Section 1.2.

Although traditional physical therapy has shown its effectiveness for the rehabilitation of physical functions of patients with movement disorders [95], a series of drawbacks can also be observed [9, 57], which are summarised as follows.

- These rehabilitation programmes are "boring" and patients are demotivated by these repetitive exercises.
- Computerised sensing techniques are not involved in these programmes, which may lead to incorrect interpretation of observed data.
- A one-to-one form of delivering rehabilitation services makes conventional rehabilitation very inefficient and costly.
- Costly equipment is required by traditional rehabilitation therapies.
- Insufficient funding for rehabilitation services results in making access to these services unaffordable.
- The workforce in the rehabilitation field is inadequate in number.
- Rehabilitation centres are usually distributed in urban areas, while a large number of people needing rehabilitation services live in rural regions.

In light of the above, in 1998, the term "telerehabilitation" was "first raised" in a scientific article, attempting to overcome the shortage in conventional rehabilitation services [303]. Co-existing with the opportunities for telerehabilitation, such as economy of scale, interactive and motivation, reduced healthcare costs, patients' privacy prevention and so on [57], a number of challenges can be seen from an engineering point of view as well. These challenges are also closely related to the aim of this book.

First of all, developing affordable, high-quality and robust hardware [300] is critical to capture human movements for further analysis. High-quality and effective hardware may provide more accurate monitoring of the movements, thereby offering better feedback to patients to correct their movements and more valuable information to therapists to make further treatment decisions. In addition, similar to the fourth issue in traditional rehabilitation mentioned above, costly equipment used in telerehabilitation services may prevent patients with low economic status from accessing them. Thus, how to develop affordable devices is critical for the development of telerehabilitation services.

Secondly, an advanced approach in representing human movement data [381] is important after capturing human motion. As pointed out in Theodoros and Russell [350], one difficulty in telerehabilitation is how to reduce information collected from sensors, thereby producing meaningful results to therapists. Therefore, an emerging challenge is to discover which features can be utilised to represent human motions so that their rehabilitation-related details can be well preserved while other irrelevant information can be reduced as much as possible.

Thirdly, developing an outcome measurement scheme to quantitatively and objectively represent the performance of patients accessing telerehabilitation services in also a challenge [382]. As previously described [10], one of the goals of RERC is to develop assessment tools to monitor the progress of patients accessing telerehabilitation services. These tools can not only be a feedback to stimulate the patients to perform more exercises but can also provide therapists with general information about their patients in terms of functional rehabilitation.

Last but not the least, enabling patients to access physical telerehabilitation services regardless of their location and time is a challenge [382]. As one purpose of rehabilitation services is for patients to recover the ability to perform ADLs, enabling them to perform rehabilitation exercises in their most familiar and natural environment is very important. Due to the different preferences between patients, developing telerehabilitation services that can be run on mobile devices is extremely useful. By doing so, patients' kinematic performance in telerehabilitation exercises and daily living can be assessed pervasively.

Because of the importance of telerehabilitation, as well as the automated kinematic performance assessment tool, a significant amount of effort has been made to improve the physical telerehabilitation, which will be discussed in Sections 1.5 and 1.6.

1.3 Sensors in Telerehabilitation

In the past few decades, various types of sensors have been considered as patient monitoring and data acquisition tools. In this section, the use of three main types of sensors, namely Kinect, RGB camera and IMU, is reviewed.

1.3.1 Opto-electronic sensing

The Vicon [7] motion capture studio is used in some clinical settings as well as in certain other human motion assessment applications in sports. A number of fixed cameras within a dedicated infrastructure are used to capture the position of markers on the moving body part to greater precision, with the resulting Vicon system being used as the benchmark for motion capture systems. Due to cost, the requirement for a fixed and dedicated infrastructure and the specialised technical knowledge, this has not been used as a standard clinical practice and remains predominantly used in research and development.

Recently, a number of non-invasive, portable and affordable optical 3D motion capture devices have emerged. These products include Leap Motion® controller, ASUS®Xtion PRO LIVE, Intel®Creative Senz3D, Microsoft Kinect®, and so on. Among them, Kinect is the most popular motion capture device for whole-body motion capture. The first version was released in 2010 with Xbox 360 and was used for gaming purposes and the second version was released with Xbox One in 2014.

The first version of Kinect utilised a depth sensor provided by a company named "PrimeSense" [400]. The appearance and components of this version of Kinect are shown in Figure 1.3.

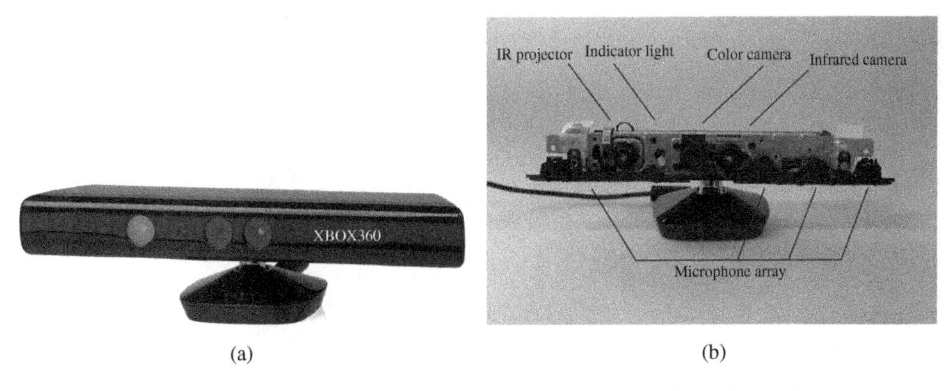

Figure 1.3 Appearance and components of Kinect version 1. Source: Evan-Amos, Image taken from https://commons.wikimedia.org/w/index.php?curid=16059165.

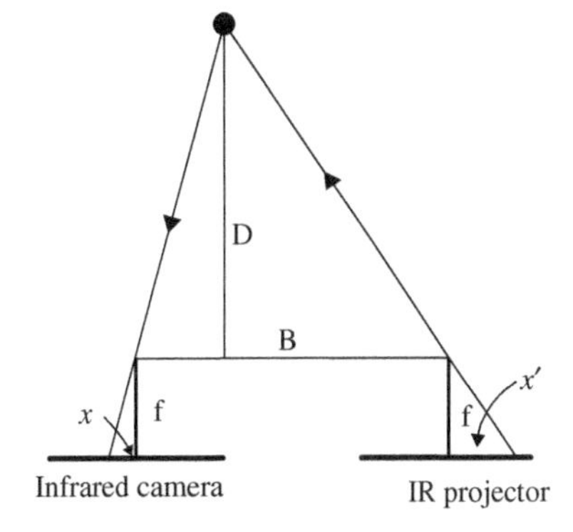

Figure 1.4 The pinhole camera model of Kinect version 1 [373]. Source: From Wang *et al.* [373]. © 2012, University of British Columbia.

The infrared projector and the corresponding camera can be modelled as in Figure 1.4.

These sensors measure the depth information via a structured light principle, which analyses a pattern (such as that in Figure 1.5) of bright spots projected on to the surface of an object [328]. In the case of Kinect, these "bright spots" are infrared light and are unobservable by human eyes directly.

Furthermore, Kinect utilised the other two techniques to further process the information to generate depth maps. These two tools include depth from focus and depth from stereo [121]. The principle of the former is that the further away the object is, the more blurred it will be [125], while the latter utilised parallax to estimate the depth information.

Different from the first version of Kinect, the second version (refer to Figure 1.6) measures the depth information with the time-of-flight (ToF) technique [189], which is stated as the distance that can be measured by knowing the speed of light and the duration the light uses to travel from the active emitter to the target. Estimated in Lachat *et al.* [189], this version of Kinect utilised the indirect time-of-flight, which measures the "phase shift between the emitted and received signal". The depth is computed as

$$d = c\frac{\Delta\phi}{4\pi f}, \tag{1.1}$$

where f is the modulation frequency, c is the light speed and $\Delta\phi$ is the determined phase shift.

Figure 1.5 An example of the projected pattern of bright spots on an object [328]. Source: Shpunt and Zalevsky [328].

Figure 1.6 Appearance of Kinect version 2. Source: Evan-Amos, Image taken from https://commons.wikimedia.org/wiki/File:Xbox-One-Kinect.jpg.

As for the accuracy of joint positions tracked by the first version of Kinect, some studies have been done. Different studies have come to different conclusions on the accuracy of skeleton joint tracking. For instance, Webster *et al.* [376] reported that the accuracy was around 0.0275 m after removing offset by resetting the alignment of the average points for each record set. Therefore, they concluded that Kinect version 1 was sufficient for clinical and in-home use. Obdrzalek *et al.* [263] evaluated the accuracy of the first version of Kinect, PhaseSpace Recap and Autodesk MotionBuilder in the environment of coaching elderly people. The result reported in their paper is that the error of the skeleton built by both Kinect and MotionBuilder is around 5 cm. However, in general postures, the accuracy is about 10 cm due to unavoidable factors, such as occlusions. Therefore, they suggested that the current skeletonisation approach enabled Kinect to measure general trends of movements, while an improved skeletonisation algorithm should be investigated if Kinect was used for quantitative estimations. Furthermore, Xu *et al.* [388] evaluated the accuracy of both the first and second versions of Kinect for static postures. From their experiment, they concluded that the accuracy of the first version Kinect varies from posture to posture. For instance, the error was only 26 mm for a shoulder centre in an upright standing posture, while it was 452 mm for the right foot joint in a sitting posture with the right leg on top of the left one. By comparison, for the second version of Kinect, when the right foot was raised, the error of the left elbow was only 26 mm, while it was 418 mm when the right leg was on the left one. As a result, it was concluded that though the resolution of the second version of Kinect had been improved significantly, its tracking accuracy of the joint centre had not improved.

The comparison of the specifications between the two versions of Kinects is shown in Table 1.3. From the comparison, it is obvious that the second version of Kinect provides a larger viewing angle, a higher resolution in both depth images and colour images, and more tracking joints.

Though Kinect was initially developed for gaming, it is widely applied in telerehabilitation as a non-invasive and affordable motion capture device. A telerehabilitation system (KiReS) using Kinect as the motion capture device has been proposed. On the patient side, two avatars were displayed to represent the motion recorded by the therapist (reference motion) and that performed by the patient. Therefore, the patient was able to see the differences between his/her motion and the reference. Eventually, the incorrect movements could be corrected over time. On the therapist side, new motions could be created to suit the patient's conditions by composing various existing movements or recording completely new ones. Luna-Oliva *et al.* [217] utilised Kinect Sports I™, Joy Ride™ and Disneyland Adventures™ to provide telerehabilitation services to children with cerebral palsy in their school. Their experimental results showed that it is feasible to use Kinect as a therapeutic tool for children with cerebral palsy and the improvements in global motor function could be the result of using this tool. Ortiz-Gutiérrez *et al.* [268] applied Kinect in providing telerehabilitation services to patients with postural control disorders. The experiment results showed an improvement over a general balance in both groups. In the experimental group, the significant differences resulted from visual preference and the contribution of vestibular information.

1.3.2 RGB camera and microphone

Apart from Kinect, conventional RGB cameras and microphones are also pervasively used, especially in the early stages of the history of telerehabilitation when virtual reality devices had not been well developed and pervasively utilised. One of the potential reasons is that they are easy to install and are cost-effective and well-developed.

In the early stages of the development of telerehabilitation, the plain old telephone system (POTS) was widely used as the infrastructure of videoconferencing, which was sufficient to provide a teleconsultation. Delaplain *et al.* [93] made a pioneering trial between two islands to conduct 59 medical tele-consultations in the form of a videoconference. In this trial, diagnostic and therapeutic decisions were made in a number of specialities, including physical therapy. This is deemed to be one of the first examples of applying videoconferencing in telerehabilitation (although, at that time, the word "telerehabilitation" had not been invented) with cameras and microphones. Its success illustrates the feasibility of using a videoconference in telerehabilitation. Later, in 2002, Clark *et al.* [72] successfully managed a teletherapy case for 17 months. In this case, a POTS was set up between the therapist's site with a desktop videophone and Mrs M's home with a traditional telephone and a television to provide post-stroke telerehabilitation services in the form of a two-way interactive videoconference. However, the lesson learnt from the case is that the use of this novel approach

Table 1.3 Comparison of basic technical specifications between two versions of Kinects. Sources: Based on Sempena *et al.* [322]; https://medium.com/@lisajamhoury/understanding-kinect-v2-joints-and-coordinate-system-4f4b90b9df16 microsoft.kinect.jointtype.aspx. Accessed: 2015-05-22.

	Version 1 [322]	**Version 2 [4]**
Viewing angle (vertical)	43°	70°
Viewing angle (horizontal)	57°	60°
Vertical tilt range	±28°	no
Frame rate (frame per second)	30	30
Depth resolution (pixels)	320 × 240	512 × 424
Colour stream resolution (pixels)	640 × 480	1920 × 1080
Skeleton	HAND_RIGHT, HEAD, SHOULDER_CENTER, HEAD_LEFT, WRIST_RIGHT, WRIST_LEFT, ELBOW_RIGHT, ELBOW_LEFT, SHOULDER_RIGHT, SHOULDER_LEFT, SPINE, HIP_CENTER, HIP_RIGHT, HIP_LEFT, KNEE_RIGHT, KNEE_LEFT, ANKLE_RIGHT, ANKLE_LEFT, FOOT_RIGHT, FOOT_LEFT [6]	HAND_TIP_RIGHT, HEAD, HAND_TIP_LEFT, THUMB_RIGHT, HAND_RIGHT, HAND_LEFT, THUMB_LEFT, WRIST_RIGHT, WRIST_LEFT, ELBOW_RIGHT, NECK, ELBOW_LEFT, SHOULDER_RIGHT, SHOULDER_LEFT, SPINE_SHOULDER, SPINE_MID, HIP_RIGHT, HIP_LEFT, SPINE_BASE, KNEE_RIGHT, KNEE_LEFT, ANKLE_RIGHT, ANKLE_LEFT, FOOT_RIGHT, FOOT_LEFT [3]
Number of tracking subjects	6 (can only display 2)	6 (can display 6)
Sensing principle		

has a number of requirements for the patients, as well as the caregiver, which may be a potential issue in developing telerehabilitation systems in the future. As a conclusion, they mentioned that although telerehabilitation cannot totally replace the conventional way to deliver rehabilitation services, it indeed contributes to traditional therapy. Furthermore, Savard *et al.* [316] reported two cases of using a videoconference to provide a teleconsultation service for neurological diagnoses. Different from the previous two studies where the videoconference systems were utilised by patients at home, the systems in Savard *et al.* [316] were installed in clinics. Therefore, patients had to visit the clinics in order to use these facilities. Although it was found that the time taken for a tele-consultation was similar to an in-person consultation, the former was more efficient as multiple parties were able to participate in the consultation simultaneously. However, completing remote tests for clinicians would be an important factor that gives telerehabilitation an advantage.

With the advancement of the Internet and computers to deliver telerehabilitation services, POTS was gradually replaced. Russell *et al.* set up an Internet-based computer system in two separate rooms in a clinic to evaluate the feasibility of using a videoconference to assess the kinematic gait [308] by evaluating the performance of measuring knee angles via the Internet, called an Internet-based goniometer (IBG), against the traditional face-to-face approach. The interface of the software is shown in Figure 1.7(a). As a conclusion of the experiment, the IBG was found to be comparable to the universal goniometer (UG) in terms of intra- and inter-rater reliability. After that, the same setup was used the following year to provide tele-medicine to patients after a total knee replacement [309]. This modern method of delivering a rehabilitation service was welcomed by both therapists and patients since it was safe, easy to use and could be integrated into daily clinical practice. More importantly, the outcome of using this approach was similar to a traditional rehabilitation method. Another example is that video cameras were used by Lemaire [204] in a tele-health program to provide tele-consultation and education services for various disorders. The majority of the patients accessing this service were satisfied with telerehabilitation. Though

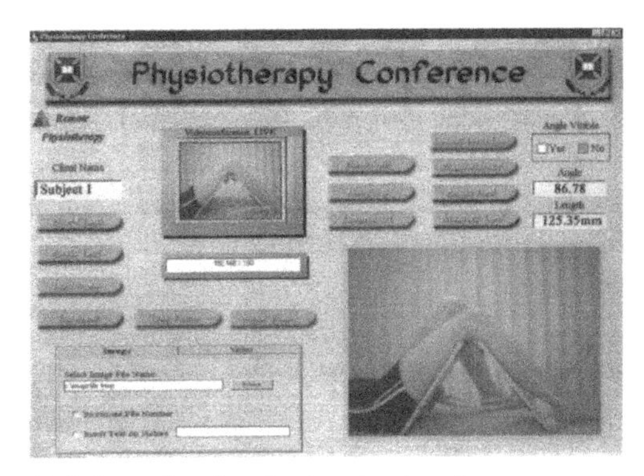

(a) Software interface developed and used to measure the angle of knee via the Internet [308].

(b) A therapist monitoring the knee angle measurement remotely through the Internet.

Figure 1.7 The physiotherapist monitoring the exercise on his patient remotely through the Internet [308]. Source: Russell *et al.* [308].

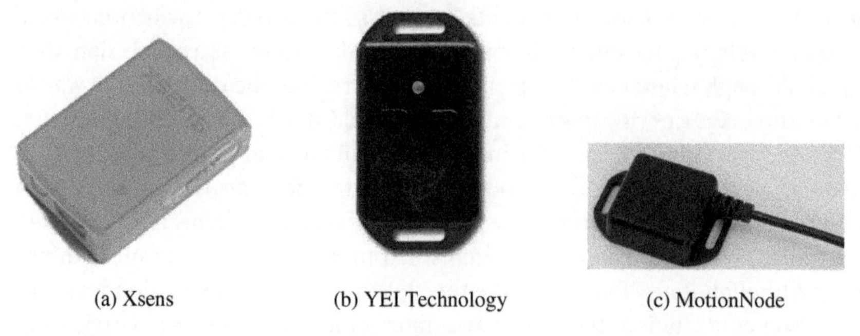

Figure 1.8 Pictures of animals. Sources: (a) Xsens; (b) Amazon; (c) MotionNode.

the study also found that the time taken for tele-consultation was similar to a traditional approach, the majority of the therapists agreed that the tele-health was easy to use and they were confident of the assessment results done with tele-health. However, as it was in 2000, not all hospitals could afford brand new computers. Therefore, it was critical to develop a low-cost telerehabilitation system so that everyone could afford it.

1.3.3 Inertial measurement unit (IMU)

Inertial measurement unit (IMU) is a device that mainly measures angular velocity, orientation, gravitational force and magnetic direction. In an early stage of the development of the IMU, a gyroscope and an accelerometer were usually utilised to provide angular velocity and inertial acceleration. Later on, integration of a magnetometer enabled an IMU to measure magnetic direction. As a result, measurements from an IMU can be more accurate [20]. As all these sensors are able to provide three-dimensional measurements, the IMU is widely utilised in movable applications, such as for aircraft navigation [398].

Recently, thanks to the advancement in a micro-electromechanical system (MEMS), IMUs can be produced in a size small enough to be worn by human beings. As a result, in recent years, more and more applications of IMUs have been seen in rehabilitation and telerehabilitation fields as human motion capture devices [20]. Currently, there have been a number of companies producing and selling IMU sensors, for example Xsens®, YEI Technology®, MotionNode® and so on (see Figure 1.8). These products can be attached on humans for motion tracking. Using 19 sensors in a full-body suit, Rokoko Smartsuit Pro™[5] captures full-body motions using IMU sensors in real time and is available as a commercial product.

In a small number of applications, a single IMU is used to monitor specific conditions (usually relating the movement of one joint) in telerehabilitation. For instance, Giansanti *et al.* [116] utilised an IMU with one three-axis accelerometer and a gyroscope to detect the risk of falling in telerehabilitation. The other example is Han *et al.* [141], who integrated a 6 degrees of freedom (6DOF, including three-axis accelerometer and gyroscope) IMU with a customised ankle foot orthosis (AFO) to provide a telerehabilitation diagnostic service for patients with two types of conditions, including those with muscle weakness because of brain injuries and those who were about to receive total knee replacement surgery due to osteoarthritis of knee joints. Experiments in the study confirmed the high sensitivity and specificity of the AFO-IMU module in measuring the flexion and extension motions of knee joints.

However, it is obvious that a single IMU is insufficient to monitor the movement involving multiple joints, such as upper extremity movements and whole body movements. Therefore, a body area network (BAN) or a body sensor network (BSN) was developed to fill this gap. For instance, Nerino *et al.* [257] proposed a BSN to provide knee telerehabilitation services to patients with anterior cruciate ligament (ACL). In the proposed system, multiple IMUs with 9 DOF (including a three-axis accelerometer, gyroscope and magnetometer) were attached to the thigh, calf and foot, thus enabling measurements to be taken of the angle of the knee and ankle. After comparing this to the Vicon system, the average angular errors measured by the BSN on the knee and ankle were 2.4° and 3.1° with a standard deviation of 1.8° and 2.4°, respectively. Horak *et al.* [145] summarised the role played by body-worn movement monitor devices in rehabilitation services for balance and gait. Cancela *et al.* [61] evaluated the wearability of a BAN-based system, named PERFORM, to monitor the symptoms of patients with PD. This system is composed of four tri-axial accelerometers locating on two legs and two arms, respectively, and a central sensor with one tri-axial accelerometer and gyroscope positioned on the waist (refer to Figure 1.9). Analysis was conducted considering comfort, biomechanical and physiological aspects of the system. According to the experiment, it was found that patients were generally satisfied to wear such a system. However, some patients were concerned about their privacy, as well as how others might think about them particularly in a public area, thereby showing a bit of anxiety and unwillingness to use this system. Furthermore, the strap made patients uncomfortable and difficult to wear when by themselves. Last, but not least, feedback was necessary during monitoring so that patients knew the system was working properly.

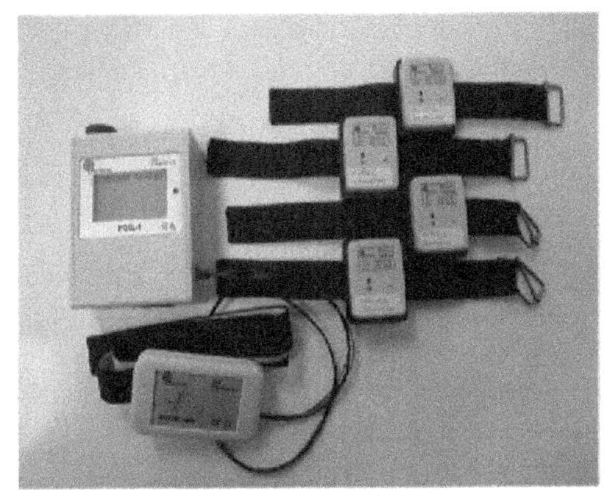
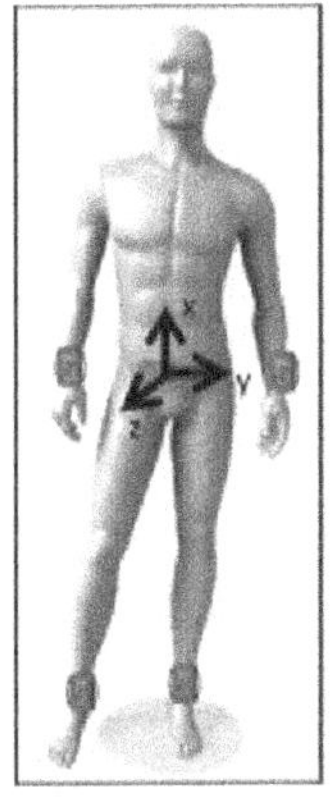

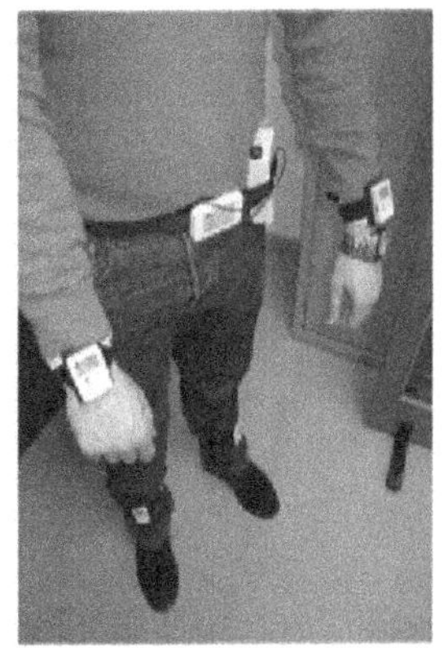

Figure 1.9 Locations of five sensors worn by a subject. Source: Cancela *et al.* [61].

1.4 Model-based State Estimation and Sensor Fusion

Kinematic model-based state estimation can often be used to estimate parameters of interest while combining different information available typically in real-time applications. Let the N sensor measurements be $y = \lfloor y_1 \cdots y_N \rfloor$ and the unknown states of the dynamic system be $x = \lfloor x_1 \cdots x_M \rfloor$. The dynamic system can be modelled in the form of a state space model,

$$\dot{x} = Ax + Bw \quad (1.2)$$

$$y = Cx + d \quad (1.3)$$

where A, B and C denote the system matrices and w is the uncertainty input.

Three filters are commonly used to achieve the goal.

First of all, a Kalman filter is used extensively as an optimal state estimator under Bayesian assumptions and is extended to cover non-linear dynamic and measurement models with stochastic, normal or Gaussian distributed noise [29, 210, 231]. This filter was introduced with the purpose of addressing the limitations of other filters in solving the Wiener problems [163]. Various versions of the Kalman filter, ranging from optimal (also called standard) to extended and various unscented versions pertaining to both linear and non-linear systems have covered a large number of application scenarios. The proposed system is linear with position, velocity and acceleration acting as state variables, enabling the implementation of an optimal Kalman filter for data fusion. Detailed information about the Kalman filter can be found in Chapter 6. Although Kalman filtering is pervasive with the underlying assumptions of Gaussian noise or system uncertainty distributions, when the uncertainties deviate from these assumptions, the performance degradation can be significant and a more generic uncertainty assumption is warranted.

Secondly, particle filters, which are a form of sequential Monte Carlo sampling, are widely used for state estimation problems with both linear and non-linear dynamic systems [172, 383, 385]. They relax the single Gaussian assumption in the state and the measurement uncertainties, allowing the handling of more complex noise situations via sampling from multiple probability densities. When this function is a single Gaussian, the particle filter essentially simplifies to the standard Kalman filter. The application of a particle filter in our linear system is described in Chapter 6.

1.4.1 Summary and challenges

From the literature, it is clear that a number of sensors have been applied in telerehabilitation. For the evaluation of the approaches introduced in this thesis, we utilised Kinect as the example of affordable OBMCDs to collect data for the following two major reasons.

First of all, Kinect is a non-contact motion capture device. Patients do not need to attach any extra item to their bodies. As a result, they may be able to perform rehabilitation exercises or ADLs with more natural poses.

Secondly, compared to conventional RGB cameras, Kinect is able to track 3D positions of 20–25 joints throughout a human body. Therefore, without making much effort in processing images, data collected from a Kinect are much more convenient for further analysis.

This is critical for automatically assessing the kinematic performance of patients in a telerehabilitation system without the presence of a professional therapist.

However, the majority of affordable OBMCDs still suffer from a number of drawbacks. One is that it may not be accurate enough to track small movements, as well as the positions of joints occluded by other body parts. Secondly, its small viewing range limits its application in lower extremity rehabilitation, which, in many cases, involves a large moving area. A solution for these limitations is discussed in Chapter 2.

1.5 Human Motion Encoding in Telerehabilitation

After capturing patients' movements, it is critical to reduce information so that the key features representing the characteristics of the movements can be selected. Therefore, before conducting an automated performance assessment, it is critical to extract these features by encoding human motions.

1.5.1 Human motion encoders in action recognition

Actually, this problem has been extensively studied in the field of human action recognition. For instance, Ren *et al.* [297] employed the silhouette of a dancer to represent his/her performance by extracting local features to control animated human characters. Wang *et al.* [372] obtained the contour of a walker from his/her silhouette to represent the walking motion. A spatio-temporal silhouette representation, the silhouette energy image (SEI), and variability action models were used by Ahmad *et al.* [19] to represent and classify human actions. In both visual-based and non-visual-based human action recognition of *differential features*, such as velocity and acceleration, motion statistics, their spectra and a variety of clustering and smoothing methods have been used to identify motion types. A two-stage dynamic model was established by Kristan *et al.* [182] to track the centre of gravity of subjects in images. Velocity was employed as one of the features by Yoon *et al.* [390] to represent the hand movement for the purpose of classification. Further, Panahandeh *et al.* [272] collected acceleration and rotation data from an inertial measurement unit (IMU) mounted on a pedestrian's chest to classify the activities with a continuous hidden Markov model. Ito [154] estimated human walking motion by monitoring the acceleration of the subject with 3D acceleration sensors. Moreover, angular features, especially the joint angle and angular velocity, have been used to monitor and reconstruct articulated rigid body models corresponding to action states and types. Zhang *et al.* [397] fused various raw data into angular velocity and orientation of the upper arm to estimate its motion. Donno *et al.* [97] collected angle and angular velocity data from a goniometer to monitor the motions of human joints. Angle was also utilised by Gu *et al.* [129] to recognise human motions to instruct a robot. Amft *et al.* [26] detected the feeding phases by constructing a hidden Markov model with the angle feature from the lower arm rotation. Apart from the above, only a few have considered a similar approach of trajectory *shape features* such as curvature and torsion. For example, Zhou *et al.* [401] extracted the trajectories of the upper limb and classified its motion by computing the similarity of these trajectories.

1.5.2 Human motion encoders in physical telerehabilitation

Although a number of encoding approaches have been investigated, not all of these approaches have been adopted in the physical telerehabilitation field, where the angles and trajectories of joints are usually utilised.

In these two representations, the angle-based approach is more widely used. This is mainly because human limbs are normally modelled as articulated rigid bodies. Additionally, some measurement devices, such as IMUs, are able to measure the orientation of limbs easily. Therefore, angles of joints and orientation of limbs can be acquired without much difficulty. Limb segments are hinged together with various degrees of freedom (DOF), which can be seen in Tables 1.1 and 1.2.

A number of examples utilising angle and its derivatives as encoders can be found. For instance, Tseng *et al.* [357] evaluated two platforms (Octopus II and ECO) to capture human motions for home rehabilitation. In two platforms, angles of joints were measured to represent the movement of limbs by the same type of compass (TDCM3 electrical compass) and different accelerometers, including the FreeScale MMA7260QT accelerometer and the Hitachi-Metal H34C accelerometer, respectively. Two types of angles were taken into consideration. One was the joint angle between limbs, such as the angle of the elbow and knee. The second type was the angle between the orientation of a sensor and gravity. Moreover, in the telerehabilitation system developed by Luo *et al.* [218], angles of joints were utilised to encode the movements of the upper extremity. In this system, the angles of the shoulder and wrist were measured by two IMUs, as these two joints were modelled with three degrees of freedom on each joint while those of the elbow and fingers were measured by an optical linear encoder (OLE) and a glove made by multiple OLEs because these joints could be modelled with one degree of freedom. Additionally, Durfee *et al.* [102] introduced two bilateral electrogoniometers in a home telerehabilitation system for post-stroke patients. These bilateral electrogoniometers were attached to the wrist and hand of a subject, respectively. The angles of flexion and extension movements in the wrist and the first MCP joints were measured to represent the movement of the wrist and hand. Two potentiometers (refer to Figure 1.10a) were utilised to calculate the angles of joints (θ) as

$$\theta = 180 - \cos^{-1}\left(\frac{c^2 + L^2 - a^2}{2cL}\right) - \cos^{-1}\left(\frac{d^2 + L^2 - b^2}{2dL}\right), \tag{1.4}$$

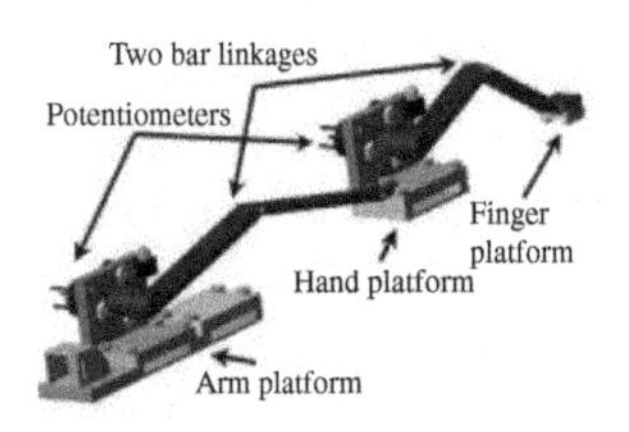

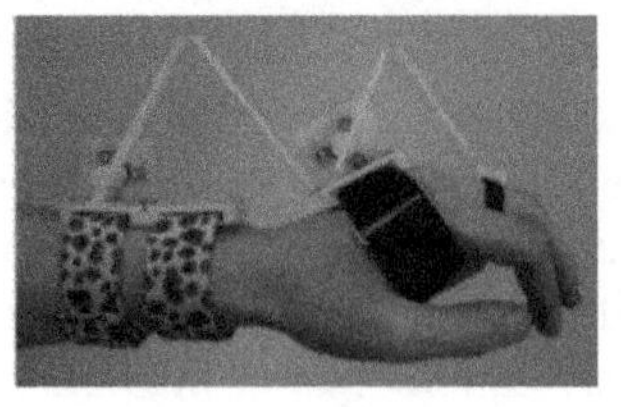

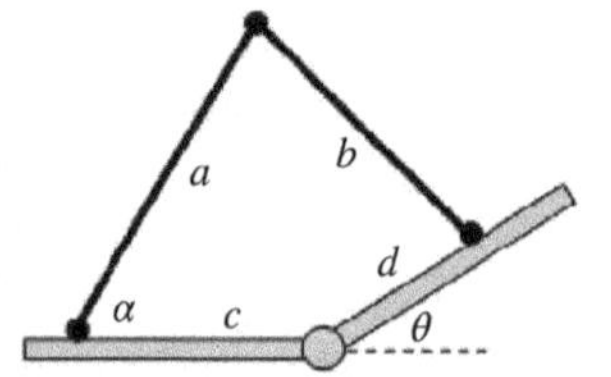

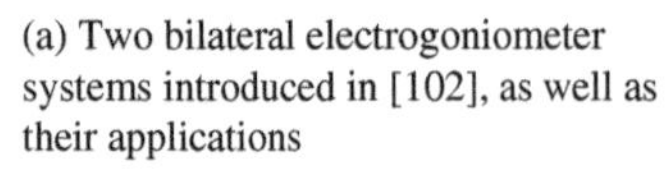
(a) Two bilateral electrogoniometer systems introduced in [102], as well as their applications

(b) Concept and parameters in an electrogoniometer system [102]

Figure 1.10 Pictures of animals. Source: Durfee *et al.* [102]. © 2009, ASME.

where the distance from the anatomic joint to the linkage joint is

$$L^2 = a^2 + c^2 - 2ac\cos\alpha. \tag{1.5}$$

Apart from the above three examples, in some studies where motion trajectories of joints are captured, angle information is still derived for encoding human movements. For example, Adams *et al.* [15] developed a virtual reality system to assess the motor function of upper extremities in daily living. To encode the movement, they used the swing angle of the shoulder joint along the Y and Z axes, the twist angle of the shoulder, the angle of the elbow, their first and second derivatives, the bone length of the collarbone, upper arm and forearm, as well as the pose (position, yaw and pitch) of the vector along the collarbone to describe the movement of the upper body. Here the collarbone is a virtual bone connecting two shoulders. These parameters were utilised in an unscented Kalman filter as state, while the positions of the shoulders, elbows and wrists reading from a Kinect formed the observation. Another example is that of Wenbing *et al.* [378], who evaluated the feasibility of using a single Kinect with a series of rules to assess the quality of movements in rehabilitation. Five movements, including hip abduction, bowling, sit to stand, can turn and toe touch, were studied in this paper. For the first four movements, angles were used as encoders. For instance, the change of angle between left and right thighs (the vector from the hip centre to the left and right knee) was used to represent the angle of hip abduction, while the dot product of two vectors (from the hip centre to the left and right shoulders) was utilised to compute the angle encoding the movement of bowling. Additionally, Olesh *et al.* [267] proposed an automated approach to assess the impairment of upper limb movements caused by stroke. To encode the movement of the upper extremities, the angle of four joints, including shoulder flexion-extension, shoulder abduction-adduction, elbow flexion-extension and wrist flexion-extension, were calculated with the 3D positions of joints measured with Kinect.

Though angles of joints, as well as their derivatives, are utilised widely in encoding human motions, trajectories of joints and their derivatives can also be observed in some rehabilitation and telerehabilitation applications.

The first example is that of Chang *et al.* [67], who developed a programme to use Kinect as the motion capture device for spinal cord injury (SCI) rehabilitation. In this programme, the trajectories of the hand, elbow and shoulder were recorded to represent the external rotation of upper extremities. Similarly, Su [339] also developed a rehabilitation system, named KHRD, to provide home-based rehabilitation services. To represent human motions, two key features were used, including trajectories of joints, as well as their speed. Additionally, Cordella *et al.* [78] modified the Kinect into a marker-based device to measure the positions of joints on a hand (refer to Figure 1.11). Markers with a dimension of 1.2 cm were attached to the joints of fingers, as well as the wrist. After detecting the centre of these markers, a robust tracking scheme was developed to track the position of each marker. Thus the movement of a hand was encoded by the trajectories of each joint on the hand, as well as the trajectory of the wrist.

1.5.3 Summary and challenge

From the literature, it is found that encoders used in human motion recognition are similar to those in physical telerehabilitation in many studies. For instance, features like

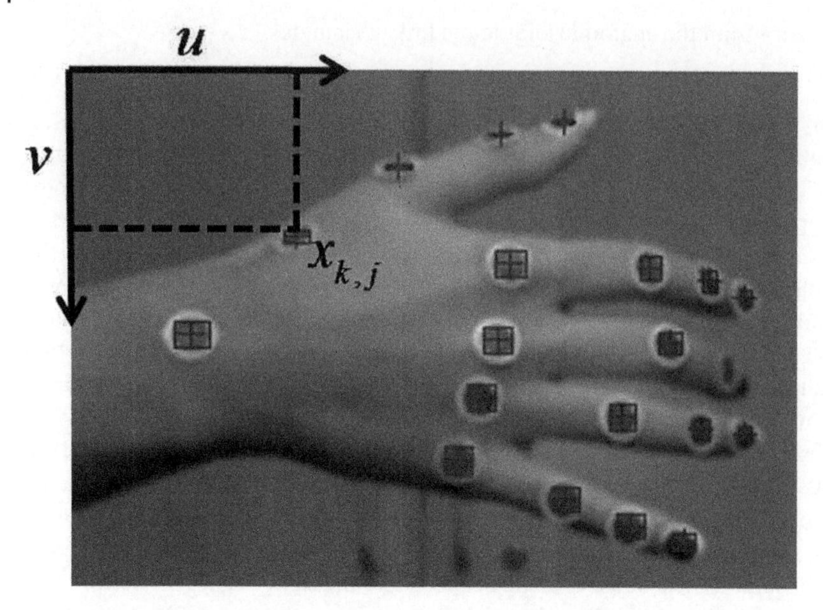

Figure 1.11 Marker-based hand tacking system. Source: Cordella *et al.* [78].

trajectories, velocity, acceleration, angle, angular velocity and angular acceleration are most commonly used in both fields. Though patients with movement disorders usually have a limited range of motions, they may be required to do certain tasks so as to evaluate their ability to perform ADLs, which usually are composed of a series of simple movements.

As a result, there remain challenges in developing formal descriptions and robust computational procedures for the automatic interpretation and representation of motions of patients. The majority of studies [92, 158] employed a variety of human motion encoders to recognise or decompose general movement, such as reaching, waving hands, jumping, walking and so on. Few of them investigated details in each general movement, for example, the even smaller atomic components included in these general movements that are of importance for syntactic and structural descriptions of human movements in detail, especially in a clinic and rehabilitation environment, where the details of movements of body parts require a form of motion language or, at least, syntax. A novel approach to encode human motion trajectories will be discussed in Chapter 3.

1.6 Patients' Performance Evaluation

In recent decades, with the advancements in telerehabilitation and associated motion capture technologies, an increasing number of research and development activities are focusing on the development of automated quantitative measures of patient performance in ADLs [136, 262]. Due to the important role played by the upper extremity in ADLs [99], an automated approach for measuring and assessing the ability of upper extremities to perform certain tasks is vital for telerehabilitation systems to deliver their full potential.

1.6.1 Questionnaire-based assessment scales

In the past few decades, a number of approaches have been proposed for assessing upper extremities, the majority of which are questionnaire-based. For musculoskeletal movement disorders of the extremities, most scales are generic. For instance, the self-reported Musculoskeletal Function Assessment (MFA) instrument [226], Short Musculoskeletal Function Assessment (SMFA) questionnaire [344] and self-administered measure of disabilities of the arm, shoulder and hand [148] were developed, but few of them are condition specific. More examples can be found in Adam *et al.* [99]. However, for neurological movement disorders, assessments are more disease-specific and rarely focus on upper extremities. For instance, Chedokee-McMaster (CM) assessment, the Fugl-Meyer (FM) assessment and the Wolf Motor Function Test (WMFT) assessment are for stroke. The Fahn-Marsden rating scale (F-M) [58], the Global Dystonia Rating Scale (GDRS) [12], the Unified Dystonia Rating Scale (UDRS) [11] and so on [22] were developed for dystonia. The Parkinson's Disease Questionnaire (PDQ-39) [284] and its shorter version (PDQ-8) [159], as well as the Parkinson's Disease Quality of Life (PDQL) questionnaire [144], the Webster [375] and the Unified Parkinsonâs Disease Rating Scale (UPDRS) [227] were developed for Parkinson's disease. More relevant to our work, Lane *et al.* [192] developed the Abnormal Involuntary Movement Scale (AIMS) to assess patients with dyskinesia. In this scale, the amplitude of involuntary movements was taken into consideration. In addition, Goetz *et al.* [119] proposed to use the Objective Dyskinesia Rating Scale (also known as the Rush Dyskinesia scale) to assess the severity of dyskinesia.

1.6.2 Automated kinematic performance assessment

Recently, with the development of sensing technologies, a number of approaches have been proposed to either automate conventional testing scales or develop new methods. Some examples are shown as follows.

Knorr *et al.* [177] automated two tasks, including reaching to the front and to the side, in the WMFT for people after stroke. Three accelerometers were attached to the hand, the corresponding forearm and the upper arm respectively to capture the characteristics of motion patterns. Eventually, two linear features (root mean square errors of acceleration and jerk) and a non-linear feature (approximate entropy of acceleration) were evaluated as the measurements of functional limitation and motor impairment. Similarly, Wade *et al.* [369] tried to automate the widely used WMFT for post-stroke patients. In their proposal, one IMU was attached to the wrist of the subject to measure the time used by the subject to finish each task in different WMFT tasks. Hester *et al.* [143] utilised 14 features to predict the clinical measurement scores of patients with stroke. These features were extracted from data collected by four accelerometers, three of which were attached to the affected hand, forearm and upper arm, while the fourth sensor was on the trunk. As for the tasks, different from the two previous studies, some tasks were not included in the WMFT. The process of predicting clinical scores from these 14 features is shown in Figure 1.12. Furthermore, Leibowitz *et al.* [202] introduced a protocol to quantitatively measure the proprioception deficit. In the protocol, the affected hand was located under a square board so that it could not be seen by

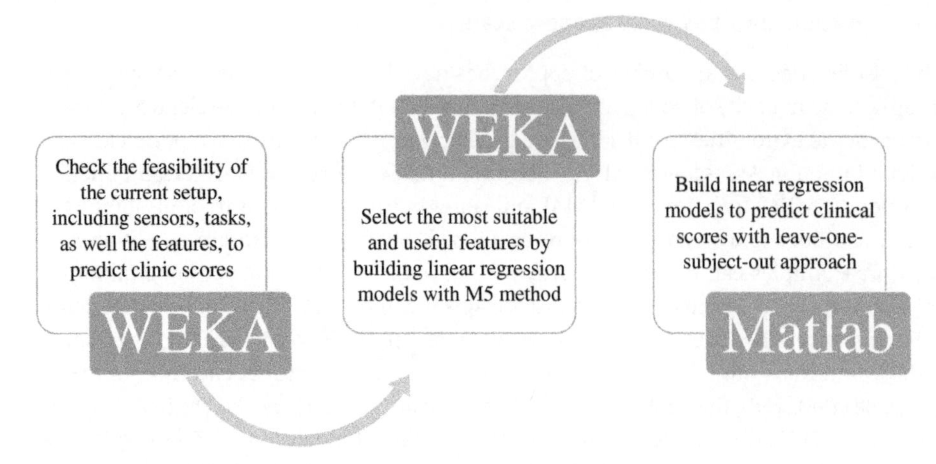

Figure 1.12 The process of predicting clinical scores from 14 features.

the subject. After passively moving the impaired hand to one of four locations, the healthy hand was required to be moved to the same location actively. Eventually, the positions of both hands were measured by a MiniBIRD500 magnetic tracking system and the positional differences between the impaired and non-impaired hands were computed as an indicator.

1.6.3 Summary and challenge

In the rehabilitation field, assessing the performance of limb functions is of importance in evaluating other medical procedures [219]. This task can be relatively easy to achieve in clinical environments. However, when it comes to telerehabilitation, it becomes difficult due to the absence of well-trained clinicians in the majority of cases. From the literature, it is found that a number of features have been used to perform automated assessment. However, the majority of them are calculated from a dynamic aspect of movements, such as velocity, acceleration and jerk. Therefore, a challenge here is whether it is possible to derive features for kinematic movement assessment based on shape information from motion trajectories. A novel approach to evaluate patients' kinematic performance by investigating both the shape and dynamics of the motion trajectories will be discussed in Chapter 2.

2

Kinematic Performance Evaluation with Non-wearable Sensors

2.1 Introduction

To accurately assess human kinematic performance, three elements are critical: high-quality raw data, proper data extraction methods and data analysis approaches. In this chapter, some examples are given for each aspect, such as data fusion algorithms for improving data quality, a motion encoding method for extracting hidden information from raw data and a data analysis approach for kinematic performance classification.

First of all, Section 2.2 investigates the robust and accurate capture of human joint poses and bio-kinematic movements for exercise monitoring in real-time tele-rehabilitation applications. Recently developed model-based estimation ideas are used to improve the accuracy, robustness and real-time characteristics considered vital for applications where affordability and domestic use are the primary focus. We use the spatial diversity of the arbitrarily positioned Microsoft Kinect© receivers to improve the reliability and promote the uptake of the concept. Skeleton-based information is fused to enhance accuracy and robustness, critical for biomedical applications. A specific version of a robust Kalman filter in a linear framework is employed to ensure superior estimator convergence and real-time use, compared to other commonly used filters.

Secondly, human actions have been widely studied in a number of areas, such as human–computer interaction (HCI) and sports. However, the majority of studies focus on the high-level representation of actions rather than basic components. In Section 2.3, we propose to utilise a syntactic two-level model to concentrate on the low-level interpretation of relatively simple motions as a preliminary study. By introducing this encoding model, more information could be extracted from the raw data.

Lastly, effective tele-rehabilitation technologies enable patients with certain physiological disabilities to engage in rehabilitative exercises in their natural environment. These recommended routines essentially constitute activities of daily living (ADLs). Therefore, training and assessment scenarios for the performance of ADLs are vital for the promotion of tele-rehabilitation. As an example, in Section 2.4 we look at quantitatively assessing the patient's ability to perform functional upper extremity reaching tasks with an automated approach based on action kinematics. The shape of the movement trajectory and the instantaneous acceleration of kinematically crucial body parts, such as wrists, are used to compute the approximate entropy of the motions to represent stability (smoothness) in addition to the duration of the activity.

Human Motion Capture and Identification for Assistive Systems Design in Rehabilitation, First Edition.
Pubudu N. Pathirana, Saiyi Li, Yee Siong Lee and Trieu Pham.

2.2 Fusion

2.2.1 Introduction

Human motion tracking and recognition have received renewed interest, particularly due to the progress made in sensing and data integration [139, 175, 178]. Numerous approaches have been pursued, employing readily available commercial products that impact many areas, ranging from computer gaming to remote rehabilitation. Inertial measurement units (IMUs) are used [404] in wearable [340] wireless systems to capture the movements in real time (MT9 in Xsens Motion Tech, G-Link). In contrast, systems with integrated vision normally operate with markers for the real-time tracking of body parts. Robot-assisted bio-kinematic motion detector systems are typically used to extract particular actions via pattern recognition or data mining means [403]. Although a number of approaches have been discussed to address various shortcomings in these systems designed for human movement detection applications, limited attempts [97] have been made to enhance the measurement accuracy for these commercial devices that are primarily designed for the computer gaming industry. For more effective use of these systems in clinical settings, reliability and robustness are two key aspects that need closer attention. Multiple cameras [60] are commonly used for improved accuracy with marker (active or passive)-based tracking [178], with distinct advantages despite the requirement of perspective imagery, occlusions and the need for relatively significant infrastructure costs. Furthermore, the need for designated clothing, multiple cameras and addressing the data association [277] problem limit the scope of the application domain.

Recently, Microsoft Kinect©, which provides 3D locations of up to 20 joints [140] (dark grey dots on joints in Figure 2.1) at an affordable cost, has come to the attention of rehabilitation researchers as a potential device for exercise monitoring [307]. Compared to the commercially available and lab-based Vicon system (which costs hundreds of thousands of dollars), a Kinect system is more suitable for home-based tele-rehabilitation as people with

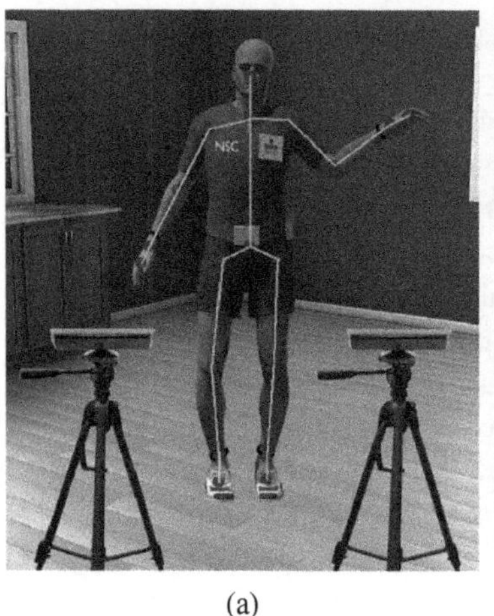

(a)

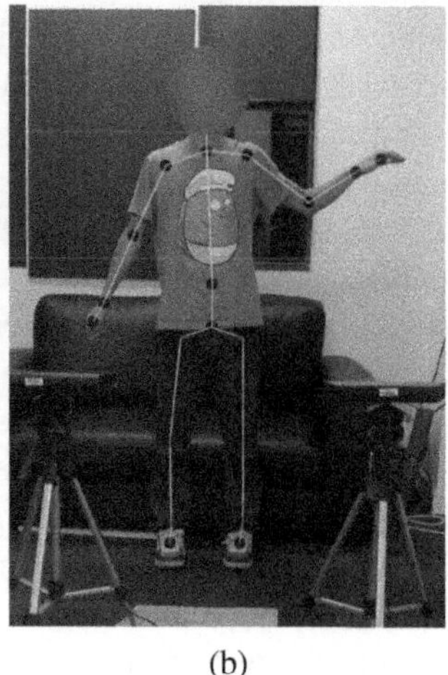

(b)

Figure 2.1 Virtual human mimicking the movements of a real human with data captured by two Kinects©. Source: Pubudu N. Pathirana.

disabilities are more likely to be financially constrained due to generally lower socioeconomic status [325]. Therefore, affordable and user-friendly sensors that can be easily set up are a core requirement, particularly in-home environments or confined clinical assessment areas. In terms of joint position measurement, the human body is segmented with randomised decision forests in a point cloud environment produced by projecting an infrared mesh on to the human body from the infrared camera on Kinect©. The positions of the joints are subsequently deduced with a local model-finding approach [327]. Primarily, considered as a motion and voice sensing input device for the Xbox video game console, Kinect© is currently the fastest-selling consumer electronics device. The introduction of this device has transformed academic research in this arena markedly as it provides the developers with data that enable them to concentrate on the higher-level architectures in integrated systems design. For example, with this device, the multi-camera vision problem traditionally considered as inherently a non-linear estimation problem [346] involving perspective projection can now be considered as a linear problem with readily available position data.

However, some issues remain with regard to the robustness, accuracy and reliability of these devices in healthcare applications, despite some attempts [139] to improve their localization accuracy. While a majority of past work [130] has primarily focused on the reconstruction of 3D object surfaces with the cloud points generated by Kinect©, some work has been reported on combining Kinect© with other devices, such as cameras [83, 139] and inertial sensors [44, 101] to improve measurement and tracking accuracy. In this work, we focus on 3D point measurements with multiple Kinects© to precisely track human movements using generic 3D point position information.

As optical tracking systems typically form a non-linear estimation problem, employing model-based filtering in such settings can cause the estimator to diverge. It is well known that linearisation of such models when implemented with extended Kalman filtering may potentially be unstable without appropriate initialisation in addition to the computational cost compared to their linear counterparts [277]. In contrast, we develop the entire formulation in a linear framework, therefore removing the need to use non-linear estimation techniques. A global coordinate system was established to map the positions of human body joints using these sensor arrays. A setup phase employing a maximum likelihood estimation was executed initially to ascertain the position and the orientation of Kinect© receivers that were positioned utilizing the empirical knowledge of the exercise routines, as this information is required for the subsequent filter implementations. Using both simulated data and real data, an optimal Kalman filter, a particle filter and a robust linear filter were used to fuse the sets of skeleton data in the local coordinate frame of each Kinect© into a common set of skeletal data in the global coordinate frame, thereby improving the accuracy of the joint positions resulting in real-time iso-skeletal positioning. Assuming a more generic uncertainty description, a robust linear filter [42, 277] was proven to perform better than the optimal Kalman filter as well as the particle filter in a number of experimental scenarios.

This section essentially looks at the problem of increasing the accuracy and reliability of a 3D vision sensor-based human motion capture aimed at tele-rehabilitation. In this context, real-time application and occlusions are intrinsic features that require closer investigation, while affordability and ease of use are vital for the uptake by healthcare providers and patients undergoing therapy [325].

2.2.2 Linear model of human motion multi-Kinect system

In this section, we employ model-based filtering for real-time motion tracking where N number of points are tracked by M number of receivers. The position of the mth Kinect© with respect to the global coordinate system is denoted as $s^m = [s_1^m, s_2^m, s_3^m]^\top \in \mathbb{R}^3$ with ⊤ denoting transposition. Here $m \in [1, \ldots, M]$ and $n \in [1, \ldots, N]$.

Assume that the nth object trajectory with respect to the global coordinate system is denoted by $X^n = [x_1^n, x_2^n, x_3^n, x_4^n, x_5^n, x_6^n, x_7^n, x_8^n, x_9^n]^\top$. Here, the three tuples correspond to position, velocity and acceleration along the Cartesian x, y and z directions, respectively.

Moreover, the observation data of the nth joint is denoted as $O^n = [o_1^n \ o_2^n \ o_3^n]^\top \in \mathbb{R}^3$ with reference to the built-in coordinate system of the mth Kinect© of the multi-Kinect© system. Since it can capture more than one object at the same time, the observation data structure of the mth Kinect© is $Y^m = [O^1 \ldots O^n \ldots O^N]^\top \in \mathbb{R}^{3N}$.

The state transaction matrix A and the input matrix B can be defined as:

$$A = \begin{bmatrix} A_1 & 0_9 & \cdots & 0_9 \\ 0_9 & A_1 & \cdots & 0_9 \\ \vdots & & \ddots & \vdots \\ 0_9 & \cdots & & A_1 \end{bmatrix} \in \mathbb{R}^{9N\times 9N}, \tag{2.1}$$

$$B = \begin{bmatrix} B_1 & 0_{9\times 3} & \cdots & 0_{9\times 3} \\ \vdots & & & \vdots \\ 0_{9\times 3} & \cdots & & B_1 \end{bmatrix} \in \mathbb{R}^{9N\times 3N}, \tag{2.2}$$

where

$$A_1 = \begin{bmatrix} I_3 & I_3\Delta t & I_3\Delta t^2/2 \\ 0_3 & I_3 & I_3\Delta t \\ 0_3 & 0_3 & I_3 \end{bmatrix} \in \mathbb{R}^{9\times 9} \tag{2.3}$$

and

$$B_1 = \begin{bmatrix} \Delta t^2/2 & 0 & 0 \\ 0 & \Delta t^2/2 & 0 \\ 0 & 0 & \Delta t^2/2 \\ \Delta t & 0 & 0 \\ 0 & \Delta t & 0 \\ 0 & 0 & \Delta t \\ 1 & 0 & 0 \\ 0 & 1 & 0 \\ 0 & 0 & 1 \end{bmatrix} \in \mathbb{R}^{9\times 3}. \tag{2.4}$$

Here Δt is the time interval between two consecutive sets of data collected from the multi-Kinect© system, and the underlying mathematical model for the human joint movement captured by a multi-Kinect© is then given by equation (1.2), where $x = [X^1 \cdots X^N]^\top \in \mathbb{R}^{9N}$ and $w \in \mathbb{R}^{9M\times 3}$ is the disturbance or uncertain biokinematic manoeuvres. The measurement matrix C is

$$C = \begin{bmatrix} C_1 \\ C_2 \\ \vdots \\ C_M \end{bmatrix} \in \mathbb{R}^{3NM\times 9N}, \tag{2.5}$$

where

$$C_m = \begin{bmatrix} S_m & 0_{3\times9} & \cdots & 0_{3\times9} \\ 0_{3\times9} & S_m & \cdots & 0_{3\times9} \\ \vdots & & & \vdots \\ 0_{3\times9} & 0_{3\times9} & \cdots & S_m \end{bmatrix} \in \mathbb{R}^{3N\times9N} \tag{2.6}$$

and $S_m = \begin{bmatrix} R_m & 0_{3\times6} \end{bmatrix} \in \mathbb{R}^{3\times9}$. $R_m, m \in \{1, 2, \dots, M\}$ is the rotation matrix of the mth Kinect© with reference to the global coordinate system. Let θ_m, ϕ_m and φ_m be the yaw, pitch and roll angles, respectively, for the mth Kinect© for counterclockwise rotation. Then the rotation matrix can be computed as

$$R_m = R_z(\varphi_m)R_y(\phi_m)R_x(\theta_m), m = 1, 2, \dots, M \tag{2.7}$$

where

$$R_x(\theta_m) = \begin{bmatrix} 1 & 0 & 0 \\ 0 & \cos\theta_m & -\sin\theta_m \\ 0 & \sin\theta_m & \cos\theta_m \end{bmatrix}, \tag{2.8}$$

$$R_y(\phi_m) = \begin{bmatrix} \cos\phi_m & 0 & \sin\phi_m \\ 0 & 1 & 0 \\ -\sin\phi_m & 0 & \cos\phi_m \end{bmatrix}, \tag{2.9}$$

and

$$R_z(\varphi_m) = \begin{bmatrix} \cos\varphi_m & -\sin\varphi_m & 0 \\ \sin\varphi_m & \cos\varphi_m & 0 \\ 0 & 0 & 1 \end{bmatrix}. \tag{2.10}$$

Therefore, the underlying measurement model for the multi-joint tracking can be stated as (1.3), where $\mathbf{y} = [Y^1 \cdots Y^M] \in \mathbb{R}^{3MN}$ and $d \in \mathbb{R}^{3MN}$ is the measurement noise. Note that the translation of each Kinect© is corrected, i.e. $\mathbf{y} = y - t_m$, where y is the actual measurement and t_m is the displacement vector of each Kinect© estimated (together with the rotational matrices pertaining to orientation), as described in Section 2.2.2.

Rotation and translation of the Kinect©

In human motion capture applications, it is vital for the user to have some flexibility in positioning the receivers (i.e. due to space restrictions, mitigating occlusions with the nature of the exercise routines). Arbitrary Kinect© positioning instigates the problem of obtaining the rotation matrix of each Kinect©, which is crucial for the physical implementation in the 3D case. This can easily be achieved by a minimum of four distinct measurements. Let the measurement be $\mathbf{y}_i$ for each $\mathbf{x}_i$ measurement, $i \in [1, \dots, N]$, $N \geq 4$. Then

$$\mathbf{y}_i = R_m\mathbf{x}_i + t_m + e_i,$$

where e_i is the residual error. We can rearrange this as

$$\hat{\mathbf{y}} = Kr + e$$

where $K = \mathrm{Diag}\{K_i\} \in \mathbb{R}^{3N\times12}$ with $K_i = \begin{bmatrix} I_3 & \mathrm{Diag}\{x_i^\top\} \end{bmatrix} \in \mathbb{R}^{3\times12}$, $\hat{\mathbf{y}} = \begin{bmatrix} \mathbf{y}_i \cdots \mathbf{y}_N \end{bmatrix}^\top$, $e = \begin{bmatrix} e_i \cdots e_N \end{bmatrix}^\top$ and $r = \begin{bmatrix} t_m^\top & r_1 & r_2 & r_3 \end{bmatrix}^\top$. Here, r_i for $i \in [1, 2, 3]$ are the rows of the R_m matrix, i.e. $R_m = \begin{bmatrix} r_1^\top & r_2^\top & r_3^\top \end{bmatrix}^\top$. Therefore, the least-squares solution $r = \left(K^\top K\right)^{-1} K\hat{\mathbf{y}}$ will provide the optimal position vector (t_m) and rotation matrix (R_m) for the mth Kinect©.

2.2.3 Model-based state estimation

The real-time position of the physical joints of the human body is estimated using the aforementioned kinematic description of the dynamics (equation 1.2) and the measurements (equation 1.3). The natural choice is to use the standard Kalman filter (refer to Section 6.2.1) under Gaussian assumptions of measurement and state noise. Particularly for the case of human joint movements during exercise or regular day-to-day movement routines, it is unrealistic to assume a Gaussian distribution for the unknown acceleration components such as w in equation (1.2) or d in equation (1.3). Therefore, similar to the implementation in other studies [42, 277], we use a robust version of the Kalman filter with a more generic assumption on the deterministic uncertainty descriptions [27, 282, 318]. We also implement two commonly used and relevant filters for real-time applications of this nature – a standard Kalman filter and a particle filter (refer to Section 6.3). We compare these with our proposed robust version of the Kalman filter, which accounts for uncertain human movements modelled in a more generic, deterministic and bounded context.

2.2.4 Fusion of information

Information from multiple Kinects© provides spatial diversity to measurements and plays a pivotal role in improving the estimation accuracy. Here we look at the advantages of employing multiple Kinects©. Denoting the state estimate with an M number of specially distributed Kinects© as $\hat{x}_M$, let the root mean squared error (RMSE) function between the M Kinect© fused trajectory and the ideal one be Φ_M, i.e. Φ_1 indicates the RMSE between the raw trajectory from the first (reference) Kinect© and the ideal trajectory. Let $\Phi_{M-1:M}$ denote the consecutive improvement with the increase (from $M-1$ to M) in the number of Kinects©. Then

$$\Phi_{M-1:M} = \frac{\Phi_{M-1} - \Phi_M}{\Phi_{M-1}} 100\% \quad \forall\, M \in [2 \cdots M_{\max}] \tag{2.11}$$

where

$$\Phi_M = \sqrt{\frac{\sum_{k=1}^{T1} (x(k) - \hat{x_M}(k))^2}{T1}}. \tag{2.12}$$

Here, $M_{\max}$ is the maximum number of Kinects© considered.

2.2.5 Mitigation of occlusions and optimised positioning

Positioning of the Kinects© can affect occlusions as well as the accuracy of the measurements. As the mitigation of occlusions in this form is heavily dependent on the actual human motion, we look at extending the linear filtering to account for missing information for the generic case. The introduction of spatial diversity in positioning can also be considered to improve the overall measurement accuracy for generic human movements.

Occlusions and incomplete information

One of the key advantages in using the skeletal information from the Kinect© is the absence of the *data association* problem as the joints are already identified. Therefore, we are able to directly address the problem of occlusions under the missing information scenario in model-based filtering. This means that $\mathbf{y}(t)$ is incomplete or not available for a certain time t. Let $\mathbf{M}(t) = [\mathcal{M}^1(t)\mathcal{M}^2(t) \cdots \mathcal{M}^{3MN}(t)]^\top$ be a given vector for $t = 1, 2, \ldots, T$ such that $\mathcal{M}^i \in \{0, 1\}$, for $i = 1, \ldots, 3MN$. Then the matrix $\mathrm{M} \triangleq [\mathbf{M}(1) \cdots \mathbf{M}(T)]^\top$ is referred as the *incomplete* matrix. With $\mathcal{M}^i$, let us define two sequences of matrices:

$$E(t) = \mathrm{Diag}[\mathcal{M}^1(t)\mathcal{M}^2(t) \cdots \mathcal{M}^{3MN}(t)],$$
$$\hat{E}(t) = [\tilde{\mathcal{M}}^1(t)\tilde{\mathcal{M}}^2(t) \cdots \tilde{\mathcal{M}}^{3MN}(t)]^\top, \tag{2.13}$$

where $\mathcal{M}^i(t) + \tilde{\mathcal{M}}^i(t) = 1$.

In Figure 2.2, four cases that lead to missing data are presented. For the first and second cases, if the joint is in these areas, it can only be tracked by one Kinect (Kinect 1 if the joint is in area 1 and Kinect 2 if the joint is in area 2) since it is out of the view of the other one. For cases 4 and 5, the joint is in the field of view of both Kinects, but occluded by an object. Therefore, the Kinect located in the same side of the object cannot track the position of the joint. The parameters of using missing information for data fusion are also shown in Figure 2.2(a).

Kinect© optimal positioning

Using a similar approach to that in Adrian *et al.* [41], the general Fisher's information matrix $I(x) = \Delta_x y(x)^\top R_y \Delta_x y(x) = C^\top C$. $\mathbf{det} I(x) = 0$, implies that the position measurement scheme has no orientation dependency in an information theoretic sense under Gaussian assumptions. This is unlike the case for the angle, distance and time delay of arrival measurement schemes. Considering the 2D orientation scenario under an equidistance exercise monitoring setting, as depicted in Figure 2.2(b), the position measurement can be considered in an angle measuring context for the purpose of an information theoretic evaluation ($x_1^n = -r \sin\alpha, x_2^n = r\cos\alpha$, where r is a constant indicating the range specified by the Kinect© to maximise the field of view). The determinant of the Fisher's information matrix (in an angle measurement context) is maximum (see Figure 2.3) when $\theta = \pi/2$. Indeed, the XY planes for the two Kinect© cases are orthogonal to each other, causing a $\pi/2$ angle subtended at the joint being tracked. For the cases of an any arbitrary number greater than two Kinects©, an extensive positioning analysis is given in Adrian *et al.* [41]. Note that this reasoning is based purely on information theoretic assumptions and as the potential occlusions depend on the exercise routine, the receiver positions may need to be adjusted accordingly. The improvements due to the suppression of occlusions are much more significant than the gain due to Kinect's optimal positioning [notice the improvement due to two Kinects© as opposed to one later in Figure 2.9 and the improvement due to missing data accounting by the robust linear filter (RLF)]. Therefore, after positioning the Kinects based on the empirical knowledge to minimise the occlusion, the position estimation approach described in Section 2.2.2 can be used to estimate the rotation matrices and translation vectors required for fusion of information.

2.2.6 Computer simulations and hardware implementation

We use a hypothetical scenario in a computer-simulated environment as well as in a practical scenario with a hardware setup to test and validate our theoretical assertions. In evaluating, the RMSE for M Kinects© (Φ_M) from equation (2.12), in computer

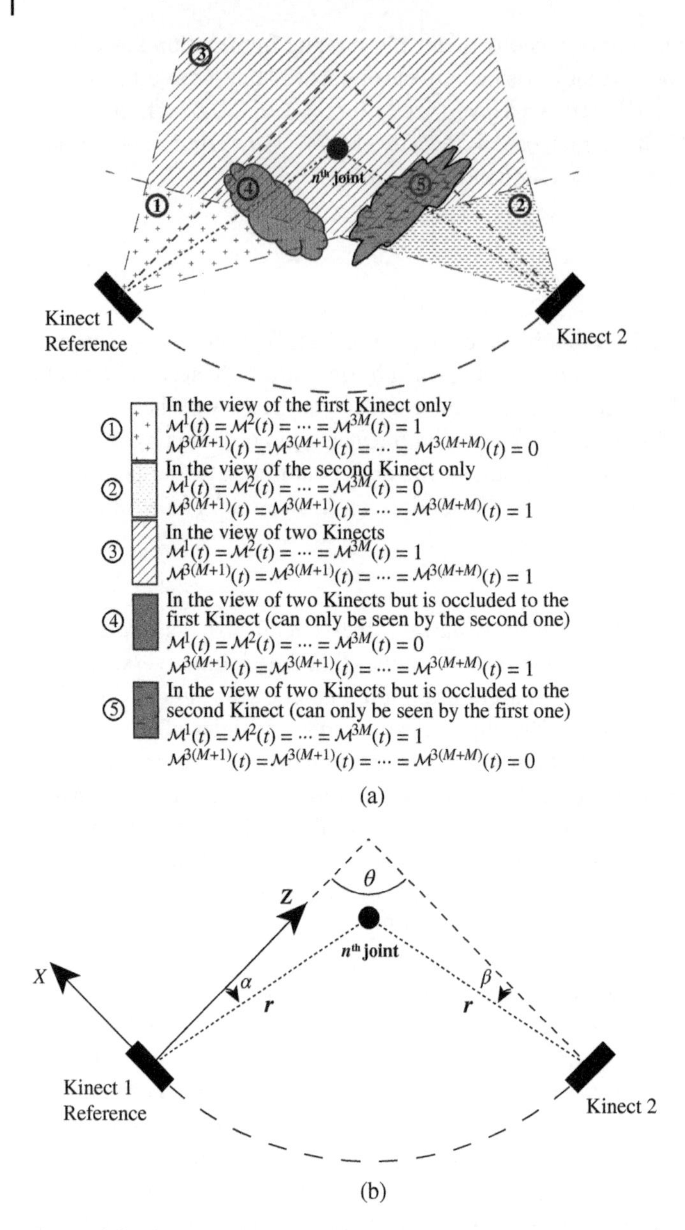

Figure 2.2 (a) Four instances where the two Kinect systems may have missing data and another instance without any occlusion. Here, the *incomplete* matrix parameters of the RLF-based two-Kinect fusion are explicitly stated. (b) The optimal positioning of two Kinects.

simulations we use the known ideal trajectory $x(k)$. In the hardware experimentation, we use the Vicon captured trajectory as the ground truth. In order to convey these ideas on multi-Kinect© fusion, we consider one Kinect© (K_1) as the reference, Φ_M is averaged over the number of runs and we denote the average RMSE as Φ_M^a ($M = 1, 2, ..., 20$ for computer simulations and $M = 1, 2, 3$ for the hardware experimentation), which in turn is used to calculate the average relative improvement percentage ($\Phi_{M-1:M}^a$). This is used to

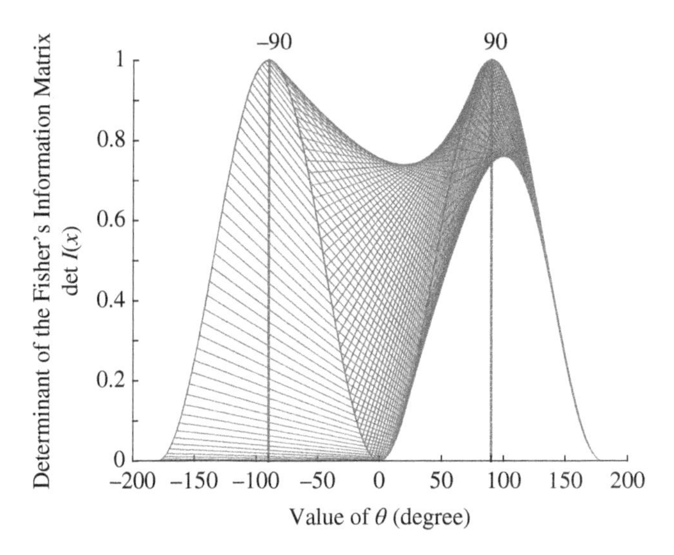

Figure 2.3 Information theoretic assessment of Kinect© orientation.

demonstrate the contribution of adding a Kinect© to the multi-Kinect© system in terms of reducing the RMSE. Note that the fused trajectories captured by the Kinect© system should be rotated and translated to the coordinate system of the Vicon system when calculating the RMSEs.

In addition to the overarching assertion of superior performance in terms of accuracy and robustness in the proposed linear filter, improvement due to multiple Kinect© sensor fusion, mitigating the effects of occlusions via extending the idea of missing information, was evaluated in simulated experiments as well as in a hardware-based practical context.

When selecting tuning parameters for various approaches considered, in a generic sense, parameters that optimised the mean squared error between the estimated states and the actual state were used. The actual state was captured from the Vicon system in the real data experiment as the ground truth. This was obtained using a common collection of training datasets. Indeed, the interval for the search was based on empirical knowledge of the underlying experiment. For real-data experiments, a smooth curve was fitted to the sample training trajectories and the standard deviations of the measurement noise with spectral densities (S_w and S_v) were obtained under Gaussian assumptions. Obviously for the simulated examples the specific noise distributions were assumed and considered as additive uncertainties to the perfect measurements. With this reasoning, we acquired an empirical understanding of the magnitude (s) of S_w, S_v and $P(0)$. A sufficiently large interval $sI_9[10^{-5}, 10^{-4}, \ldots, 10^4, 10^5]$ was chosen to test each combination of S_w, S_v and $P(0)$ in computing the error (the difference between the fused and the ideal trajectory) and the combination that provided the smallest RMSE was selected. This was further optimised by employing the expectation minimisation (EM) algorithm using all the training datasets. Indeed, the underlying assumption here is similar noise and uncertainty characteristics across the training and testing exercises. We assumed the same noise distribution for all the filters, with the standard Kalman filter being observed as a special case of the RLF with the interpretation of probabilistically bounded Gaussian noise. For the particle filter, when w_t

in equation (6.21) and $p(Y_{t+1} - CP^t_{i+1})$ in equation (6.22) were computed, the standard deviation of w and v (S_w and S_v) could be obtained. As the number of particles increase, a better approximation is reached. However, due to significant computational cost, we tested from 1000 to 10 000 with a step size of 1000. In robust linear filtering, we considered the combinations of N and Q, which were in the range of $[10^{-10}, 10^{-9}, \ldots, 10^{9}, 10^{10}]$ in order to obtain the combination that gave the smallest RMSE. Further, if $N = 10^n$ and $Q = 10^q$ gave the smallest RMSE, we refined the range of $[10^{n-1}, 2 \times 10^{n-1}, \ldots, 10^n, 2 \times 10^n, \ldots, 10^{n+1}]$ for N (and similarly for Q) and repeated the process until the improvement in RMSE was not sufficiently significant. In the simulation-based example, since we knew the standard deviation of the noise, we directly used that information to determine parameters for the Kalman filter and particle filter, while the same approach described above was used to determine the parameters for the robust linear filter. Indeed, all the data in the real experiment were collected from the same Kinects© positioned in the same environment and hence our assumption of a similar distribution over different filters and training and testing scenarios was justified.

Computer simulations

Improvement due to fusion A helical trajectory with 1000 samples and a total length of 6 metres was generated to represent the motion of a joint in the human body as the ground truth or the ideal trajectory of our experiment. Using this trajectory, another 20 trajectories with bounded noise were generated to represent data captured from 20 different Kinects© with known positions and orientations. To illustrate the robustness of these algorithms, three signal-to-noise ratios (SNRs), 20, 25 and 30, were considered.

The noisy trajectories collected from the Kinects© in each round of the experiment would be fused through each of the filters. For example, the trajectories gained from Kinects© were fused into one trajectory as per our basic postulation, leading to the model with all the filter implementations; this would be conducted for up to 20 Kinects©. The RMSEs (Φ_M, $M = 1, 2, \ldots, 20$) were computed between these 20 trajectories and the ground truth to compare the performance of the multi-Kinect© systems. Furthermore, the relative improvement percentages between $M - 1$ Kinect© fused trajectory and M Kinect© fused trajectory were computed as described in Section 2.2.4 (see equation 2.11).

The RMSE and relative improvement percentage diagrams are shown in Figure 2.4. It can be seen that RMSEs between fused and model trajectories gradually decrease with the increasing number of the Kinects© irrespective of the noise level considered. For instance, in Figure 2.4(a), the RMSEs of fused trajectories corresponding to Kalman filter (KF), particle filter (PF) and RLF drop from 0.168 m, 0.153 m and 0.122 m to 0.039 m, 0.037 m and 0.034 m with the increase in the number of Kinects© from 1 to 20. Alternatively, the increased number of Kinects© enabled countering the noise effects in the fused trajectories to be countered. Moreover, it is obvious that the robust linear filter outperformed the others by generating fused trajectories with the smallest RMSE, demonstrating its superiority over the other filters considered.

From Figures 2.4(b), 2.4(d) and 2.4(f), while RMSEs of fused trajectories decrease with the increase in the number of Kinects©, relative improvement percentages ($\Phi_{M-1:M}$) decayed almost in an exponential fashion. For example, in the three cases with different SNRs, the relative improvement percentages of RLF with two Kinects© are always more than 25%, while this number drops to around 15% when the number of Kinects© rises to three and a similar behaviour can be expected from the other two state estimators.

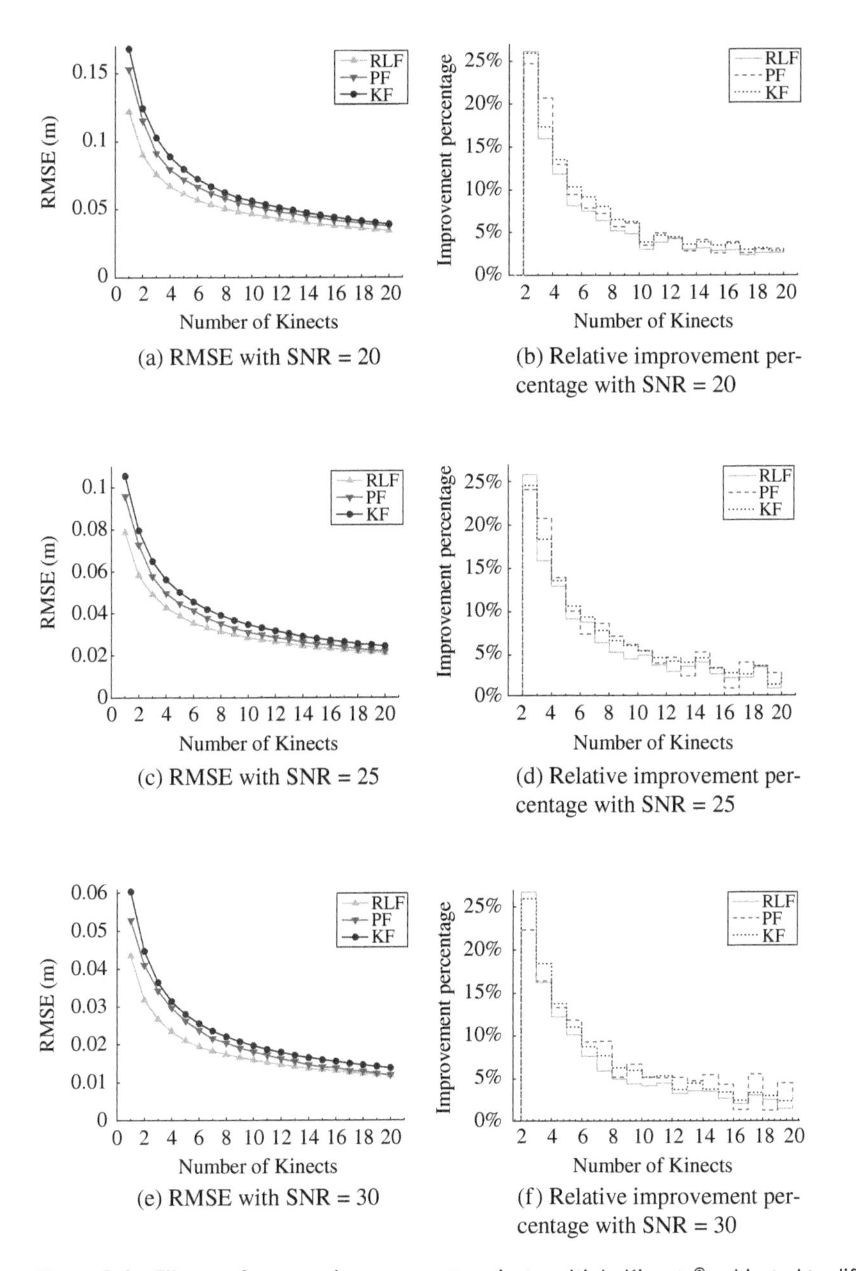

(a) RMSE with SNR = 20

(b) Relative improvement percentage with SNR = 20

(c) RMSE with SNR = 25

(d) Relative improvement percentage with SNR = 25

(e) RMSE with SNR = 30

(f) Relative improvement percentage with SNR = 30

Figure 2.4 Filter performance improvement against multiple Kinects© subjected to different uncertainty levels.

System cost and complexity in applied tele-rehabilitation Additional Kinects© indeed improve the accuracy of human kinematic estimations and also incur added cost. Assuming the cost of a Kinect© is C, the combined cost function optimisation is

$$\min_{M\in\mathbf{N}^{+}}\Psi(M) \tag{2.14}$$

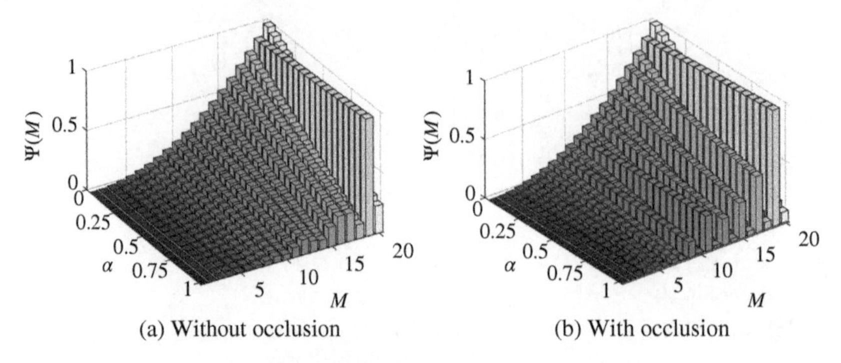

Figure 2.5 The relationship between $\Psi(M)$, α and M in the cases with and without occlusion. Robust linear filtering was utilised for data fusion.

where

$$\Psi(M) = \alpha\overline{\Phi_{M-1:M}}^2 + (1-\alpha)\overline{CM}^2.$$

Additionally, $\overline{\Phi_{M-1:M}} \in [0,1]$ and $\overline{CM} \in [0,1]$ are the normalised values of $1/(\Phi_{M-1:M})$ and CM over M. Here $\alpha \in [0,1]$ is the relative weighting parameter. To illustrate the relationship between $\Psi(M)$, α and M, a simulation experiment was conducted for the case of robust linear filtering under the two different scenarios of interest: occluded and non-occluded cases. For the former, we assumed no missing data, while for the latter, we intentionally set the second half of some datasets with missing data.

As illustrated in Figure 2.5, for both cases (with and without occlusion), the smaller M is associated with a smaller $\Psi(M)$ for a range of α. Therefore, in order to ensure a compromise in the cost and system complexity, it is reasonable to select a small M for applications in tele-rehabilitation. It is noteworthy that the change of α exerts little influence on $\Psi(M)$ because $\overline{\Phi_{M-1:M}} \approx \overline{CM}$ for $M = 19$.

Hardware experiment

Improvement due to multiple Kinects© The hardware experiment was conducted with three non-linearly placed Kinects© (30 Hz sampling rate and labelled as K_1, K_2 and K_3) and a Vicon system (250 Hz sampling rate) set up as depicted in Figure 2.6. The distance between the subject and each Kinect© was approximately 3 metres to ensure that each Kinect© could capture and generate the complete skeleton of the subject. The orientation of each Kinect© is placed with empirical knowledge to minimise occlusion. At the same time, a marker was placed on the right wrist of the subject so that the positions of the right wrist could be captured by the Vicon system simultaneously.

During the experiment, three healthy male subjects conducted three types of repetitive movements to simulate a relatively longer period of exercise: swinging (S) the right arm for approximately 1 minute, reaching (R) forward with the right arm and drawing circles (C) in front of the body with the right arm for approximately 30 seconds. The detail of these gestures, such as the total trajectory length for each type of movement performed by the three subjects, and the corresponding RMSE of the trajectories from individual Kinects© are shown in Table 2.1.

Figure 2.6 Experimental setup: Vicon and Multi-Kinect© system.
Source: Pubudu N. Pathirana.

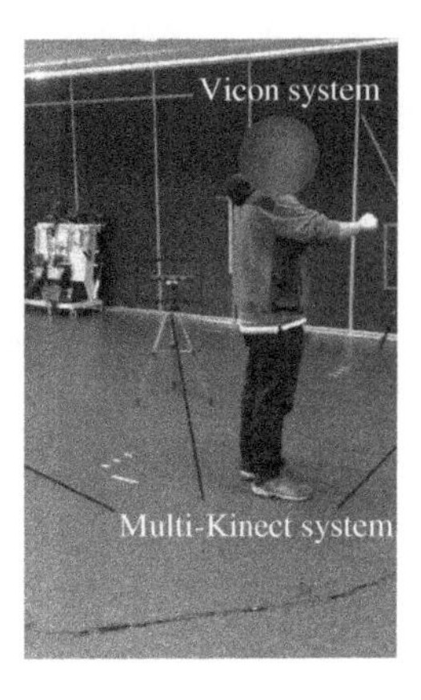

Table 2.1 Example human movements.

Motion type	Length (m)	RMSE (Φ_1, cm)
Circular	89.850	6.13
Reaching	73.536	6.60
Swinging	210.699	5.75

Since the multi-Kinect© system and the Vicon system are two independent standalone systems, certain implementation-related clarifications are needed. Firstly, although it is possible to trigger the two systems simultaneously, in this experiment, the data captured from the two systems were synchronised manually using the kinematic event itself. More specifically, when two systems began capturing data, the subjects remained steady for a few seconds prior to commencing the exercises for synchronisation purposes. This was purely for comparison (with Vicon) purposes and was not required in the actual application. The starting point of the movements can be clearly observed from the captured data, which was used to synchronise the multi-Kinect© system and the Vicon system. Secondly, data from the multi-Kinect© system were resampled at 250 Hz, again for comparison purposes with Vicon. Thirdly, during the evaluation phase, trajectories captured from the multi-Kinect© system were compared to those from the Vicon system, which were used for calibrating the two systems. In our experiment, we utilised the algorithm introduced in Section 2.2.2 to compute the rotation matrix and translation vector between each Kinect© and also with the Vicon system.

The time used to compute the fusion of each frame using different approaches in the computer with Intel® Core™ i5 (3.0 Hz) and 8 GB memory was also recorded for evaluation as a real-time system.

The results of the hardware experiments are shown in Figures 2.7 and 2.8. The legends in both figures are indicated in the form of *(number of Kinects©) – (filter name)*. For example, *3K – KF* denotes the three Kinects© fused by the Kalman filter.

Figure 2.7 illustrates the root mean square errors due to multi-Kinect© fusion. Generally, the RMSEs of three-Kinect© fusion (with lighter colours) with all the combinations of motion types and fusion approaches are smaller than that of two-Kinect© fusion (with

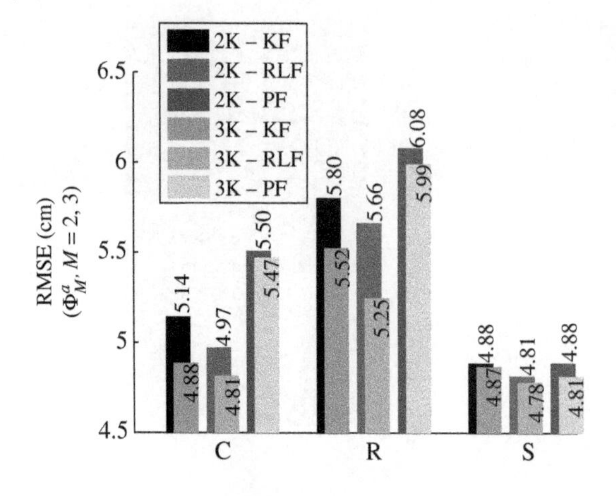

Figure 2.7 Averaged RMSEs (Φ^a_M, $M = 2, 3$) over the same type of exercises conducted by the three subjects.

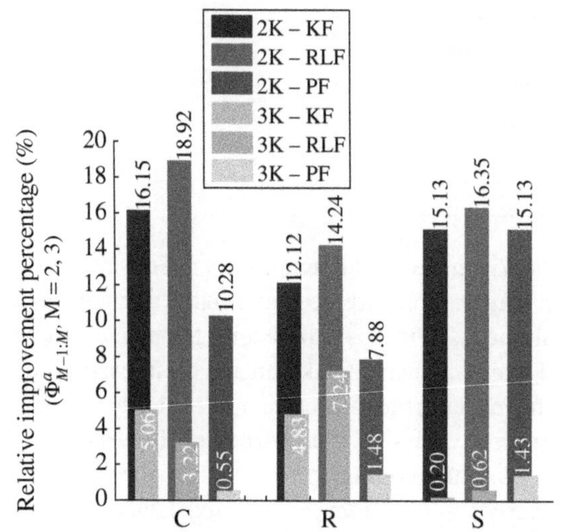

Figure 2.8 Averaged relative improvement percentages ($\Phi^a_{M-1:M}$, $M = 2, 3$) of multi-Kinect© fusions.

darker colours). Here the smaller RMSE indicates the proximity between the fused trajectory and the captured trajectory from the Vicon system. Alternatively, the fused trajectory with three Kinects© makes less noise than is the case for two. In other words, fusion of more Kinects© counter noise, resulting in smaller RMSEs with respect to trajectories captured from the Vicon system. This result corresponds to Figures 2.4(a), 2.4(c) and 2.4(e) in computer simulations. In addition, robust linear filtering generates the smallest root mean square errors in all the cases and varying numbers of Kinects©, while the particle filter gives the largest ones except for fusion of swing motion with three Kinects© (0.06 cm smaller in this case compared to the Kalman filter), which confirms the superior performance of robust linear filtering.

Furthermore, Figure 2.8 depicts the relative improvement percentages of two (darker colours) and three (lighter colours) Kinect© fusions. It is evident that the two-Kinect© fusion generates much higher relative improvement percentages (ranging from 10.28% to 18.92%) than the latter (between 0.20% and 7.24%), at least for the movement types and fusion methods considered.

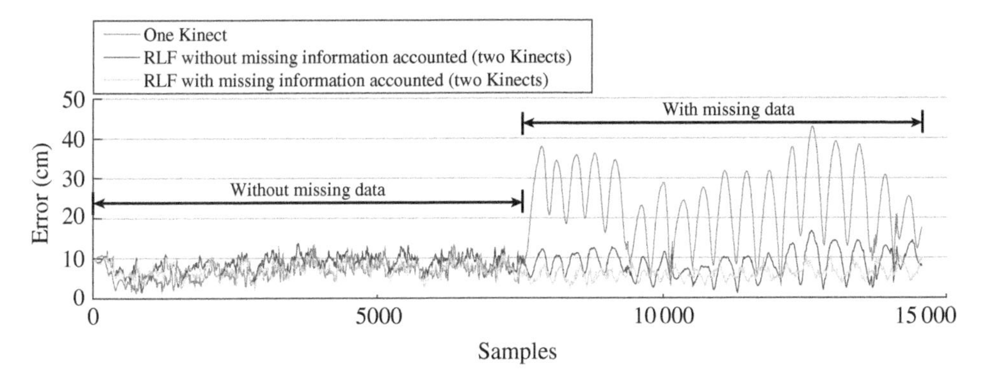

Figure 2.9 Errors of two-Kinect© fusion with missing data. The temporal interval between consecutive samples is 1/240 second.

Considering the average computational time of 4.2123e-5 seconds per frame (SPF) for RLF, compared to 3.016e-4 SPF and 0.348 SPF for KF and PF respectively, it is clear that the RLF is more suitable for real-time application. The particle filter needs rigorous optimisation prior to any consideration of real-time use.

Robustness due to occlusions A real-data experiment has been conducted to illustrate the capabilities of the proposed approach. This deals with missing information due to occlusions of body parts in monitoring human movements, which is considered as one of the major limitations in using a single Kinect© as a motion tracking device. In this experiment, we not only recorded the positions of a joint during a time frame, but the tracking status of this joint was recorded from the Kinect© SDK. Three subjects were required to perform a swing motion repeatedly in around a minute. During the first half of the experiment, the monitored joint (right wrist) was within the field of view of both Kinects©, ensuring that there were no *missing data* during this period. In contrast, during the second half of the exercise routine, all the subjects were required to conceal their right arm behind their body to ensure only the Kinect© on their right side could capture the wrist position information, with the Kinect© on their left side experiencing missing data. Data collected from both Kinects© were fused by the robust linear filtering with and without missing information accounted and then compared to the trajectory captured from the Vicon system. Due to very low computational efficiency, we have not performed the data fusion with the particle filter.

The results of this experiment are shown in Figure 2.9 and the comparison between our proposed approach and those introduced in other studies [28, 315, 389] is shown in Table 2.2. Figure 2.9 indicates the error when using one Kinect© and also when using two Kinects© fused by robust linear filtering with (Section 2.2.5) and without missing information accounted for. The error was computed as the Euclidean distance between the trajectories deduced from the Kinect© system and the Vicon system. It is evident that in the first half of the trial (with no missing data), the position estimation error for the cases of one Kinect© and RLF with and without missing information accounted are relatively similar as the Kinects© could track the monitored joint. However, in the second half, the position estimation error for the case of one Kinect© is increased significantly as it was no longer able to track the joint due to occlusions. As RLF without accounting for missing information

Table 2.2 Comparison (in the presence of missing information) of accuracy and execution times between the proposed RLF and the ones introduced in the references shown. Sources: Saputra *et al.* [315]; Yeung *et al.* [389]; Asteriadis *et al.* [28].

	Subject 1 (cm)	Subject 2 (cm)	Subject 3 (cm)	Speed (SPF)
RLF	7.77	8	10.16	4.21E-05
[315]	15.7	13.49	17.48	1.03E-07
[389]	9.14	10.12	11.82	0.518903
[28]	12.46	15.94	16.14	0.0686

can still utilise the information from the second Kinect©, it is able to track the joint with the errors still under approximately 0.2 metres. Indeed, the RLF with missing information accounted further improves the estimation accuracy by restricting the error to less than 0.1 metres. Table 2.2 depicts the average RMSE and processing time per frame for the three approaches. It is evident that RLF with missing information accounted for always has the smallest RMSE compared to other relevant approaches using skeleton information. When comparing the execution times, all these approaches were implemented in Matlab 2014b® and executed under the same conditions. The approach in Saputra *et al.* [315] required 1.03e-7 seconds, which was faster compared to RLF. The approaches in Yeung *et al.* [389] and Asteriadis *et al.* [28] required 0.519 and 0.0686 seconds per frame, respectively. As Kinect© can only provide one skeletal frame in 0.33 seconds, our approach and the ones in [28, 315] can be considered for real-time implementations.

Discussion

The computational cost in using particle filtering or a point cloud-based approach is significant in comparison to the model-based filtering, as Licong *et al.* [211] uses 0.5 of CPU time and also our simulations indicate an inferior performance with particle filtering. Skeleton-based measurement schemes such as ours require a more sophisticated model-based filtering method, unlike the simple averaging based approach used in Saputra *et al.* [315], which reports a lesser accuracy of 0.13 m compared to the 0.08 m accuracy in our proposed method. Although in computer simulations the particle filter outperforms the Kalman filter, we observe the converse in the hardware experiment. This is due to the fact that 10 000 particles used in the hardware experiment are not sufficient for this multi-Kinect© fusion. Indeed, the computational cost and the resource consumption have already made the particle filter unsuitable for real-time applications of this nature.

Particularly in an exercise monitoring scenario, the occlusions can predominantly be caused by other parts of the patient's body. We have considered occlusions as a missing information scenario and the positioning of the Kinects© need to ensure these effects are minimal. Nevertheless, some form of occlusions are likely to exist and the severity of these are dependent on the actual exercise routine. The optimal positioning of the Kinects© can be achieved based on the information theoretic reasoning utilising the Kramer Rao bound concepts, as in [41]. Optimal positioning of receivers subjected to missing information is considered as an optimal angular positioning exercise with a fixed range to maximise the field of view. Even so, the occlusions are significantly dependant on the actual nature of the exercise

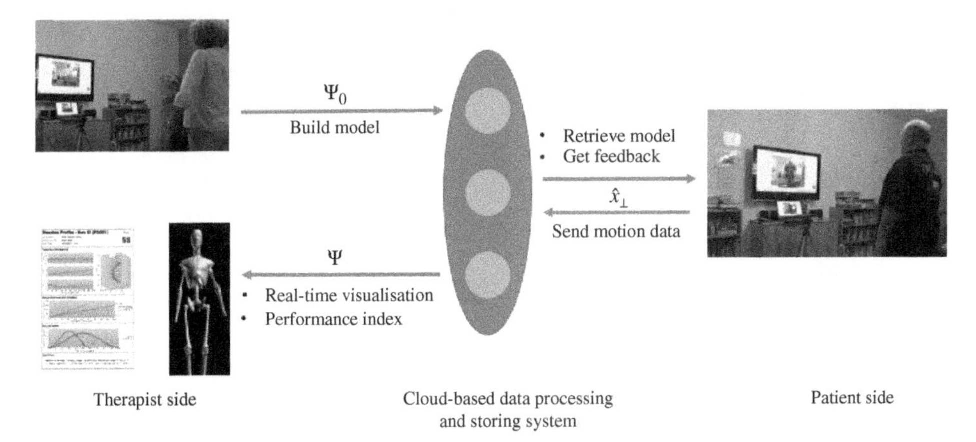

Figure 2.10 Cloud-based exercise monitoring and performance assessment with biofeedback. Source: Pubudu N. Pathirana.

movements. The resulting positioning outcomes are based on the exercise dynamics and hence the positioning needs some flexibility to cater for the nature of the exercises. For practical use, arbitrary positioning of receivers provides clinicians with greater flexibility while minimising occlusions. Robust linear filtering with missing information is more suitable for multi-Kinect© fusion for its lower RMSEs and shorter computational time. The particle filter exhibits shortcomings despite being widely used in the multi-sensor fusion arena. From both the simulation and the real-data experiment, it is obvious that increasing the number of Kinects© improves the measurement accuracy of the overall multi-Kinect© system, albeit increasing the cost as well as the complexity in setting up and managing the system. Indeed, the number of Kinects© used is a compromise between counteracting factors, such as the cost, system complexity, overall accuracy and complexity in the exercise routine. The significant improvement in increasing to two Kinects© is deemed as a notable aspect to consider for certain applications.

Application in tele-rehabilitation

An ideal good application of multi-Kinect fusion is improving the measurement and assessment accuracy in tele-rehabilitation. Based on the aforementioned theoretical assertions, a novel cloud-based remote rehabilitation exercise monitoring system with biofeedback has been implemented and is currently being deployed for relevant formal clinical trials. The conceptual architecture is depicted in Figure 2.10, where the crucial aspects are enhanced accuracy and biofeedback with performance measurements, which are intrinsic requirements in providing cost-effective remote rehabilitation without jeopardising the quality of care. Indeed, we endeavour to capture human motions more effectively using multiple Kinect© fusion to develop a more clinically effective and commercially viable system.

Let the normalised trajectory of the prescribed exercise routine captured from the clinician be denoted by $\mathscr{T}$. Then the patient exercise performance measure (Ψ) can be denoted in the mean square sense as

$$\Psi = \frac{\Psi_0 - \sum_{k=0}^{T-1} \left(\mathscr{T}(k) - \hat{x}_\perp(k)\right)^{\mathsf{T}} \left(\mathscr{T}(k) - \hat{x}_\perp(k)\right)}{\Psi_0} \tag{2.15}$$

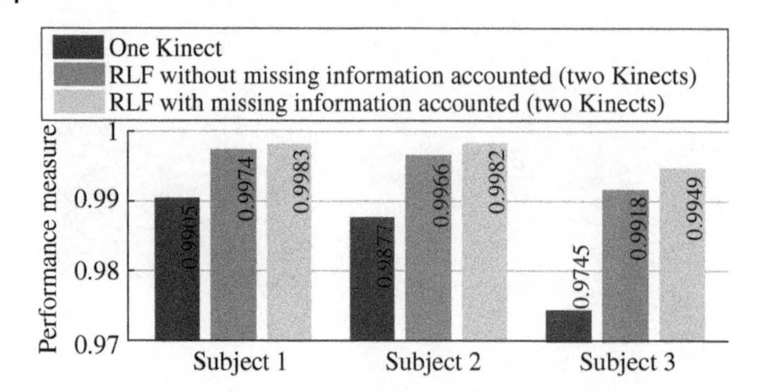

Figure 2.11 Average RMSE and performance measures of two Kinect-fusions with missing data.

where $\Psi_0 = \sum_{k=0}^{T-1} \mathcal{I}(k)^\top \mathcal{I}(k)$ and $\hat{x}_\perp$ denotes the normalised state estimate. The ideal exercise trajectory $\mathcal{I}(k)$ captured at the therapist's is used as the benchmark in the patient system at home as well as to guide the patient's exercise motion. The performance index Ψ deduced from the patient trajectory $\hat{x}_\perp$ is the quantitative measure the therapists use to monitor the improvements due to the exercises being performed remotely.

We illustrated the advantages of using multi-Kinect© fusion for tele-rehabilitation. We assumed that the motion trajectory from a professional clinician captured by the Vicon system is the model motion and the patient is provided with this trajectory to follow remotely. In our experiment, since the motions of three subjects were captured by the Vicon and multi-Kinect system simultaneously, all these subjects can attempt to gain 100% in the performance measure. From Figure 2.11, RLF with missing information accounted for gave performances closest to 100%, with 99.83%, 99.82% and 99.49%, while RLF without accounting for missing information provided 99.74%, 99.66% and 99.18%.

2.3 Encoder

2.3.1 Introduction

Actions as much as language can express ideas and communication not only in humans but also in various other species to the extent that recent work suggests there are strong links between motor and language areas of the human brain [293]. While there are a few studies that have shown the application of the language of action in various areas [62, 312, 313, 379], it is also a key to apply pervasive motion-sensing technologies to clinical kinematics, which is rarely studied.

However, there still remain challenges in developing formal descriptions and robust computational procedures for the automatic interpretation and representation of human actions. The majority of studies employed a variety of human motion encoders to recognise or decompose general movement, such as reaching, waving hands, jumping, walking and so on. A few of them investigate details in each general movement, for example, the even smaller atomic components included in these general movements that are of importance for syntactic and structural descriptions of human movements in detail, especially in clinical and rehabilitation environments, where the details of movements of body parts require a form of motion language or, at least, syntax.

A detailed example of current work is the POETICA system of Aloimonos *et al.* [8], which integrates formal generative action language with sensor (IMU) velocity and acceleration measures. Specifically, Guerra-Filho *et al.* [131] used velocity and acceleration to represent and divide human actions into atomic motions as the phonemes for a human action language (HAL). They also extracted atomic components using velocity, acceleration and joint angle – motion "phonemes" in [132]. Sant'Anna *et al.* [313] analysed gait motion in the elderly, which was performed using a motion language based on accelerometers. Moreover, joint angles have also been used pervasively to find primitives of human actions [157, 298].

It is interesting to note that in some cases motion primitives belonging to the same type of action may share the same motion trajectory with various dynamics and orientations, such as shaking hands, while sometimes, although the trajectories of two motions are similar, they are two different motions due to various dynamics, like walking and running. Therefore it is most important to separate the *shape* from the *dynamics* of human actions. Further, it is important to have motion encoders that are capable of uniquely encoding the shape and dynamics invariant to the limb's absolute position and pose.

More generally, this work is embedded in a more general three-level description framework (see Figure 2.12) involving the L_1 sensor, the L_2 limb and the L_3 complete action descriptions – and their relations – that fit within specific contexts. In the present context, this is clinical kinematics, such as rehabilitation and patient monitoring, albeit the main focus here is on the development of robust computational methods for computing unique, invariant and useful sensor "atomic states", or motion primitives for the sensor (L_1) level and how they apply to other levels. Here is a simple example. Consider the right elbow bend to straighten exercise involving the *planar* movement of the right arm, as described in PhysioAdvisor [2].

Bend and straighten your elbow as far as possible pain-free.

This implies that the arm limbs have required states as well as the sensors measuring the displacement in ways that would enable an automated valuation as to whether the exercise is performed correctly. For example, it is more or less independent of speed, smoothness, etc.

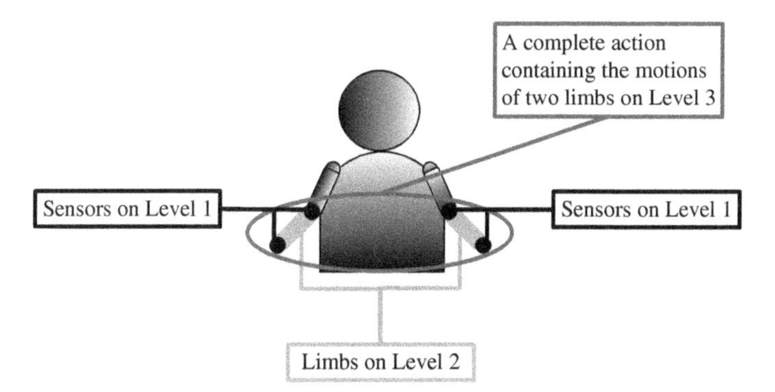

Figure 2.12 Three-level syntactic description framework for building language for human action.

2.3.2 The two-component encoder theory

In differential geometry the shape of a curve is described by the Frenet-Serret equations [112, 323], a system of linear differential equations. Let p be a point in R^3, with position vector $\mathbf{r} = (x, y, z)$, moving along a trajectory γ; the simplest form of the Frenet-Serret equations arises when γ is parametrized by arc length s, whose infinitesimal element ds is defined as

$$ds = \left(dx^2 + dy^2 + dz^2\right)^{1/2}. \tag{2.16}$$

In this parameterisation the tangent vector

$$\mathbf{u} = \frac{d\mathbf{r}(s)}{ds} = \left(\frac{dx(s)}{ds}, \frac{dy(s)}{ds}, \frac{dz(s)}{ds}\right) \tag{2.17}$$

is the unit vector and its derivative is a vector normal to $\mathbf{u}$ whose norm is the curvature κ. This result is the first Frenet-Serret equation

$$\frac{d\mathbf{u}}{ds} = \kappa(s) \cdot \mathbf{n}, \tag{2.18}$$

where $\mathbf{n}$, the unit vector, corresponds to the principal normal vector. Curvature thus measures the amount of arc rate of change of $\mathbf{u}$ and can be computed via the formula

$$\kappa(s) = \left\|\frac{d\mathbf{u}}{ds}\right\|. \tag{2.19}$$

Vectors $\mathbf{u}$ and $\mathbf{n}$ thus define a plane, the osculating plane, and the orientation of γ in space can be now completely specified once a third unit vector is given, the binormal vector, defined as

$$\mathbf{b} = \mathbf{u} \times \mathbf{n}. \tag{2.20}$$

The system of Frenet-Serret equations can now be completed:

$$\frac{d\mathbf{n}}{ds} = -\kappa(s)\mathbf{u} + \tau \cdot \mathbf{b}, \tag{2.21}$$

$$\frac{d\mathbf{b}}{ds} = -\tau(s)\mathbf{n}, \tag{2.22}$$

where the torsion τ measures how the curve winds out of the plane,

$$\tau(s) = -\frac{d\mathbf{b}}{ds} \cdot \mathbf{n}. \tag{2.23}$$

These equations are the basis for the fundamental theorem of curves, which proves that a curve parameterised by the arc length s is uniquely defined up to rigid motion and that κ, τ encode the curve in a way invariant (or blind) to rotations and translations. This is because the position vector $\mathbf{r}$ can be computed at each s by first integrating the Frenet-Serret equations to obtain the tangent, normal and binormal vectors given an initial orientation (thus losing the invariance to rotations) and next by integration of equation (2.17); the resulting vector, $\mathbf{r}$, depends on the initial position, i.e. it is not invariant to translations.

The role of k and τ can be better understood if one considers the Taylor expansion around, say, $s = 0$, retaining just the first four terms:

$$\mathbf{r}(s) = \mathbf{r}(0) + s\mathbf{u}(0) + \kappa(0)\frac{s^2}{2}\mathbf{n}(0) + \kappa(0)\tau(0)\frac{s^3}{6}\mathbf{b}(0) + O(4). \tag{2.24}$$

The first two terms provide the best linear approximation of the curve near $\mathbf{r}(0)$ and the curvature at $s = 0$ modulates the departure from linearity in the osculating plane. Finally, the torsion appears in the last and smallest term of the expansion and controls the deviation of $\mathbf{r}$ from the osculating plane. Note that, in a neighbourhood of $s = 0$, the largest variation is along $\mathbf{u}$ followed by lesser variations along $\mathbf{n}$ and $\mathbf{b}$, respectively. Suppose we sample the curve from a set of experimental data. If the neighbourhood is small enough then $\mathbf{u}$, $\mathbf{n}$, $\mathbf{b}$ are, respectively, the first, second and third principal directions of the sample distribution.

Curvature and torsion encode solely the shape of the curve and provide nothing about the speed of the point p, since the arc length parameterisation uses velocity $\mathbf{v}$ as a unit vector. In order to have a complete, invariant description of the motion encoding both the *shape and the kinematic dynamics* of point p, we need to know the speed υ at each position, noting that υ is a scalar and, as such, is invariant under rigid motion. In other words, we need the relation between time t and s, provided by the equation

$$\upsilon = \left| \frac{ds}{dt} \right| . \tag{2.25}$$

This becomes more explicit when considering the temporal parametrisation of γ.

2.3.3 Encoding methods

It is possible to compute the trajectory shape and dynamics (kinematics) using two different indexing schemes for γ: one with respect to discrete time sampling (δt) or with respect to discrete spatial/distance (δs), as follows.

Temporal index

This method assumes that each position (x, y, z) is indexed by time, resulting in a position vector [59]:

$$\mathbf{r}(t) = ((x(t), y(t), z(t))) , \tag{2.26}$$

where $t = k\delta t, k = 0, \ldots, T$ for a constant δt while the corresponding $\delta s = (\delta x^2 + \delta y^2 + \delta z^2)^{1/2}$ is variable.

From $\mathbf{r}(t)$, quantities such as $\{\kappa(t), \tau(t)\}$ and υ can be computed, via finite differences schemes, as follows. Compute first

$$\mathbf{v} = \left(\upsilon_x, \upsilon_y, \upsilon_z\right) \approx \left(\frac{\delta x}{\delta t}, \frac{\delta y}{\delta t}, \frac{\delta z}{\delta t}\right) . \tag{2.27}$$

Then

$$\mathbf{a} = \left(a_x, a_y, a_z\right) \approx \left(\frac{\delta \upsilon_x}{\delta t}, \frac{\delta \upsilon_y}{\delta t}, \frac{\delta \upsilon_z}{\delta t}\right) \tag{2.28}$$

and

$$\frac{d\mathbf{a}}{dt} \approx \left(\frac{\delta a_x}{\delta t}, \frac{\delta a_y}{\delta t}, \frac{\delta a_z}{\delta t}\right) . \tag{2.29}$$

Thus υ can be obtained from

$$\upsilon = \left[\left(\frac{dx}{dt}\right)^2 + \left(\frac{dy}{dt}\right)^2 + \left(\frac{dz}{dt}\right)^2 \right]^{1/2} \tag{2.30}$$

and $\kappa(t), \tau(t)$ from

$$\kappa(t) = \frac{\|\mathbf{v} \times \mathbf{a}\|}{v^3}, \qquad \tau = \frac{(\mathbf{v} \times \mathbf{a}) \cdot d\mathbf{a}/dt}{\|\mathbf{v} \times \mathbf{a}\|^2}, \tag{2.31}$$

respectively. However, computations of equations (2.30) and (2.31) require performing up to third-order numerical differentiation and so filtering must be performed to deal with noise.

Spatial index

Here we separate the trajectory shape from its dynamics right from the initial encoding by considering $\mathbf{r}$ as a function of s. The coordinates

$$\mathbf{r}(s) = (x(s), y(s), z(s)) \tag{2.32}$$

are sampled at values of $s = k\delta s, k = 0, \dots, S$ with δs constant, and so δt becomes a function of s and not vice versa as in the temporal indexing scheme.

In this case computations are as follows.

First

$$\mathbf{u} \approx \frac{\left(\frac{\delta x}{\delta s}, \frac{\delta y}{\delta s}, \frac{\delta z}{\delta s}\right)}{\left[\left(\frac{\delta x}{\delta s}\right)^2 + \left(\frac{\delta y}{\delta s}\right)^2 + \left(\frac{\delta z}{\delta s}\right)^2\right]^{1/2}} = \frac{(\delta x, \delta y, \delta z)}{\left[(\delta x)^2 + (\delta y)^2 + (\delta z)^2\right]^{1/2}}, \tag{2.33}$$

where the division ensures $\|\mathbf{u}\| = 1$.

Then

$$\frac{d\mathbf{u}}{ds} \approx \left(\frac{\delta u_x}{\delta s}, \frac{\delta u_y}{\delta s}, \frac{\delta u_z}{\delta s}\right) \tag{2.34}$$

and

$$\mathbf{n} \approx \frac{\left(\frac{\delta u_x}{\delta s}, \frac{\delta u_y}{\delta s}, \frac{\delta u_z}{\delta s}\right)}{\left\|\frac{d\mathbf{u}}{ds}\right\|} \tag{2.35}$$

giving

$$\mathbf{b} = \frac{\mathbf{u} \times \mathbf{n}}{\|\mathbf{u} \times \mathbf{n}\|}. \tag{2.36}$$

Note that derivatives of $\mathbf{n}$ and $\mathbf{b}$ can be computed for $\mathbf{u}$, and application of equations (2.19) and (2.23) provide values of $\kappa(s)$ and $\tau(s)$, respectively.

The speed v can be computed by making use of equation (2.25), which can then be indexed as $v(s)$; alternatively, since the interval δs is fixed for any s, we can encode the speed by using the time intervals $\delta t(s)$ variable with s.

We have shown that an action (trajectory shape and dynamics) can be uniquely encoded in a way invariant to rigid motion in the space (κ, τ, v) and we can define simple and complex single sensor actions as contours in this domain. It is sometimes convenient to encode the trajectory shape separate from dynamics in a 2D κ-τ space and plot the dynamics, v,

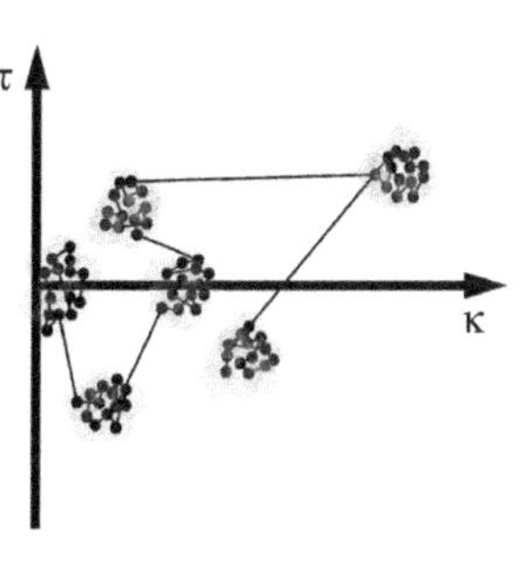

Figure 2.13 Shape model. Here the locus of points in the 2D κ-τ space corresponds to 3D trajectory positions.

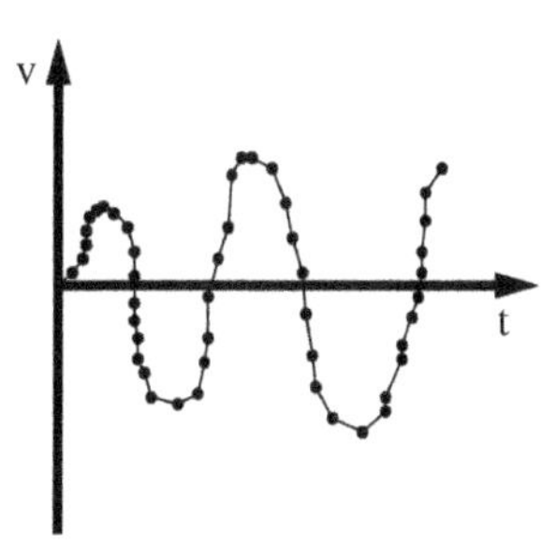

Figure 2.14 Dynamic model. Here the speed along the trajectory, υ, is indexed over time, t.

separately, as shown in Figures 2.13 and 2.14. However, there are a few *canonical trajectory shapes* worth noting, including linear ($\kappa = 0, \tau = 0$), planar ($\kappa = const, \tau = 0$), screw ($\kappa = const, \tau = const$) and spiral ($\kappa \neq 0, \tau \neq 0$) actions. Accordingly, a uniform rectilinear motion is encoded by the point $(0, 0, \upsilon_0)$, a circular uniform motion by $(\kappa_0, 0, \upsilon_0)$ and a motion with constant speed along a helix by a point $(\kappa_0, \tau_0, \upsilon_0)$. Obviously, in general, curves are mapped to curves in the κ, τ, υ space.

2.3.4 Dealing with noise

One of the major problems with collecting data from kinematic sensors is noise, particularly when the relevant information involves differential operators of different orders. Precisely because the trajectories correspond to actions, the ideal filter is one that retains important amplitudes while removing small variations due to noise. In the following we explored and compared least-square Gaussian (LS), Savitski-Golay (SG) and optimal Kalman filters (KF) with this in mind.

Least-squares Gaussian filter

The least-squares Gaussian filter is a non-linear least-squares fitting regime minimising the noise in a raw dataset and smoothing the curve by using one or a mixture of Gaussians. The fitting process is updated by minimising the sum of the squared errors between the raw dataset and the fitting equation. Here, the raw dataset is a set of paired data defined by $\{Y_k, X_k\}$, $k = 1, 2, \ldots, K$ as the indexing number [13]. The estimated values of Y_k is notated as

$$\hat{Y}_k = \sum_{n=1}^{N} \left(a_n e^{-(\frac{X_k - b_n}{c_n})^2} \right), \tag{2.37}$$

where $k = 1, 2, \ldots, K$ is the indexing number and N is the order of the filter (the number of Gaussians); a_n, b_n and c_n, $n = 1, 2, \ldots, N$ are the parameters chosen to minimise ϵ [13]:

$$\varepsilon = \sum_{k=1}^{K} (Y_k - \hat{Y}_k)^2, k = 1, 2, \ldots, K \tag{2.38}$$

and

$$\begin{aligned} \min_{a_i,b_i,c_i} \varepsilon &= \min_{a_i,b_i,c_i} \sum_{k=1}^{K} (Y_k - \hat{Y}_k)^2 \\ &= \min_{a_i,b_i,c_i} \sum_{k=1}^{K} \left[Y_k - \sum_{n=1}^{N} \left(a_i e^{(-\frac{X_k - b_i}{c_i})^2} \right) \right]^2 . \end{aligned} \tag{2.39}$$

Here, k is the indexing number and N is the order of the fitting equation.

A solution introduced in Refs [13, 253] is to compute the partial derivatives of the error and set the result to 0:

$$\nabla \varepsilon = 0, \tag{2.40}$$

where $\nabla = [\frac{\partial}{\partial a_n}, \frac{\partial}{\partial b_n}, \frac{\partial}{\partial c_n}]^T$ and $n = 1, 2, \ldots, N$ is the order. This solution can be used in both linear and non-linear least-squares methods (see Abdi [13] for more details).

However, least-squares filters have known disadvantages. Firstly, they are particularly sensitive to outliers, especially the extreme values [13] and, secondly, they are essentially smoothing filters that do not attempt to preserve steep gradients or important discontinuities so critical in kinematics. To this end, higher-order moments need to be included via filters such as the Savitzky-Golay (SG) least-squares polynomial filters.

The Savitzky-Golay filter

The SG filter is designed to preserve such higher-order moments by approximating the underlying function using a best-fitting polynomial moving window. Since the process uses least-squares fitting, it involves matrix inversion to derive the coefficients of a fitted polynomial that form a convolution kernel defined by their sampling in (window size) and polynomial order [317]. The result is a set of least-squares polynomial filters that can be applied to each of the $x(t), y(t), z(t)$ recordings of the form:

$$X(t) = SG(n, m) * x(t), \tag{2.41}$$

$$Y(t) = SG(n, m) * y(t), \tag{2.42}$$

$$Z(t) = SG(n, m) * z(t), \tag{2.43}$$

where $*$ denotes convolution, n and m refer to the window size and order of the polynomial, respectively.

The most important benefit of such polynomial approximations is that higher-order derivatives can be determined algebraically from the derived polynomial coefficients. The norms of velocity, $V(t)$, and acceleration, $A(t)$, can then be computed using these coefficients for each position parameter $(X(t), Y(t), Z(t))$, as

$$V(t) = \sqrt{X_t(t)^2 + Y_t(t)^2 + Z_t(t)^2}, \tag{2.44}$$

$$A(t) = \sqrt{X_{tt}(t)^2 + Y_{tt}(t)^2 + Z_{tt}(t)^2}. \tag{2.45}$$

V and A provide an invariant (to absolute pose and position) description of the dynamics of a curve, but not its shape. Again, the reason for separating dynamics from shape features is that a known action can have the same shape but different dynamics, even when repeated by the same person.

An experiment was done (refer to Section 2.3.7) to compare the performance of SG and LS filters to show why and how the SG filter outperforms the other one.

Kalman filter

The third type of filter examined was the class of Kalman filters (KF), varying from the normal linear to extended and unscented versions to accommodate for non-linearities in the data. Although the extended KF linearises inherent non-linearities and the unscented one allows for more robust sampling and estimation [196], we have found that the standard (optimal) KF performed as well as the others so we have compared this with the previous two filters. The normal KF smoothes the data by adapting the Kalman gain online in order to minimise the error in fitting a single Gaussian error model. The detailed information about the forward process of a Kalman filter can be found in Section 6.2.1.

A series of experiments (stated in Section 2.3.7) was then performed to determine the best combination of indexing and noise-filtering schemes for computing shape and kinematics for a given sensor.

2.3.5 Complex motion decomposition using switching continuous hidden Markov models

To syntactically describe complex motions, it is of importance to segment and classify them into smaller atomic units. In this work, switching continuous hidden Markov models [294] will be investigated to perform motion segmentation and classification simultaneously that use atomic motion components to identify complex actions.

Building model for atomic motions

Since a hidden Markov model (HMM) involves two major sets of variables, hidden states and observations, it is essential to determine these variables. For states, because they are hidden, it is more critical to know the number of states rather than what they are, while for observations, a Gaussian mixture model (GMM) is used, in the continuous HMM case, to model the observations where each observation symbol corresponds to a Gaussian component. The GMM is the weighted sum of M component Gaussian densities, defined by

$$p(\mathbf{x}|\lambda) = \sum_{i=1}^{M} \omega_i g(\mathbf{x}|\mu_i, \mathbf{\Sigma}_i), \tag{2.46}$$

where ω_i, μ_i and $\mathbf{\Sigma}_i$ correspond to the weight, mean and covariance of the ith Gaussian model and $\mathbf{x}$ is a multi-dimensional vector. The expectation-maximisation (EM) algorithm is then used to estimate these GMM parameters ω, μ and $\mathbf{\Sigma}$.

As is typical of an HMM estimation, each HMM atomic model uses, again, the EM algorithm with randomly set initial guesses and updated using training data till an optimal (MAP) solution is obtained [361], thereby establishing an HMM notated as $\lambda = \{A, B, \pi\}$. Here, A is a state transition probability distribution with $A = \{a_{s_i s_j}\}, i, j \in [1, N]$ (where s_i is the ith state at time t, s_j is the jth state at time $t+1$ and N is the total number of states),

B is the observation symbol probability distribution with $B = \{b_{s_i}^{o_k}\}, i \in [1, N], k \in [1, M]$ (where s_i is the ith state at time t and o_k is kth observation symbol) and π is initial state distribution with $\pi = \{\pi_i\}, i \in [1, N]$. This version of the EM model is implemented efficiently using the forward-backward algorithm [33].

Complex motion decomposition

In the previous section, the approach for building an HMM for one atomic motion is introduced. Since a complex motion may contain Q types of atomic motion, it is essential to build an HMM for each type of atomic motion; the collection of these HMMs is notated as $\Lambda = \{\lambda_1, \lambda_2, \ldots, \lambda_Q\}$.

To decompose this complex motion based on its shape, it should be represented by $\{\kappa_t, \tau_t\}, t = 1, 2, \ldots, T$. Decomposing the motion is actually solving the following problem using the forward algorithm, defined by

$$\underset{q}{\operatorname{argmax}}\, p(\kappa_t, \tau_t | \lambda q), \tag{2.47}$$

where $t = 1, 2, \ldots, T$ and $q = 1, 2, \ldots, Q$. The change of q means the change of atomic motions, thereby segmenting and classifying the complex motion into pre-defined atomic motions.

2.3.6 Canonical actions and the action alphabet

Though the above derivations describe two components (trajectory shape and dynamics) of complex human motions, approaches for symbolic representation or the classification of these components are not provided, which will be more practical for those who need to monitor, analyse or describe these motions [59]. For example, in situations where it is important to describe how a patient should perform specific rehabilitation exercises or even how a specific limb is not performing normally, it is necessary to have a description of the action that fits with current understanding of motor programmes and processes underlying normal action productions. In fact, the challenge to construct generative languages for actions has become a focus in recent years in both human kinematics and robotics via the POETICA project [276].

To this end we have developed a "point screw decomposition" to generate basic "atomic" action descriptions, at least for the sensor movements. This method is related, but is not identical, to screw kinematics for rigid bodies as developed over the past two centuries for mechanical motions encoding in terms of the rotations and translations of rigid bodies about a given screw axis (see, for example, McCarthy [232] for a treatment in spherical kinematics terms). It is not aimed at the descriptions of actions at the next level of specific limb movement types and their relations in kinesiology but it is intended to underpin the inference of such actions.

It must first be noted that the curvature and torsion of a helical (screw) motion is constant [96]. For this reason, then, any point at time t in $\kappa\tau$ space $[\kappa(t), \tau(t)]$ can be interpreted as corresponding to a screw action such that, if the motion remains at a "fixed point" (does not move) in the $\kappa\tau$ plane then, during this time period, its 3D trajectory corresponds to a *constant point screw* whose shape is defined by the κ and τ values at that point. Specifically, for a helix defined by

$$\mathbf{r}(t) = (a\cos(t), a\sin(t), bt), \tag{2.48}$$

it is easy to show that

$$\kappa = \frac{a}{a^2 + b^2}, \qquad \tau = \frac{b}{a^2 + b^2}. \tag{2.49}$$

Similarly, the radii of first (curvature-type) and second (torsion-type) curvatures, a, b, are defined by

$$a = \frac{\kappa}{\kappa^2 + \tau^2}, \qquad b = \frac{\tau}{\kappa^2 + \tau^2}. \tag{2.50}$$

These latter relations in equation (2.50) illustrate the relationship between κ, τ values and the types of point screw actions (such as "left-handed" and "right-handed") [59], which can be used to estimate an atomic motion. Equally, one single point screw action or helix can be utilised to approximate temporally contiguous points that are close in a $\kappa\tau$ space or curve. *Contiguity in local shape and time* is the basis of our method to encoding such motions. Consequently, in the following section, our aim is [59]:

- to describe atomic motions with a sequence of point screws (specific κ, τ values) "states";
- to determine how to relate the types of point screws and their dynamics, given the variabilities that occur in the execution of such actions by humans;
- to explore the approaches to compile efficient criteria for the encoding, prediction and recognition of complex single sensor actions via their point screw decompositions.

Again, it must be emphasised that we have chosen to encode sensor motions by two sets of invariant descriptors: $\kappa\tau$ and speed (v) signatures.

Nevertheless, as a result of this formulation we propose a few basic (canonical) motion types, as shown in Table 2.3. As for direction each such motion can be, for planar: forward/backward; planar: clockwise/anticlockwise, helical: left-handed/right-handed screw actions.

2.3.7 Experiments and results

In this section, four experiments were done to compare filters and indexing schemes, as well as to illustrate the application of a two-component encoding model. To quantise the

Table 2.3 Sensor motion types.

Type	Curvature	Torsion
Linear	0	0
Curved planar	> 0	0
Non-planar screw/helix	> 0	$\neq 0$

Table 2.4 Comparison between SG and LS.

Filter	SG	LS
Order	2	2
Window size	27	-
STD (κ) (m^{-2})	**9.3953**	33.9046
STD (τ) (m^{-2})	**62.4270**	333.0902
Corr/MSE (m^2)	**0.9995/0.0015**	0.8979/0.0153
Time (s)	**0.0163**	1.1324

performance, standard deviation (STD), correlation coefficient (Corr) and mean square error (MSE) methods were used in the following subsections. First of all, as shown in Section 2.3.6, curvatures and torsions for linear, planar and helical motion should be constants, and standard deviations of the noise (outliers) in curvatures and torsions were employed to show how computed curvatures and torsions related to expected values. Moreover, correlation coefficient and mean square error methods were used in the first experiment (refer to the comparison between the Savitzky-Golay filter and the least-squares Gaussian filter) to show the distortion between filtered trajectories and the raw one. Moreover, in these experiments, all the parameters, such as the spatial intervals, orders of the least-squares Gaussian filter, orders and window sizes of the Savitzky-Golay filter and so on, were selected by minimising the standard deviations of curvatures and torsions of related trajectories.

In terms of the abbreviations, NF means that no filter was applied (the results were computed based on raw data), while SG, LS and KF are for the Savitzky-Golay filter, the least-squares Gaussian filter and the Kalman filter.

Comparison between the Savitzky-Golay filter and the least-squares Gaussian filter

The first experiment was done to illustrate how the SG and LS smooth the trajectory of a helical motion collected from a Microsoft Kinect with a sampling rate of 30 frames per second (FPS), which was composed of the 3D positions of the right wrist with a total of 118 samples (indexed by a temporal scheme) performed by a healthy subject.

From Table 2.4, it can be seen that the SG filter results in a smaller standard deviation of curvature and torsion with less distortion (higher correlation coefficient and lower MSE with respect to the raw trajectory) and a shorter computing time. By comparison, the LS filter yields curvature and torsion with standard deviations much higher than SG with lower values of the correlation coefficient and higher MSE than those of SG; moreover, LS has a longer computational time. All in all, when suitable parameters are selected, the SG filter appears to outperform LS in generating curvature and torsion with minimised errors (standard deviations).

The computability of the two-component model

In this experiment, these two components were computed for a variety of actions including two linear and two planar motions (with 121, 121, 215 and 277 samples indexed by

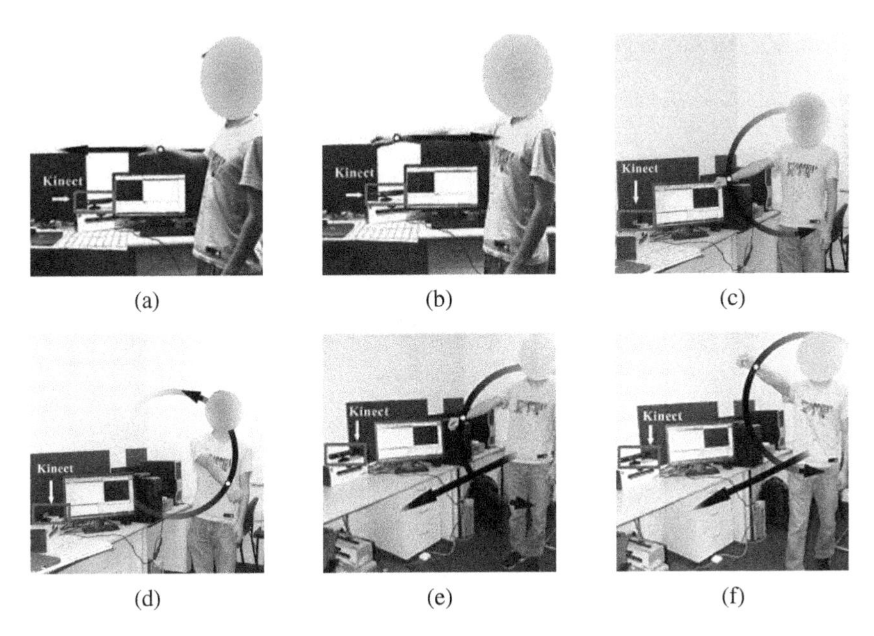

Figure 2.15 Three motions used in experiments of this work. (a, b) Linear motion; (c, d) planar motion; (e, f) helical motion. Source: Saiyi Li.

time) of the right wrist by using the Kinect®, sampling 3D positions of joints at 30 Hz. Figure 2.15 shows how these motions were captured, as it was with all the experiments in this section.

In Table 2.5 the first column shows the 3D action trajectories while the second shape column is divided into two, corresponding to the κ, τ values as a function of time. The third dynamics column shows the speed υ over time. The rows refer to different motions and only trajectories with zero torsion have been considered. Here spatial indexing and SG filtering have been used in the calculations.

For linear motions, the shape column reflects the shape of the motion trajectories independent of their dynamics. In these cases curvature and torsion for linear motion should be 0 as there was no rotation or twist. Graphs of curvature and torsion as functions of time show this to be the case for most of the time, with points located away from 0 occurring mainly at the beginning and at the end of the motion, that is, at points of temporal discontinuity. The shape computation results were consistent with this, with points mostly clustered around the origin of the κ, τ plane. In terms of the dynamic computation results, the speed was used, rather than velocity vectors because, as noted in Section 2.3.3, they can be recovered from κ, τ, i.e. the shape component of this model. Graphs of speed versus time show two different kinds of motion. In the first upward motion an initial increase in speed was followed by an interval in which υ varied a little around a constant value and decreased when the motion stopped. In the second downward motion υ increased linearly, uniformly accelerating up to the end and then rapidly decreasing to zero. Thus, although the motions had different locations and directions, they had very similar shapes (in the sense of κ, τ values) while exhibiting different dynamics. The third and fourth rows in Table 2.5 show results for two planar motions forming closed curves with different directions (tangent vectors) and

Table 2.5 Shape and dynamics for four motions. In each left figure, the colour changes from black to grey with time.

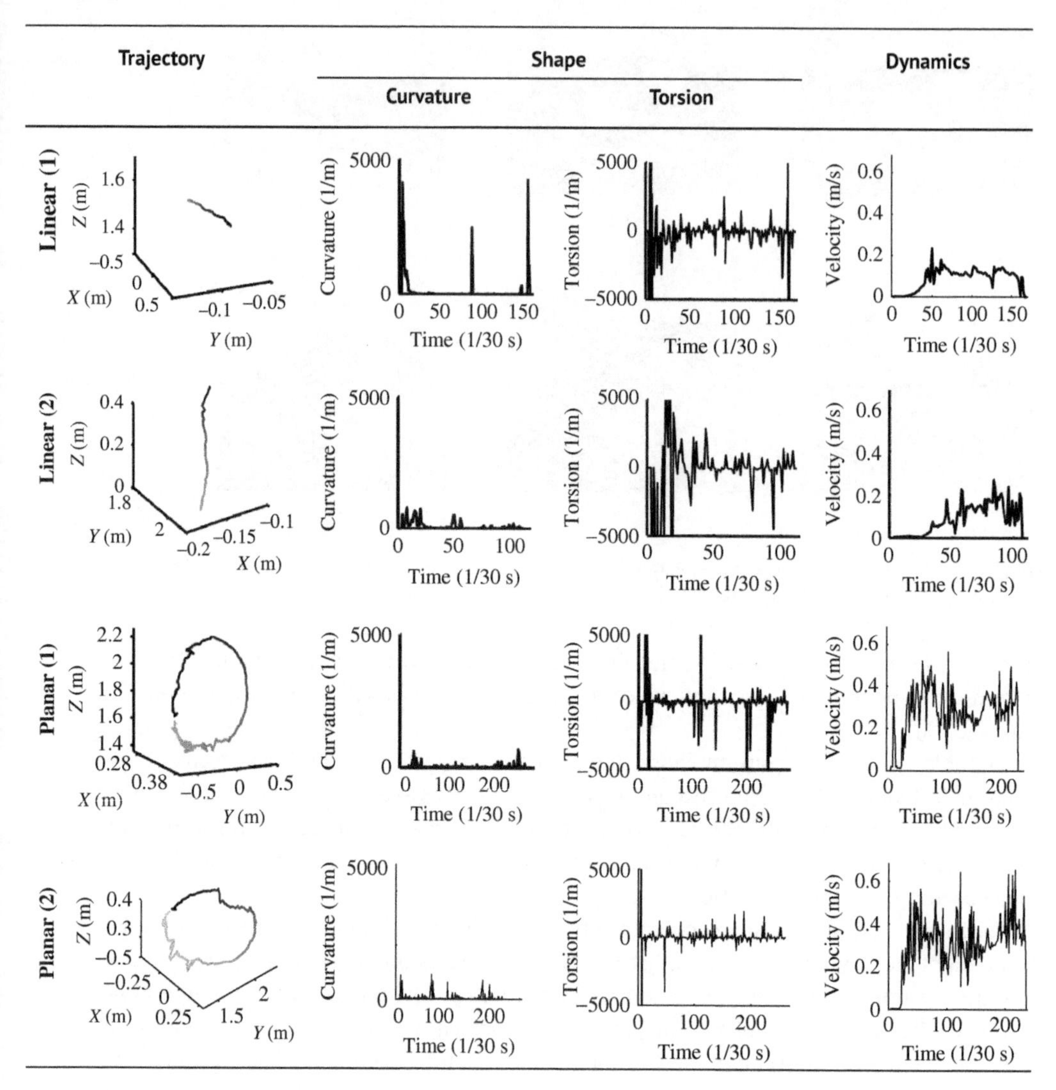

locations. In both cases the curvature is relatively small for large portions of the paths while the torsion should be zero since there was no twist in these motions.

The results were affected by the fact that curvature and torsion are very sensitive to noise, so that values κ show a spread in the range from 0 to 200 m^{-1}. Although values of τ are greater than those of κ, they fluctuate around 0 m^{-1}, which generally follows the expectation shown in Table 2.3.

However, most of the significant outliers occurred at the beginning and end of the motion or at points where sharp cusps occurred in the trajectory as, for instance, where there were two peaks in the bottom part of the left planar motion 1 or in the left-hand side of motion 2.

Finally, computed υ values showed very similar dynamics (independently of their shape); here, not surprisingly, υ values were more scattered than in the linear case due to the inaccuracy of the Kinect and the difficulty of the subject accurately outlining the specified action. In spite of this, this first experiment demonstrates that the system can encode the shape and dynamics of an action reliably, albeit somewhat noisily, at the beginning or end of an action, as expected from the nature of the filtering computations involved.

In addition, a simulation was implemented to illustrate the independence between a shape model and a dynamic model. In this part, two helical trajectories with the same shape but different dynamics were simulated with Matlab®. For the first trajectory, the speed was constant, while in the second trajectory, the speed varied. In the first, second, third and fourth quarters of the second trajectory, the speeds were 10, 30, five and 40 times as much as those in the first trajectory, respectively. These two trajectories are shown as Figure 2.16.

The two trajectories are the same and then they share the same curvature and torsion (shape model), as shown in Figure 2.17. However, the dynamics of these two trajectories were different, as can be proved by investigating their dynamic models, as shown in Figure 2.18.

From Figure 2.18(a), we can see that the average speed of the trajectory is around 0.1 m/s. In Figure 2.18(b), the speed for the first quarter (from 1 s to 50 s) is approximately 1 m/s,

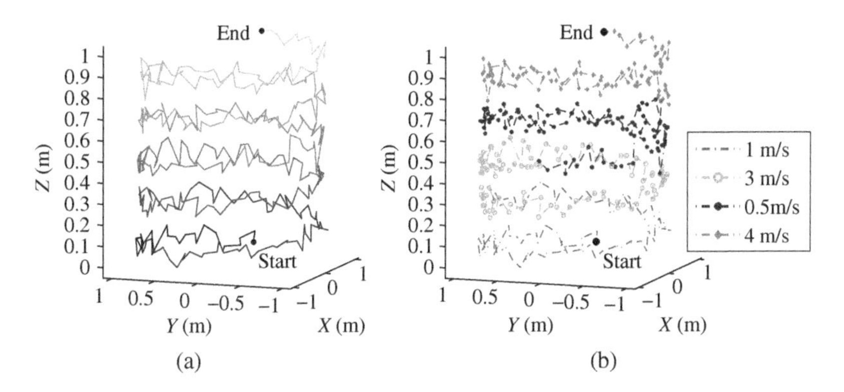

Figure 2.16 Two helical trajectories with the same orientation and shape, but two different dynamics. The colour of the trajectory in (a) with constant speed (roughly 0.1 m/s) changes from black to grey with time, and the different line styles and colours of the trajectory in (b) indicate the change of velocities. The legend on the right side of (b) shows the approximate speed of different parts of the trajectory.

Figure 2.17 Shape models: (a) for the trajectory in Figure 2.16(a) and (b) for the trajectory in Figure 2.16(b).

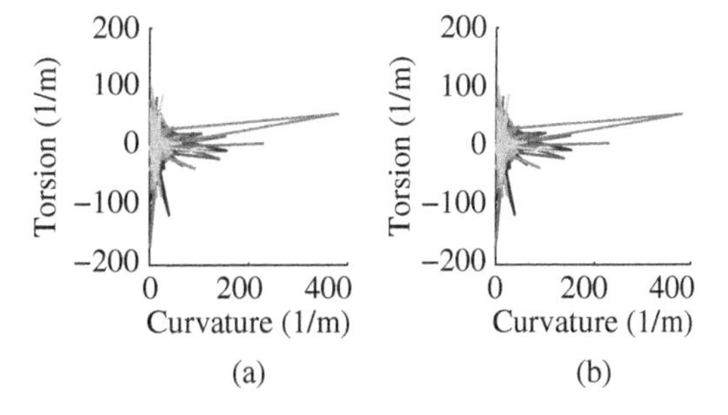

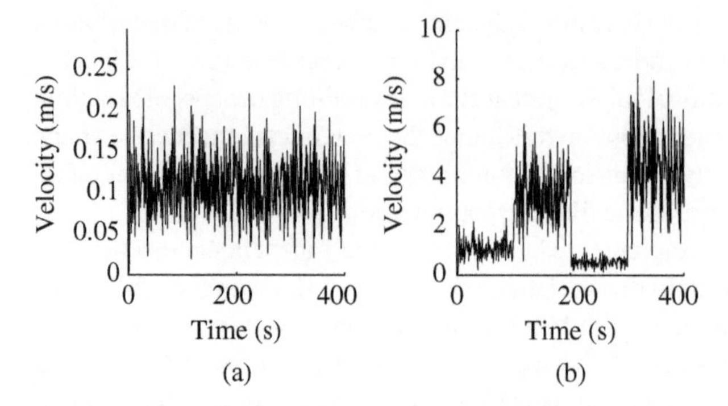

Figure 2.18 Dynamic models: (a) for the trajectory in Figure 2.16(a), (b) for the trajectory in Figure 2.16(b).

which is 10 times as much as that of the average of the first trajectory. Similarly, it is clear that in the second quarter (50–100 s), the third quarter (100–150 s) and the fourth quarter (150–200 s), the speeds are 3, 0.5, and 4 m/s separately. This result corresponds to the expectation in the beginning of the simulation.

Comparisons between the combinations of various indexing schemes and filtering approaches

In this experiment we have carefully examined the performance of the indexing and filtering methods for the linear and planar actions mentioned in the previous experiment, again with the purpose of determining the best way to minimise noise. The indexing schemes (temporal and spatial) were used in conjunction with three filtering methods (LS, SG, KF) where, in each case, optimal parameters were used for comparison purposes. At the same time, sets of data generated from trajectories without applying any filter (NF) were used for comparison. Here, we select one linear and one planar motion for a demonstration of the results.

Table 2.6 is a summary of the parameters used by filters to process trajectories with temporal and spatial indexing schemes, respectively. As can be seen, the orders and window sizes for the Savitzky-Golay smoothing filter change with time for two motions indexed with the temporal indexing scheme, while keeping almost the same when the spatial indexing scheme was applied. Moreover, the reason order 8 was used for the LS filter is that, although the lower orders gave a smoother trajectory, it is distorted, like that in Table 2.4. In terms of the KF the optimal iteration number varied considerably.

Table 2.7 illustrates the standard deviations of curvatures and torsions of two trajectories, namely linear (L) and planar (P), indexed with the temporal and spatial indexing scheme and denoised with three filters. By comparing the performances between the temporal and spatial indexing schemes (top four rows and bottom four rows), it can be seen that the spatial indexing scheme outperforms the temporal one by generating curvatures and torsions with smaller standard deviations. Moreover, SG outperforms LS and KF since the trajectory in each row filtered by SG has the smallest standard deviations of curvatures and torsions. It is also noteworthy that some numbers, such as the torsion for linear motion without applying filters, are extremely large, which is caused by some outliers in these trajectories.

Table 2.6 Parameters used by filters to process trajectories with the temporal and spatial indexing scheme. Here, L and P are for the linear and planar motions, respectively (the same in the rest tables).

		Spatial interval	SG		LS	KF
			Order	Window size	Order	Iterate number
Temporal	**L**	-	6	121	8	101
	P	-	5	181	8	576
Spatial	**L**	0.01 m	5	45	8	118
	P	0.05 m	5	59	8	566

Table 2.7 The comparison of curvatures and torsions for two trajectories indexed by temporal and spatial indexing schemes and filtered by SG, LS and KF. Here C is for curvature and T is for torsion.

			NF	SG	LS	KF
Temporal	**C**	**L**	71.56	**25.12**	101.79	161.83
		P	32.49	**4.98**	6.12	7.43
	T	**L**	1.16E+7	**7.75**	42.77	63.09
		P	70621.72	**2.16**	40.76	2.22
Spatial	**C**	**L**	4.06	**2.64**	3.84	4.03
		P	2.88	**0.87**	1.14	2.74
	T	**L**	7.49	**1.85**	5.41	7.45
		P	3.23	**0.27**	1.14	2.74

Motion decomposition using switching continuous hidden Markov model (SCHMM)

Based on the conclusion of Experiment 1, it is clear that our two-component model can be used for motion decomposition. With the help of the result of the second experiment, the noise in raw data collected by a typically noisy sensor like the Kinect® can be significantly reduced. Therefore, in the third experiment we explored how the two-component model can be used to segment action data into such components using a standard hidden Markov model (HMM) formulation [294]. In this experiment, a continuous Gaussian hidden Markov model (CHMM) [294] was implemented in Matlab® based on a Bayes Net Toolbox [248], where three HMMs were constructed for linear, planar and helical motions, respectively, and the segmentation into these three motion classes was determined by MAP criteria, that is, at any given time, the HMM was selected that produced the highest Viterbi score [294]. For testing, four complex motions consisting of various combinations of linear, planar and helical motions were analysed – all based purely on their curvature and torsion values and having different dynamics.

Three linear motions ([89, 94, 87] samples indexed by a spatial indexing scheme with a spatial interval of 0.015 m), three planar motions ([113, 122, 116] samples indexed by a spatial indexing scheme with a spatial interval of 0.04 m) and two helical motions ([142, 147] samples indexed by a spatial indexing scheme with a spatial interval of 0.04 m) were selected as training datasets to construct three HMMs for linear, planar and helical motions respectively. Already these can define quite complex and different actions depending on how the motion components are sequenced. Six Gaussian mixtures, clusters, were used as six observation symbols in each HMM along with six (latent) states. These parameters resulted in excellent recognition rates compared with other values that were explored.

After building HMMs for trajectories of various atomic motions, four complex motions combining these atomic motions were used to illustrate the performance of trajectory decomposition with a two-component encoding model, spatial indexing scheme and SG filter. Before decomposing these complex motions automatically with the proposed approach, we manually segmented them into different atomic motions and counted the samples in each segment as a reference for computing the recognition rate. The trajectory of these four complex motions and the examples of decomposition graphs are shown as follows.

Table 2.8 illustrates the performance of the SCHMM with shape models as features. The first complex motion consisted of a linear motion and a planar motion and is indexed by a spatial indexing scheme with a spatial interval of 0.007 m. From the classification/segmentation results, it is easy to see that the first 0.14 m of this motion is linear, while the remainder corresponds to a planar motion. For motion 2, the first 0.154 m was classified as planar, followed by planar motion of around 0.161 m, linear motion of about 0.35 m and planar motion of approximately 0.14 m. In this diagram, the change from linear to planar motion after about 0.8 m is caused by the small planar motion included in the linear motion. After all, it is very hard for a human being to do an exact linear motion. Moreover, the third complex motion is compounded with one helical motion and a linear motion. This sequence can be easily seen from the classification diagram, while the unexpected change in the last part is caused by the same reason as the previous motion. The last motion was combined by a helical, a planar and a linear motion, which could be reflected from the classification graph.

At the end of the experiment, 70 different complex motions combining linear, planar and helical motions conducted by 10 healthy people in a lab environment were tested in order to gather the classification rate. Each person not only performed four motions listed in Table 2.8, but also did three more trajectories comprising linear, planar and helical atomic motions while different from four pre-defined motions. The average classification rate for linear motions was around 89.2%, for planar motions it was roughly 90.6%, while for helical motions it was approximately 92.2%. There are two major factors that may lead to this result. First of all, the noise coming from testing subjects and Kinect®, which is unlikely to be eliminated completely by filters, may negatively influence the classification rate and, next, it is very hard for a person to make an exact linear, planar or helical motion. Therefore, it is reasonable to expect some parts of linear motion to be classified as a planar part, and analogously, parts of planar trajectory to be classed as helical, especially when one motion changes to another, as well as at the beginning and end of complex motions, such as cases 2, 3 and 4 in Table 2.8.

Table 2.8 Trajectories and decomposition of complex motions. In each top figure, the colour changes from black to grey with time in the trajectories.

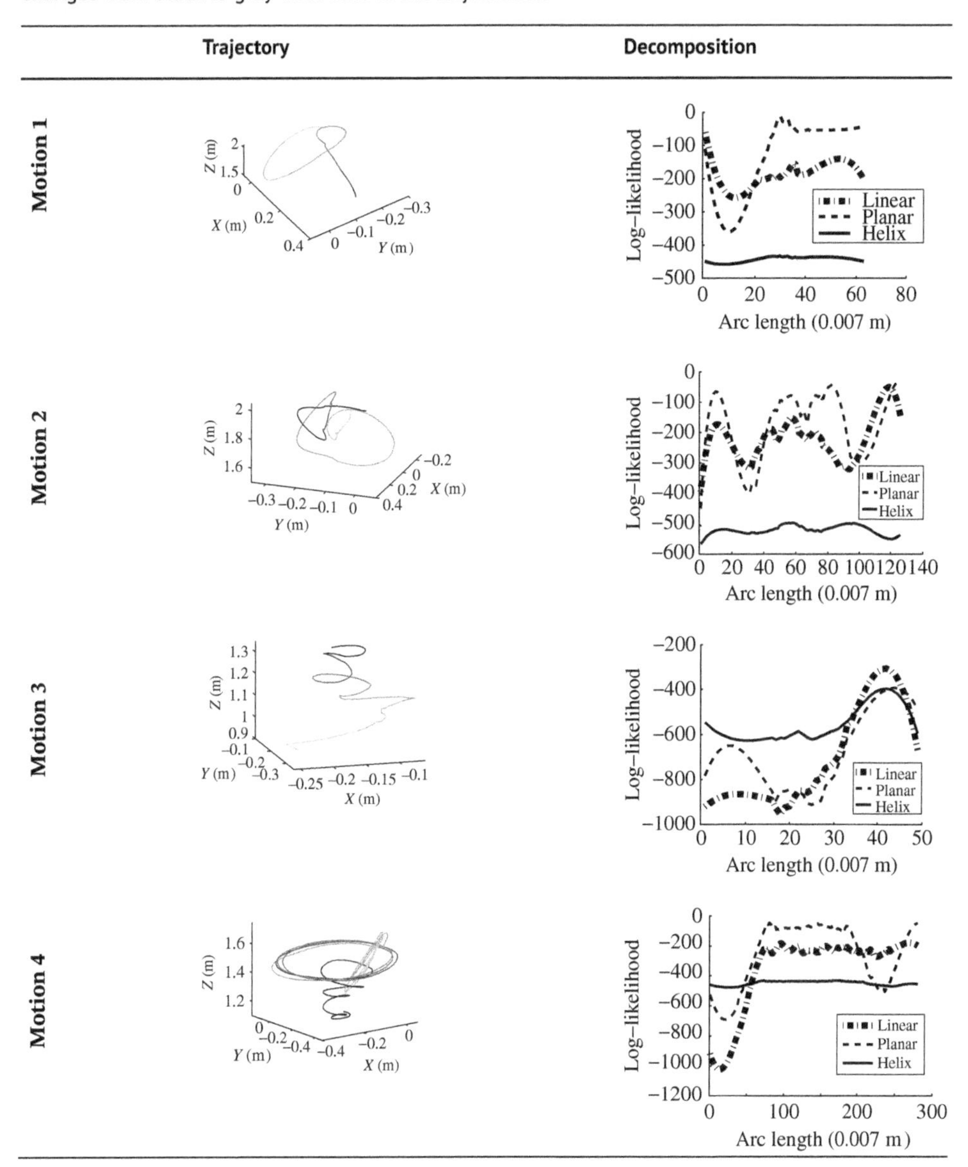

2.4 ADL Kinematic Performance Evaluation

2.4.1 Introduction

In recent decades, with the advancements in tele-rehabilitation and associated motion capture technologies, an increasing number of research and development activities are focusing on the development of automated quantitative measures of patient performance in activities of daily living (ADLs) [136, 262, 290]. Due to the important role played by the upper extremity in ADLa [94, 99], an automated approach for measuring and assessing the ability of an upper extremity to perform certain tasks is vital for tele-rehabilitation systems to deliver their full potential.

Although questionnaire-based tools have been utilised pervasively by clinicians, they are not suitable for the tele-rehabilitation environment. One reason is that the use of the majority of these tools requires clinicians, who are usually absent in the tele-rehabilitation sessions. Furthermore, some self-report questionnaires may lead to biased results. Therefore, in tele-rehabilitation, it is critical to develop an automated approach to objectively assess the ability of patients in order to assist therapists to make further clinical decisions.

In this section, we conducted a preliminary investigation of the feasibility of utilising an automatic approach to assess the ability of patients suffering from dyskinesia to perform an upper extremity reaching task in their daily living. This is assessed by measuring the smoothness of motion trajectories and the duration to finish the task. As pointed out by Daneault *et al.* [85], dyskinesia is one of the factors that adversely influence voluntary movement since some involuntary movements would be performed. Therefore, they proposed that dyskinesia in Parkinson's disease could be seen as a factor in the signal-to-noise ratio (SNR) equation with voluntary movements as the numerator (motion input) and dyskinesia as one element in the denominator. Furthermore, to assess the severity of dyskinesia, the amplitude of involuntary movements is one of the important elements that has been utilised in Abnormal Involuntary Movement Scale (AIMS) [85, 192]. In addition, dyskinesia may be associated with some degree of jerk in the extremities [321, 332]. Due to inaccurate motion trajectories, patients with dyskinesia are more likely to reduce their speed and take a longer time to finish a task in comparison to healthy subjects [109]. Therefore, it is reasonable to infer that by looking at the sub-movements and jerks (smoothness) in motion trajectories, as well as the motion duration, we are able to evaluate the ability of the subject to perform reaching tasks in daily life from an action kinematic standpoint. In this work, we used sub-movements and jerks to infer involuntary movements with large and small amplitudes, which are defined in Table 2.9 for the experiments.

In line with our work, a number of studies have been conducted to evaluate automated performance measurements or kinematics relating to upper extremities [91] (some commonly utilised features are given in Figure 2.19). One of the most obvious critical factors negatively impacting on the quality of upper extremity movements is the smoothness of

Table 2.9 The definition of three kinematic severity levels of involuntary movements and jerks, as well as their corresponding abilities in performing reaching tasks in daily living.

Level	**1**	**2**	**3**
Number of sub-movements	0	≤ 3	> 3
Amplitude of sub-movements	0	≤ 0.3 m	> 0.3 m
Number of jerks	0	≤ 3	> 3
Amplitude of jerks	0	≤ 0.03 m	> 0.03 m
Duration	≤ 5 s	5 s ~ 10 s	> 10 s
Ability in performing reaching task	High	Medium	Low

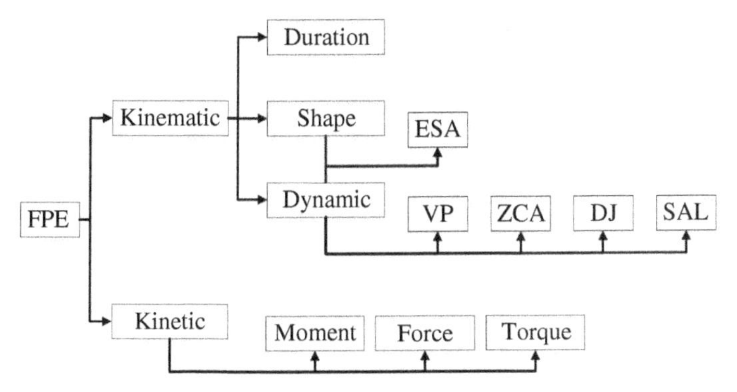

Figure 2.19 Examples of commonly used techniques with features considered for ADL performance measurement. Feature-based performance evaluation (FPE) is primarily based on kinematic or kinetic measurements. Dynamic measurements such as number of velocity peaks [25], number of zero-crossing tangential accelerations [355], dimensionless jerk metrics [304] and spectral arc-length [31] are used in techniques such as VP, ZCA, DJ and SAL, respectively, in the literature. Our proposed entropy of shape and instantaneous acceleration (ESA) introduced in this section uses both shape and kinematic-based measurements. Kinetic-related features, such as moment [160], force [115] and torque [281] have also been investigated. Sources: Alt Murphy *et al.* [25]; Trombly and Wu [355]; Rohrer *et al.* [304]; Balasubramanian *et al.* [31]; Jevsevar *et al.* [160]; Pentland and Twomey [115]; Frossard *et al.* [281].

motion trajectories. Zariffa *et al.* [395] considered directional changes, mean velocity, ratio of mean and maximum velocity and mean jerk to measure the smoothness of trajectories in patients' upper limb movements collected from a robotic rehabilitation device. In addition, Rohrer *et al.* [304] compared five features, namely jerk, speed, mean arrest period ratio, number of peaks in speed and "tent" metric, to evaluate the smoothness of an arm motion trajectory performed by stroke patients. Moreover, Lum *et al.* [216] counted the number of times the tangential acceleration of a hand passed zero to measure the smoothness of upper extremity movements in stroke patients. Apart from smoothness in motion trajectories, the duration of a specific task is also important for an upper extremity performance evaluation. Murphy *et al.* [24] took the duration of drinking into account to analyse the kinematic aspect of drinking with a cohort of healthy subjects. Similarly, the duration was also considered as a factor of an upper extremity movement assessment in two studies [31, 368]. In addition, Balasubramanian *et al.* [31] highlighted the disadvantages in some existing approaches and proposed utilising the spectral arc-length metric of the movement speed profile's Fourier magnitude spectrum to evaluate the smoothness of the movements.

2.4.2 Methodology

Severity levels definition

In order to perform computer simulations as well as to obtain data from healthy subjects mimicking the underlying involuntary movements, it is important to precisely specify the severity levels of involuntary movements in a kinematic standpoint. Since the frequency and amplitude of involuntary movements, in addition to the duration of the specific task, are important factors in assessments [192, 194, 332], we defined three severity levels of involuntary movements by assessing the kinematic performances of the upper extremity in a reaching task, as indicated in Table 2.9. As this work is a preliminary study to investigate the feasibility of using kinematic measurements to evaluate the severity of involuntary movements, a more focused exercise to describe each level and a larger set of levels can be used [192]. Nevertheless, our proposed approach can simply be used with an improved distinction of severity levels and hence we consider this aspect as not the primary focus of this work.

Feature extraction

In order to quantitatively evaluate the ability to perform the experimental task, various features need to be extracted from a raw 3D trajectory ($\gamma(k) = [x(k), y(k), z(k)]^{\top}$, where $x(k)$, $y(k)$ and $z(k)$ are the joint position in a Cartesian coordinate frame at time $k \in [1, 2, \cdots, K]$ with a temporal interval of δk) captured from the Kinect ($\delta k = 30$ ms). In this work, we considered the concept of approximate entropy of motion trajectory and the duration associated with the motion. The approximate entropy of the trajectory is related to the shape and the dynamics (i.e. instantaneous acceleration) computed from the trajectory. The feature extraction process is depicted as follows:

- *Shape model*
  Apart from the dynamics in a trajectory, its shape is also taken into account since the involuntary movements are usually associated with randomly moving joint positions, which are represented as unknown uncertainties in the shape. Therefore, it is important to extract the shape of the trajectory for evaluation [209]. The method used to model the shape is introduced in Section 2.3.3.
  Since the normal approach to compute numerical differentiation is very sensitive to noise, we utilised the approach introduced in Lewiner *et al.* [207] to estimate V, A and J, which will be used to calculate the shape model.
- *Instantaneous acceleration*
  The instantaneous acceleration is the magnitude of the acceleration throughout the trajectory computed as

$$A_k^i = \|A_k\|, \tag{2.51}$$

  where $k = 1, 2, \ldots, K$ and A_k was estimated in the previous step.
- *Approximate entropy*
  Since the uncertainty in the features, including acceleration and shape model, is implicitly captured to determine how smoothly or not a person performs a task, approximate entropy [289] can be computed based on the previous features.
  To compute the approximate entropy of a variable, i.e. instantaneous acceleration, we denote it as

$$A = [a_1, a_2, \ldots, a_K] \tag{2.52}$$

  for a trajectory with the length K. By defining a constant $m \in \mathbb{Z}$ for the length of the captured sequence of data, the vector format is given as

$$B(i) = [a_1, a_2, \ldots, a_{i+m-1}], \tag{2.53}$$

  where $i = 1, 2, \ldots, L - m + 1$.

A given constant $r > 0$ indicates the filtering level and for each i, $C_i^m(r) \in \mathbf{R}$ is calculated by finding the number of a_js that satisfies the condition $d[a_i, a_j] \leq r$ and dividing it by $K - m + 1$, where $j = 1, 2, \ldots, L - m + 1$ and $d[a_i, a_j] = \max_{p=1,2,\ldots,m}(|a_{i+p-1} - a_{j+p-1}|)$. Then the approximate entropy of the variable in equation (2.52) can be computed as

$$H_A = \lim_{r\to 0}\lim_{m\to+\infty}\lim_{K\to+\infty}[\Phi^m(r) - \Phi^{m+1}(r)], \tag{2.54}$$

where

$$\Phi^m(r) = (K - m + 1)^{-1}\sum_{i=1}^{K-m+1}(log_2 C_i^m(r)). \tag{2.55}$$

However, for a 3D trajectory shape, there are two variables of significance: curvature (κ) and torsion (τ). Here we compute the approximate entropy of the trajectory considering joint approximate entropy of κ and τ as $H_{\kappa,\tau}$. However, the following remark is vital for computational simplicity.

Remark: Curvature and torsion are independent variables ($\kappa \perp\!\!\!\perp \tau$).

According to the Frenet-Serret formulas [113, 324],

$$\begin{bmatrix} dT \\ dN \\ dB \end{bmatrix} = \begin{bmatrix} 0 & \kappa & 0 \\ \kappa & 0 & \tau \\ 0 & -\tau & 0 \end{bmatrix}\begin{bmatrix} T \\ N \\ B \end{bmatrix} \tag{2.56}$$

where κ and τ describe the relationship between T, N and B, where $T \perp\!\!\!\perp N \perp\!\!\!\perp B$. From

$$\frac{dT}{dt} = \kappa N, \tag{2.57}$$

we can see that κ is the amplitude of the projection of the change of tangent vector on the normal vector. Similarly,

$$\frac{dB}{dt} = -\tau N \tag{2.58}$$

shows that τ is the amplitude of the projection of the change of binormal vector on the normal vector. Since κ and τ indicate the change in two independent vectors, they are independent of each other. Therefore, $H_{\kappa,\tau} = H_\kappa + H_\tau$.

2.4.3 Experiment setup

Simulation data collection

The simulations were conducted with Matlab 2013a® to ensure that the proposed approach for smoothness measurement met the consistency, sensitivity and robustness requirements given in Balasubramanian *et al.* [31].

To simulate the reaching movement, we utilised the following process:

1. *Voluntary movement*
   A noiseless, free reaching involuntary movement is simulated as a smooth arc with a duration K,

$$\gamma(k) = \{\cos(\pi k/(3 \times K) + \pi/3), \sin(\pi k/(3 \times K) + \pi/3), 0\}, \tag{2.59}$$

   where $k = 1, 2, \ldots, K$.

2. *Sub-movements*
   The sub-movements are generated by the sum of multiple Gaussian models for each axis (X, Y and Z) as

$$I(t) = \sum_{n_i=1}^{N_i} A_{n_i} \exp\left(-\frac{(t-\mu_{n_i})^2}{2\sigma_{n_i}^2}\right), \tag{2.60}$$

   where N_i is the number of involuntary movements, and A_{n_i}, μ_{n_i} and σ_{n_i} are the amplitude, mean time and standard deviation of the duration of the n_ith involuntary movement. These variables can be different for various axes. By adding the $I(t)^x$, $I(t)^y$ and $I(t)^z$ to γ_t, we are able to create a trajectory with involuntary movements.
3. *Jerk*
   Jerks are simulated by adding Gaussian noise with normal distribution and various amplitudes into the motion trajectories. There are four parameters to determine the jerk, including the number of jerks (N_j), starting time (S_{n_j}), duration (D_{n_j}) and amplitude of jerks (A_{n_j}).

To ascertain the consistency and sensitivity of the proposed approach, 50 trajectories were generated to simulate a reaching movement with various numbers and amplitudes of involuntary movements and jerks. The specifications for these trajectories are shown in Table 2.10.

To simplify the simulation without losing the effects, we only simulated the trajectories in levels two and three. Furthermore, we assumed the numbers and amplitudes of sub-movements in three axes of the trajectories are the same, which means $I(t)^x = I(t)^y = I(t)^z$. It is the same for jerks.

In addition, to evaluate the robustness of different approaches, we generated 100 noiseless trajectories in level two and the numbers and amplitudes of involuntary movements were randomly generated in the given range (refer to Table 2.10). For each trajectory, we generated 100 noisy trajectories by adding Gaussian noise with zero mean and $0.33v_{\text{peak}}$ standard deviation following the experiment in Balasubramanian *et al.* [31].

Real-data experiment data collection

For a real-data experiment, no film recordings of subjects were made in this study. The Kinect camera provided numerical data that directly related to arm movements. Only de-identified numerical data, representing motion vectors, were stored in the database. Volunteers were researchers and students at Deakin University. Ethics for conducting the experiments in this study were approved by Deakin University.

The real-data experiment was conducted with four healthy subjects mimicking three severity levels of involuntary movements (refer to Table 2.9) while performing an upper limb task, i.e. moving a book from one location to another and bringing it back to the original location. Before recording the data, subjects were required to practise the tasks to make sure that their movements for different levels involved the required involuntary movements and durations.

In the experiment, a table, a chair, a book, a second version of Kinect and a laptop were used (refer to Figures 2.20 and 2.21). The chair had no arms and its height was adjustable to suit the subjects. During the experiment, the subject was about 20 cm away from the front

Table 2.10 Parameters used to simulate two groups of trajectories. These two groups of trajectories correspond to two severity levels of involuntary movements. The first 25 trajectories belong to the second level with two involuntary movements and two jerks. The last 25 trajectories are in the third level with four involuntary movements and jerks. To simulate the severity in one level, the amplitudes of involuntary movements and jerks increase with the index.

		Level 2	Level 3
Sub-movements	Index	1, 2,..., 25	26, 27,..., 50
	duration	4.5 seconds (135 frames)	7.5 seconds (210 frames)
	Ni	2	4
	μ_{n_i}/σ_{n_i}	45/50	40/50
		90/50	90/60
		–	140/60
		–	190/30
	$A_{n_i}(m)$	Index × 0.009	Index × 0.009
		Index × 0.01	Index × 0.01
		–	Index × 0.011
		–	Index × 0.012
Jerks	N_j	2	4
	$A_{n_j}(m)$	Index × 0.001	Index × 0.001
		Index × 0.0011	Index × 0.0011
		–	Index × 0.0012
		–	Index × 0.0013

of the table so that the book could easily be reached with pure arm movements (without moving his/her trunk). The dimensions of the book were 23.5 × 15.5 × 1 cm and it weighed 0.25 kg. At the same time, the book was placed near the edge of the table so that the subject could hold the book easily.

To accurately track the involuntary movements and jerk, we attached a small infra-red reflective marker to the tracking joint (wrist) and made sure the marker always faced the Kinect so that the Kinect could capture the position of the wrist. The data collection program was written with Kinect SDK v2.0-1409 with C# under Windows®8.1. Although there was no precision evaluation on the second version of Kinect, according to Breuer *et al.* [53], the precision of the new version of Kinect was close to its predecessor (less than 5 mm for the distance between the Kinect and the object less than 1.5 m [280]).

Four healthy subjects were recruited for the experiment. Their demographic data can be seen in Table 2.11. They were required to perform the task of taking a book from one pre-defined location to another location and then returning it to the original location. The original and target locations were marked on the table in advance and the subjects were required to place the book at the designated position accurately. Further, the subjects were required to mimic three different severities of involuntary movements with various

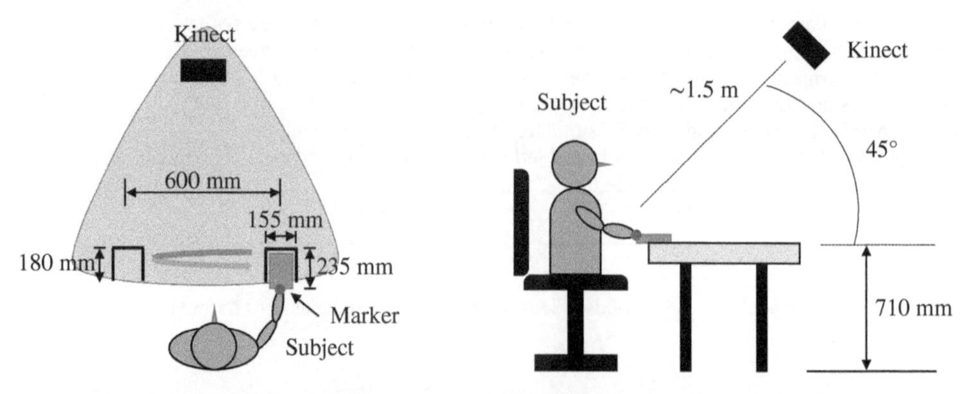

Figure 2.20 A diagrammatic view of the experimental setup.

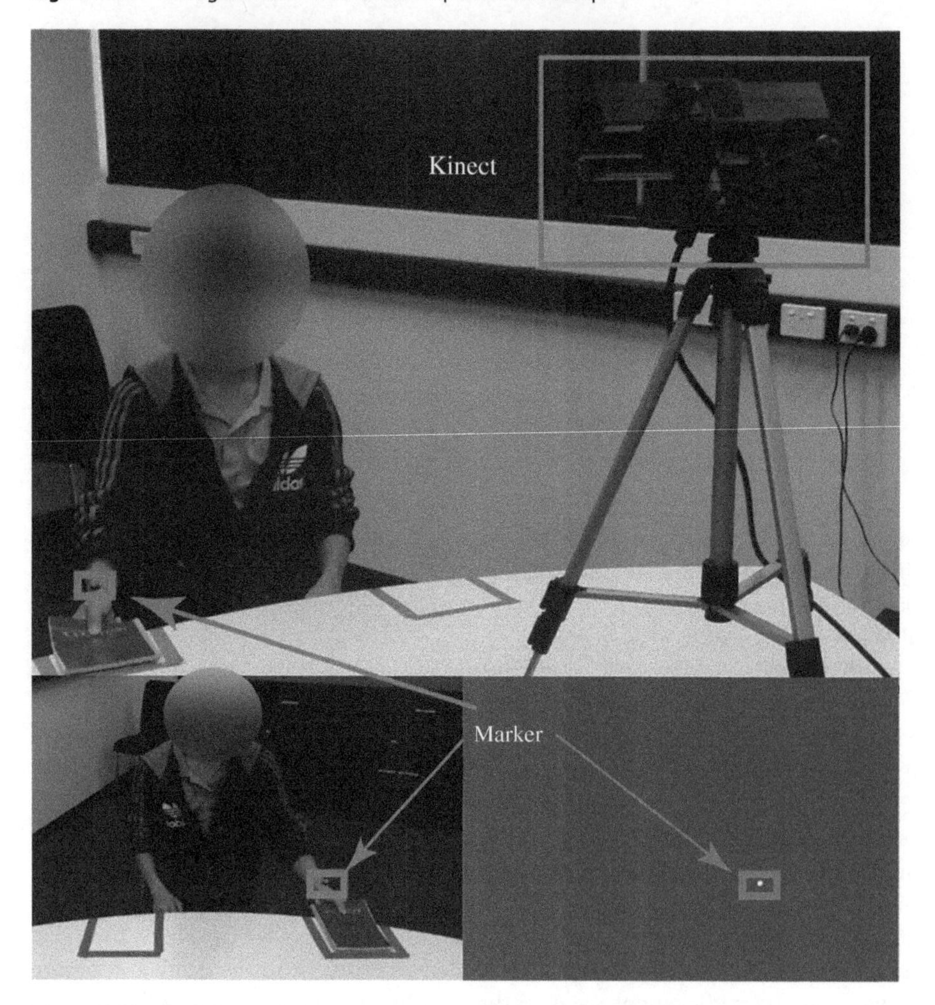

Figure 2.21 Real-data experiment setup image. The top image shows the setup of the Kinect and the subject. The bottom left and right are the RGB and depth images taken from the Kinect. Note that the marker was on the right wrist of the subject (the depth and RGB camera in the Kinect reversed the image). Source: Saiyi Li.

Table 2.11 Demographic data of subjects.

	Age	Weight (kg)	Height (cm)	Gender
Subject 1	28	55	172	Male
Subject 2	29	70	175	Male
Subject 3	27	60	173	Male
Subject 4	22	58	160	Female

durations (10 iterations for each subject). These involuntary movements were mimicking patients experiencing dyskinesia so that the motion trajectories involved jerkiness and uncontrollable sub-movements. The numbers and amplitudes of involuntary movements generally followed the specifications in Table 2.9. In addition, we played music with three different durations (4, 7 and 15 seconds for levels one to three) so that the motion duration of the subjects could be generally controlled on three levels. Eventually, we expected a deterioration in smoothness from the first to the third level and increased duration. Therefore, the ability to perform reaching tasks decreased. These criteria would be used by a human observer to classify the ability to perform the reaching task into three levels manually for the purpose of validation.

During the experiment, the subject initially held the book with his/her dominant hand (right hand for all subjects) and kept it steady. At the same time, the system operator checked whether the Kinect could capture the marker. If the Kinect could capture the marker and the subject was ready, the system operator gave the subject an instruction to start moving the book and played the music. In the meantime, the Kinect system started recording the position information of the marker into a database for offline analysis. As soon as the subject finished the task (replacing the book in the original position), the system operator stopped the system. Apart from the system operator and the subject, another researcher classified the task (one of the three levels). The manual classification criteria include the duration of finishing the task and numbers and amplitudes of involuntary movements listed in Table 2.9.

Each subject was required to perform the task at least 30 times in total to ensure that there were at least 10 trials at each level. All of the 30 trials were conducted over 3 days with 10–15 trials per day depending on the subject availability. Between each trial, the subject could rest for 30 seconds.

2.4.4 Data analysis and results

Computer simulation data analysis

Three simulations were conducted to assess the performance of the proposed approach in terms of motion smoothness, which was evaluated using five approaches, namely, the number of tangential velocity peaks (VPs) [25], the number of zero-crossing tangential accelerations (ZCAs) [355], dimensionless jerk metrics (DJs) [304], spectral arc length (SAL) [31] and entropy of shape model and instantaneous acceleration (ESA). Except for SAL, the increase of metrics computed by the other approaches illustrates the deterioration of smoothness. Simulation results were analysed from three aspects, namely consistency, sensitivity and robustness. For consistency, the raw metrics computed with these five approaches were compared to see if the metrics could keep the same trend with the change of smoothness of trajectories. Secondly, the sensitivity of these approaches with respect to the change of smoothness was analysed. The improvements are computed as

$$Q_i = (M_i - M_1)/M_1, \tag{2.61}$$

where M_i is the metric value of trajectories with index of $i = 1, 2, \ldots, 50$. The approach that had the most significant improvement with respect to the trajectory index was the most sensitive to the change of smoothness since the smoothness deteriorates with the increase of the trajectory index (refer to Figure 2.23). Eventually, the robustness of these approaches was analysed. Firstly, the noisy trajectories were processed by a low-pass filter with a cut-off frequency of 14 Hz since the simulated frequency of the trajectories was 30 Hz. The normalised difference metrics were computed between the noisy and noiseless trajectories with approach a (a can be VP, ZCA, DJ SAL and ESA) as follows:

$$(d_i^j)_a = \left(\frac{M_i^j - m_i}{\max(m) - \min(m)} \right)_a, \tag{2.62}$$

where m is the collection of the metrics (we call them ground truths in the rest of this section) of the 100 noiseless trajectories and m_i is the metric of the ith ($i = 1, 2, \ldots, 100$) noiseless trajectory. In addition, M_i^j is the metric of the jth ($j = 1, 2, \ldots, 100$) noisy trajectories generated based on ith noiseless trajectory. Moreover, $\max(\cdot)$ and $\min(\cdot)$ selected the maximum and minimum data from the collection m. Lastly, the probability density functions of d (the collection of the normalised differences) were estimated using the *ksdensity* function in Matlab. The whole evaluation process followed that introduced in Balasubramanian *et al.* [31].

Computer simulation result

Examples of generated trajectories are depicted in Figure 2.22. Before comparing the metrics from various approaches, the change in smoothness with respect to trajectory indices is shown in Figure 2.23.

Figure 2.24 illustrates the consistency characteristics of different approaches. Since the simulated trajectories were classified into two levels of severity of involuntary movements, the trend in individual levels were analysed first. For the results in level one (in light), VP, SAL and ESA showed a consistent trend with the metrics where VP and ESA kept increasing from around 5 and 0.013 to approximately 17 and 0.1, while SAL decreased from −4 to around −5. The fluctuations were caused by the randomly generated values in the simulation (the same as follows). However, the consistent trend was hardly observed in ZCA and DJ. The metrics of the former fluctuate between 10 and 15, while those of the latter decreased from −25 to around −75 and then increased to approximately −10. For the metrics of trajectories belonging to the third level, DJ, SAL and ESA show a consistent trend, where DJ and ESA increased from about −25 and 0.2 to approximately −5 and 0.3, while SAL decreased from around −5.5 for all of these approaches to −6. As for the other two approaches, the consistency was not obvious. Lastly, according to Section 2.4.2, the trajectories in level three were less smooth than those in level two. Therefore, metrics from these two levels should show the differences, which can be observed in the result for all these

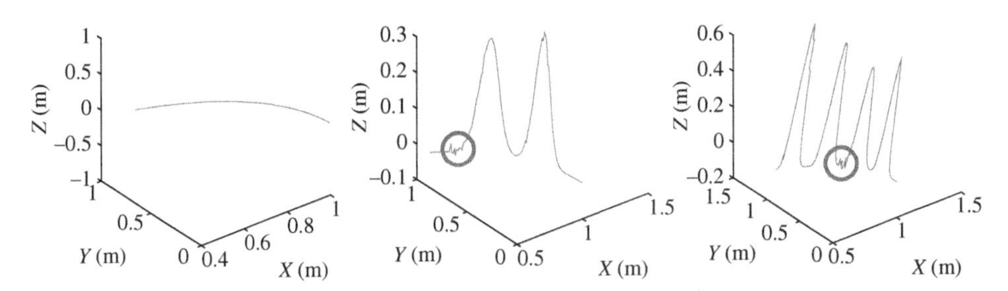

Figure 2.22 These three graphs show trajectories in three levels. The left one is in level one for natural movements without involuntary movements and jerks. The middle trajectory is in the second level with two involuntary movements (with amplitudes of 0.225 and 0.25 m) and jerks (with amplitudes of 0.025 and 0.0275 m). The last one is in the third level with four involuntary movements (with amplitudes of 0.45, 0.5, 0.55 and 0.6 m) and jerks (with amplitudes of 0.05, 0.055, 0.06 and 0.065 m). The dark circles are examples of jerks in the second and third level.

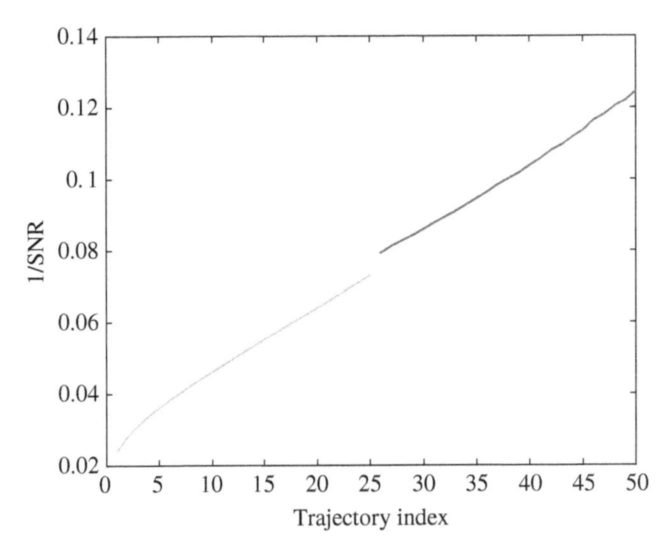

Figure 2.23 The smoothness level of trajectories, which is represented by one over the signal-to-noise ratio (SNR) since the SNR is non-linear with respect to the linear change of smoothness, while 1/SNR is linear. It is easier for the following comparison.

approaches. For example, the average metrics from VP, ZCA, DJ and ESA in level three were higher than the average metrics in the second level, while SAL showed lower metrics in level three than in level two. All in all, SAL and ESA outperformed VP, ZCA and DJ in terms of consistency.

The second aspect was sensitivity (refer to Figure 2.25), which analysed the change rate of metrics from various methods with respect to the change of smoothness. From the result, it is not hard to observe that ESA was the most sensitive approach since the improvement percentages changed from 0% to around 1000% for the second level and from 2000% to about 2300% for the third level. By comparison, the metrics of SAL increased from 0% to around 50% from trajectory indices of 1 to 50. As for VP, it was quite sensitive to the change of the

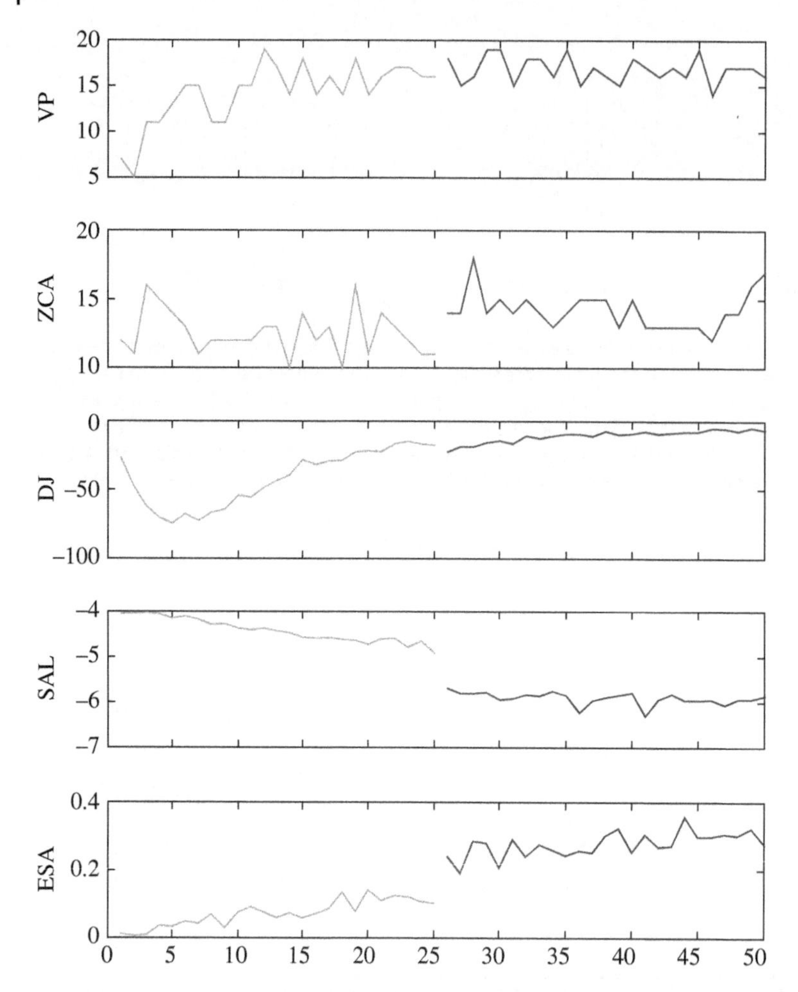

Figure 2.24 The metric given by these approaches tends to illustrate the consistency characteristics in various approaches used to evaluate the smoothness of trajectories in two severity levels. With the increase in numbers and amplitudes of both involuntary movements and jerks, the smoothness of the trajectories deteriorates. The first half (with light from 1 to 25) is in the second level and the last half (with dark from 26 to 50) is in the third level. The VP was computed with a threshold value of 0.01 m/s and the temporal gap between two consecutive peaks was 100 ms.

severity in the first 15 trajectories (increased from 0% to about 100%), but less sensitive for the rest. For ZCA, the improvement percentages were very minimal. Although DJ was very sensitive in the second level, the sensitivity gradually reduced in the third level. In addition, since the smoothness decreased linearly with respect to the trajectory index, we fitted a line (regression lines) to the metrics generated by each method and computed the gradient of each line indicating the general sensitivity. This confirmed our conclusion that ESA was the most sensitive with the highest gradient.

Lastly, the robustness of the proposed approach was investigated. From Figure 2.26, the performance of ESA was very close to SAL as they had a similar value (around 0.05) with

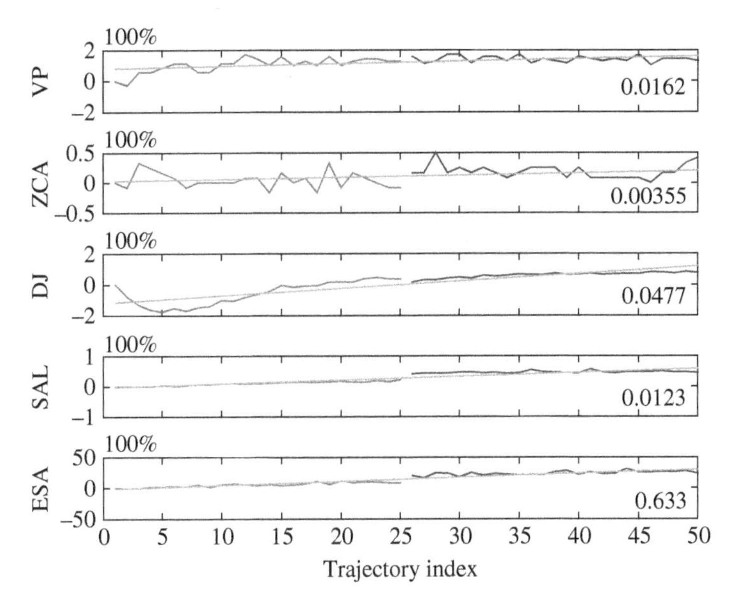

Figure 2.25 Sensitivity comparison of the five approaches with respect to the change in the severity of involuntary movements. A better evaluation technique is preferred in order to be sensitive to the small changes in the severity of involuntary movements and the change rate metric should be proportional to the change rate of severity. The two jittery lines show the improvement of metrics of the second and third severity levels with respect to the metric of the first trajectory computed using various approaches (refer to equation 2.61). The regression line of each corresponding metric shows the approximate improvement rate of the method. The value at the bottom right of each graph is the gradient of the regression line.

the highest density. However, the range of ESA (0 to 0.75) was a little bigger than SAL (−0.1 to 0.5). By comparison, VP and DJ were not able to maintain the metrics with the influence of measurement noise as they had a large metrics range and a large difference from the ground truth. Although the metric for ZCA was in a reasonable range (from 0.1 to 0.5), the offset from the ground truth was large (around 0.3).

Eventually, ESA outperformed SAL as it was more sensitive to the change of smoothness and also met the requirement of dimensionless, consistency and robustness.

Healthy subjects simulation data analysis

In the real-data experiment, firstly, the smoothness of all the trajectories was evaluated using the same approaches considered in the computer simulation section. Additionally, by taking the duration into consideration, all these trials were classified into three levels of ability to perform the task by using three commonly used clustering methods, namely k-means clustering, the Gaussian mixture model (GMM) and fuzzy clustering [37], in order to determine which clustering method suits the purpose of classifying motion trajectories into different levels of ability to perform upper extremity reaching tasks (severity levels of involuntary movements). Since the trials had also been classified by a human observer, a Cohen's kappa correlation coefficient was computed between the human observer and the computer to indicate the assessment agreement. Higher coefficient values indicate a greater

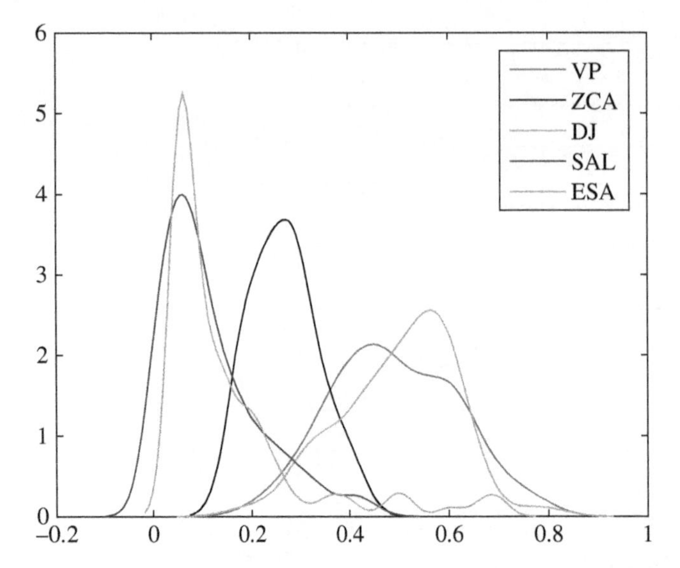

Figure 2.26 Robustness comparison of the five approaches with respect to the change in the severity of involuntary movements. A better evaluation technique needs to be limited to a smaller range (on the horizontal axis) in terms of the differences in metrics between the noisy trajectories and their corresponding noiseless ones, and the value with the highest amplitude in vertical axis should be close to zero.

level of agreement between the utilised approach and the human observer. Except for the proposed method, the other four approaches, including VP, ZCA, DJ and SAL, were used for comparison.

Healthy subjects simulation result

Here we present the results of our preliminary real-data experiment with healthy subjects mimicking different severity levels of involuntary movements with their upper extremities while seated.

First of all, some examples of simulated trajectories and features are shown in Figure 2.27. The first three rows are the trajectories in three axes, thereby showing the sub-movements (with circles) and jerks (with rectangles) more clearly. The third and fourth rows were shape models of these trajectories with curvature and torsion, while the last row is the instantaneous acceleration. As for the columns, levels one to three (refer to Section 2.4.2) are in the first to the third columns. As can be observed, from the first to the third level, the number of sub-movements and jerks increased from 0 to 4 and 5, respectively. Correspondingly, the curvature, torsion and instantaneous acceleration were increasingly noisy.

Secondly, the distributions of various features, including the duration of the task, as well as metrics computed with various approaches, are shown in Figure 2.28. The first box plot was the distribution of durations for three severity levels of involuntary movements. As can be seen, although there was some overlap between two consecutive levels, the interquartile ranges followed the definition of severity levels (refer to Table 2.9). The remaining five graphs show the metrics computed with five approaches. First of all, it is obvious that, in

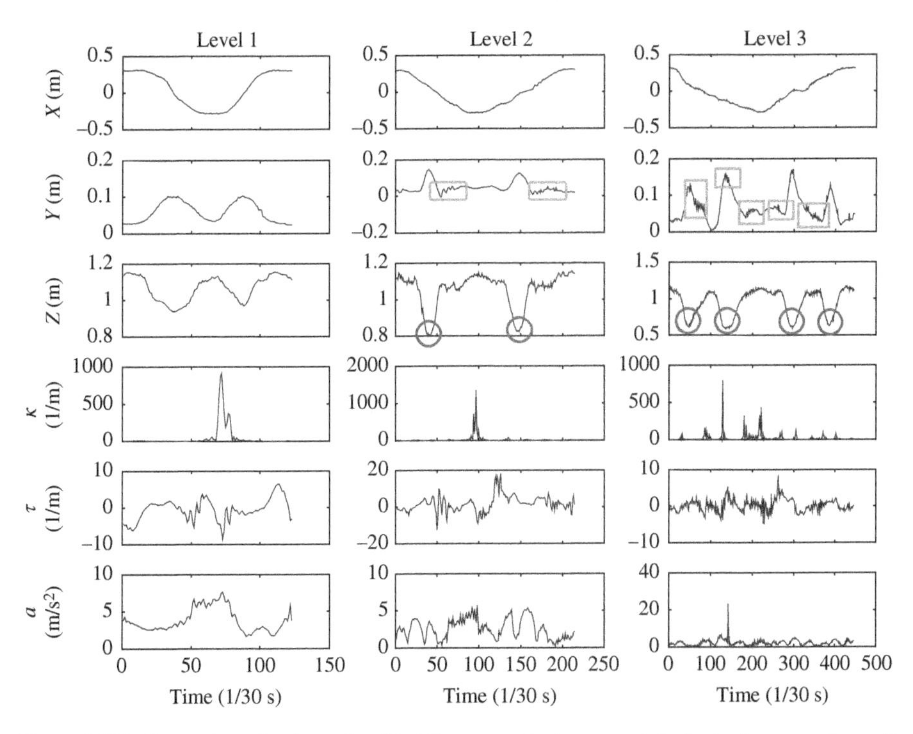

Figure 2.27 Examples of trajectories (first three rows), shape models, including curvatures (fourth row) and torsions (fifth row), and instantaneous accelerations (sixth row) are illustrated for three levels of the ability to perform an upper extremity reaching task (columns one to three corresponding to levels one to three of the severity of involuntary movements). The circles show examples of sub-movements and rectangles are examples of jerks.

VP, ZCA, DJ and ESA, the more severe involuntary movements were associated with higher values, while it was the opposite in SAL (the more severe involuntary movements had lower metrics). However, this trend between the first and the second levels in DJ was very small. Secondly, although every approach showed a certain degree of overlap between two consecutive levels, DJ showed the worst situation because the metrics of its first and second levels were very similar. As for the other four approaches, VP and ZCA showed a larger overlap between the first and second levels than SAL and ESA, and VP had the largest overlap between the second and third levels. Generally speaking, the differences between these three levels of ESA were more obvious than in the other four approaches.

Table 2.12 shows the Cohen's kappa between five automated approaches and the human observer generated by three clustering approaches. Generally speaking, ESA always gave the highest agreement (0.8250 for k-means and GMM, with 0.85 for fuzzy clustering). In addition, SAL and ZCA had a similar performance with kappa values slightly smaller than 0.8 in the majority of cases. By comparison, VP was a little better than DJ. But these two approaches were worse than the other three methods. Moreover, the comparison of these three clustering approaches suggested that the fuzzy clustering was more suitable for automated assessment of the severity of involuntary movements, thereby evaluating the ability to perform reaching tasks in daily living in terms of kinematics.

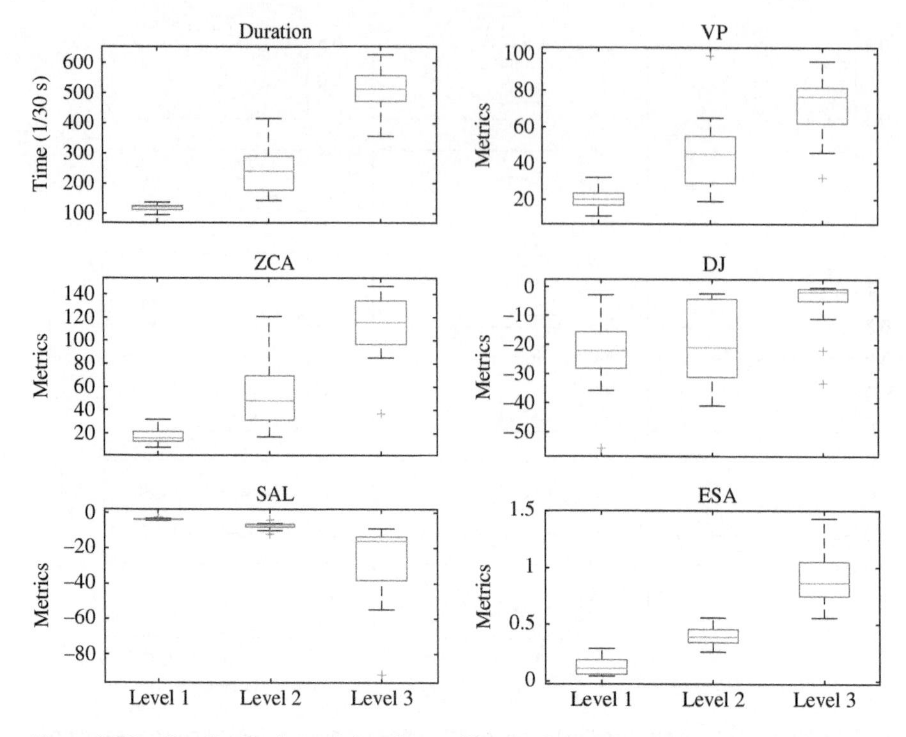

Figure 2.28 The distributions of durations utilised to finish the reaching task, as well as metrics computed by five approaches, for three severity levels. The threshold utilised to compute VP was 0.25 m/s and the temporal gap was 100 ms for all subjects.

Table 2.12 Cohen's kappa ($p < 0.05$) between various automated approaches and the human observer.

Approach	k-means	GMM	Fuzzy clustering	Average
VP	0.6875	0.6926	0.7250	0.7017
ZCA	0.7625	0.8000	0.7500	0.7708
DJ	0.5500	0.5631	0.5500	0.5544
SAL	0.7875	0.7875	0.7500	0.7750
ESA	0.8250	0.8250	0.8500	0.8333

Discussion and conclusion

From the simulation and real-data experiment, the proposed approach has shown its superior performance in terms of capturing non-smooth movement patterns due to involuntary movements during the performance of a specific upper extremity task. The reasons for this are three-fold. Firstly, a dimensionless and duration-independent entropy was utilised as a metric for evaluation. It is well known that entropy can be utilised to analyse the regularity of variables, which makes it suitable for analysing the ability to perform a task in terms of motion smoothness, because motions involving involuntary movements tend to have an irregular trajectory shape. Secondly, apart from considering the dynamics (instantaneous acceleration) of the motion trajectory, the shape of the trajectory was also taken into account, thereby not only meeting the requirement of dimensionless, consistency and robustness, but also more sensitivity than other approaches compared in the computer simulation (refer to Section 2.4.4). Thirdly, the shape model, including the curvature and torsion, is sensitive to noise in trajectories. Therefore, the involuntary movements in motion trajectories significantly introduce noise, thereby increasing the entropy of the shape model. At the same time, by considering the instantaneous acceleration, the dynamics of the trajectories were utilised. Eventually, the trajectories containing involuntary movements were evaluated comprehensively.

However, since this was a preliminary study, there are some areas that require further attention. Firstly, being an affordable device, the Microsoft Kinect is not as accurate as other more expensive commercially available products, such as Vicon. Therefore, the lower resolution (especially in the Z axis) hinders the Kinect in identifying small movements of human joints. In other words, jitters or tremors with a minimum amplitude may not be captured by the Kinect. Therefore, at present, it can only be utilised to detect involuntary movements with relatively large amplitude. In real applications, the Kinect should be optimised so that it can detect involuntary movement of a few millimetres. However, this study illustrates the ability of the proposed approach to evaluate a person performing reaching tasks, particularly in a non-clinical environment. It is noteworthy that although the real-data experiment consisted of a seated activity involving the arms, the proposed approach is not limited to upper body movements. Other applications will be further investigated in future work. Secondly, healthy subjects, instead of patients with involuntary movements, were evaluated. Therefore, the proposed approach should be further validated and analysed with people who have involuntary movements and other movement impairments.

This section presents a novel approach for the quantitative evaluation of the ability of individuals with involuntary movements to perform reaching tasks involving the upper extremity. We based our approach on the smoothness of the movement trajectory and also the duration of the activity. In particular, the entropy of the shape model and the instantaneous acceleration were used to capture the appropriate performance indices. Experiments with computer simulation and professional role-playing mimicking involuntary movements were conducted to provide preliminary validation of the feasibility and performance of using the proposed approach with an affordable Microsoft Kinect. The computer simulation showed the effectiveness of using entropy of the shape and instantaneous acceleration for motion smoothness evaluation in terms of their consistency, sensitivity and robustness. The real-data experiment results showed that agreement (Cohen's kappa correlation coefficient) between a human observer and the proposed automated approach with fuzzy clustering in the experiment was 0.8500, compared to 0.7250, 0.7500, 0.5500 and 0.7500 using the number of tangential velocity peaks, the number of zero-crossing tangential accelerations, dimensionless jerks and spectral arc-lengths, respectively. Further studies involving patients with movement disorders will be conducted in future to validate the feasibility of the proposed method.

2.5 Summary

In this chapter, we introduced the key aspects of analysing a person's kinematic performance in ADLs.

More specifically, three algorithms are compared to fuse data from a number of motion capture devices, such as Kinects©, to improve the quality of data from the very beginning. We have employed linear versions of the Kalman filter, a particle filter and robust linear filtering to demonstrate the benefits of using multiple Kinects© in an information fusion context to increase the accuracy of human joint position estimations. We have demonstrated the value of using strong and effective model-based linear filtering through which we confirmed the use of readily available commercial products in biokinematic measurement system applications, such as rehabilitation and physiotherapy. With guaranteed performance and accuracy, these types of affordable implementations are destined to resolve a number of issues of practical significance, i.e facilitation and monitoring of home-based prescribed exercise routines that can significantly reduce the need for patients to travel to regional centres.

After that, a novel encoding approach is introduced to extract the curvature and torsion information from the raw data. Human actions have been widely studied for their potential application in various areas such as sports, pervasive patient monitoring and rehabilitation. However, challenges still persist pertaining to determining the most useful ways to describe human actions at the sensor, then limb and complete action levels of representation and deriving important relations between these levels each involving their own atomic components. In this chapter, we report on a motion encoder developed for the sensor level based on the need to distinguish between the shape of the sensor's trajectory and its temporal characteristics during execution. This distinction is critical as it provides a different encoding scheme than the usual velocity and acceleration measures that confound these two attributes of any motion.

Lastly, we also present an approach for the quantitative evaluation of the ability of individuals with involuntary movements to perform reaching tasks involving the upper extremity. We based our approach on the smoothness of the movement trajectory and also the duration of the activity. In particular, the entropy of the shape model and the instantaneous acceleration were used to capture the appropriate performance indices. Further studies involving patients with movement disorders will be conducted in future to validate the feasibility of the proposed method.

3

Biokinematic Measurement with Wearable Sensors

3.1 Introduction

Inertial sensors measure the acceleration, the earth's magnetic field and the rotational rates experienced by the sensor – all with respect to the local coordinate frame of the sensor unit. By positioning the sensor unit on a moving object, it is possible to estimate the orientation of the object with respect to the global coordinate frame. This has been used in numerous attempts to capture the orientation of human limbs using wearable sensors. This poses fundamental questions about estimating the orientation of a movable object using observation vectors measured by the inertial measurement unit (IMU) sensors in the moving object. The underlying problem has a historical relevance in the aerospace arena and was first formulated as minimising a loss function by Grace Wahba in 1965 [370]. The loss of one degree of freedom of a rotating body in a multi-dimensional space is called a Gimbal lock and was first described by the Greek inventor Philo of Byzantium in 200–220 BCE [241]. In the literature, numerous studies were conducted to avoid singularities, which are defined as relative orientations where output torque along particular directions are unusable due to the Gimbal lock. Particularly in many spacecraft attitude estimation systems, the underlying problem is the deduction of the attitude of the craft using the observation vectors, which can typically be the unit vectors of the stars, sun or the earth's magnetic field. This underlying theory and concepts are widely applied to inertial navigation of ships, rocket engines and aircraft.

3.2 Introduction to Quaternions

Quaternions are used to represent orientation with improved computational efficiency while avoiding singularities. Consider complex quantities i, j and k such that

$$ij = k, kk = -1 \implies i^2 = j^2 = k^2 = ijk = -1. \tag{3.1}$$

Consider two coordinate frames – earth frame E and the frame of reference attached to the dynamic rigid body B. The unit quaternion is the mathematical representation of the

Human Motion Capture and Identification for Assistive Systems Design in Rehabilitation, First Edition.
Pubudu N. Pathirana, Saiyi Li, Yee Siong Lee and Trieu Pham.

rigid body attitude between the frames E and B. For $i \in \{0, 1, 2, 3\}, q_i \in \mathbf{R} : \sum_{i=0}^{3} q_i^2 = 1$ is denoted by the hypercomplex number of rank 4 such that

$$\mathbf{q} = q_0 + q_1 i + q_2 j + q_3 k. \tag{3.2}$$

Considering q_0 as the scalar part and $\begin{bmatrix} q_1 & q_2 & q_3 \end{bmatrix}^{\mathsf{T}}$ as the vector part,

$$q = \begin{bmatrix} \cos\frac{\alpha}{2} & \mathbf{u}\sin\frac{\alpha}{2} \end{bmatrix}^{\mathsf{T}} = \begin{bmatrix} q_0 \\ \mathbf{q} \end{bmatrix} \tag{3.3}$$

where $\mathbf{u} = 1/\sin\alpha/2\begin{bmatrix} q_1 & q_2 & q_3 \end{bmatrix}^{\mathsf{T}}$ is a vector in the Euler coordinate frame. Therefore, this indeed represents a coordinate frame rotated around $\mathbf{u}$ by an $\alpha = 2\cos^{-1}(q_0)$ angle around the reference axis. Quaternion multiplications can be performed using the definition provided in Section 3.1 and the complex conjugate of a quaternion $\mathbf{q}$ is $\mathbf{q}^* = q_0 - q_1 i - q_2 j - q_3 k$. Therefore, considering the normalised case, $\mathbf{q}\mathbf{q}^* = q_0^2 + q_1^2 + q_2^2 + q_3^2 = 1$ and the inverse of a quaternion is the same as the complex conjugate, i.e. $\mathbf{q}^{-1}\mathbf{q} = \mathbf{q}\mathbf{q}^{-1} = 1$ and $\mathbf{q}^{-1} = \mathbf{q}^*$.

Any vector $\mathbf{v} = \begin{bmatrix} v_1 & v_2 & v_3 \end{bmatrix}$ can be rotated (left hand) around $\mathbf{u}$ by an angle α and the resulting vector $\mathbf{v}_r$ is given by

$$\mathbf{v}_r = q\mathbf{v}_\mathbf{p}q^{-1}.$$

Here, v_p is the *pure quaternion* defined as $v_p = \begin{bmatrix} 0 & v_1 & v_2 & v_3 \end{bmatrix}$. For the case of a vector rotated (right hand), this can be given in the matrix form

$$\mathbf{v}_r = \mathbf{R}^\mathbf{q}\mathbf{v}, \tag{3.4}$$

where $\mathbf{R}^\mathbf{q}$ is the rotation matrix given as

$$\mathbf{R}^\mathbf{q} = \begin{bmatrix} 2q_0^2 - 1 + 2q_1^2 & 2q_1q_2 - 2q_0q_3 & 2q_1q_3 + 2q_0q_2 \\ 2q_1q_2 + 2q_0q_3 & 2q_0^2 - 1 + 2q_2^2 & 2q_2q_3 - 2q_0q_1 \\ 2q_1q_3 - 2q_0q_2 & 2q_2q_3 + 2q_0q_1 & 2q_0^2 - 1 + 2q_3^2 \end{bmatrix}.$$

3.3 Wahba's Problem

Here the attitude (a special orthogonal matrix called a rotation matrix) of a second coordinate frame with respect to a first coordinate frame (usually a global drama) is sought with a set of (weighted) vector observations. Particularly in the case of IMUs, these are magnetometer and accelerometer readings captured from the sensors in the second frame. Almost all algorithms for estimating attitude from vector measurements are based on minimising a loss function proposed by Grace Wahba in 1965. Given the following cost functional,

$$L(A) \equiv \frac{1}{2}\sum_i a_i|b_i - Ar_i|^2, \tag{3.5}$$

the problem is to find the orthogonal matrix A with determinant +1 that minimises the loss function [223, 225], i.e,

$$A : min_{A\in\mathrm{R}^{3\times3}}\ L(A). \tag{3.6}$$

Here $\{b_i\}$ is a set of unit vectors measured in a local body, $\{r_i\}$ are the corresponding unit vectors in a reference frame and $\{a_i\}$ are non-negative weights. The convenient form of loss function [223] is as follows:

$$L(A) \equiv \lambda_0 - tr(AB^T), \tag{3.7}$$

where $\lambda_0 = \sum_i a_i$ and $B = \sum_i a_i b_i r_i^{\mathrm{T}}$, while this transforms the minimisation of $L(A)$ into maximisation of $tr(AB^T)$.

Essentially the the same physical quantity is measured in the two frames simultaneously and in this case it can be either the earth's magnetic field and/or the gravity vector.

1. The magnetic measurements are subjected to large uncertainties due to magnetic disturbances from equipment such as electric motors.
2. When the object is moving at a constant velocity in a linear path, the resultant (measured) acceleration is gravity; hence the acceleration vector is common to two frames.

In Mirzaei and Roumeliotis [239] an extended Kalman filter-based non-linear approach was used to determine the unknown transformation between a camera and an IMU in a vision-aided inertial navigation system (V-INS). Wahba's problem can be stated as follows.

3.3.1 Solutions to the Wahba problem

TRIAD

The TRIAD method is one of the earliest and simplest methods used to determine attitude in spacecraft science [30, 43]. It was introduced by Harold Black when he was serving at the US Navy's Transit satellite system at Johns Hopkins Applied Physics Laboratory prior to the formulation of the Wahba problem and its subsequent optimal solutions. The algorithm is referred as the triad algorithm as it is simply constructing two triads of orthonormal unit vectors using the vector information that we have. The two triads are in the same reference frame, frame 1 and 2. Considering a rotation matrix A from frame 1 to 2,

$$b_1 = Ar_1, \quad b_2 = Ar_2.$$

Assuming non-co-linear measurements in each frame, the two triads are

$$t_1 = \frac{b_1}{||b_1||}, \quad t_2 = \frac{b_1 \times b_2}{||b_1 \times b_2||} \quad t_3 = t_1 \times t_2,$$
$$u_1 = \frac{r_1}{||r_1||}, \quad u_2 = \frac{r_1 \times r_2}{||r_1 \times r_2||} \quad u_3 = u_1 \times u_2.$$

Define two matrices with the orthogonal vectors

$$T_1 = \begin{bmatrix} t_1 & t_2 & t_3 \end{bmatrix}, \quad U_1 = \begin{bmatrix} u_1 & u_2 & u_3 \end{bmatrix}. \tag{3.8}$$

Considering the vector by vector mapping,

$$T_1 = AU_1.$$

Observing that T_1 and U_1 are both orthogonal, i.e. $T_1^{-1} = T_1^{\mathrm{T}}$ and $U_1^{-1} = U_1^{\mathrm{T}}$,

$$A = T_1 U_1^{\mathrm{T}}.$$

The TRIAD algorithm is a single frame deterministic method for solving Wahba's problem without considering any noise in the measurements. If the measurements of the gravity and the magnetic field are ordered as described earlier, the cross-products that are used to calculate s_2 and r_2 eliminate any contribution of the magnetic measurements relative to the vertical axis. Thus, pitch and roll components of orientation are determined using only the accelerometer measurements.

Singular value decomposition method

Singular value decomposition is generally used to decompose real or complex matrices. However, this approach has not been widely used in practice due to the computational cost. The approach provides a valuable insight [224, 225] into an analytical context. As in equation (3.2), it is required to maximise $tr(AB^T)$. Since U and V are orthonormal vectors, the matrix B has the singular value decomposition [8]

$$B = USV^T = B = U^T SV.$$

Assuming that s_1, s_2 and s_3 are the singular values of B (i.e diagonal elements of S), and noticing that U and V are orthonormal vectors,

$$B = U\text{diag}\begin{bmatrix} s_1 & s_2 & s_3 \end{bmatrix} V^\top \tag{3.9}$$

and $s_1 \geq s_2 \geq s_3 \geq 0$.

Noticing that $tr(X^\top Y) = tr(YX^\top)$ and substituting B into $tr(AB^T)$ gives

$$tr(AB^\top) = tr(AV\text{diag}\begin{bmatrix} s_1 & s_2 & s_3 \end{bmatrix} U^\top) = tr(U^\top AV\text{diag}\begin{bmatrix} s_1 & s_2 & s_3 \end{bmatrix}). \tag{3.10}$$

Considering the constraint $det(A) = 1$, $tr(AB^T)$ is maximised with an optimal attitude matrix, A_{opt} that can be stated as follows:

$$A_{\text{opt}} = U\text{diag}\begin{bmatrix} 1 & 1 & \dfrac{\det(U)}{\det(V)} \end{bmatrix} V^\top \tag{3.11}$$

It is defined in Markley and Mortari [225] that

$$s_1 \equiv \sum_{11}, s_2 \equiv \sum_{22}, s_3 \equiv (\det U)(\det V) \sum_{33}, \tag{3.12}$$

where s_1, s_2 and s_3 are singular values, although the third singular value of B is actually $|s_3|$. Further, the rotation angle error vector Φ_{err} can be defined as

$$\exp[\Phi\times] = A_{\text{true}} A_{\text{opt}}^T \tag{3.13}$$

The covariance of the rotation angle error vector can be written as

$$P = U\text{diag}[(s_2 + s_3)^{-}1(s_3 + s_1)^{-}1)(s_1 + s_2)^{-}1]U^T \tag{3.14}$$

3.3.2 Davenport's q method

This method is mainly based on the concept of quaternion and is considered to be the first useful solution to the aircraft attitude estimation problem. Under this method, the attitude matrix can be written as a unit quaternion q[223, 331]:

$$q = \begin{bmatrix} q_0 \\ \mathbf{q} \end{bmatrix} = \begin{bmatrix} \cos\dfrac{\alpha}{2} \\ \mathbf{q}\sin\dfrac{\alpha}{2} \end{bmatrix}. \tag{3.15}$$

Considering the maximisation in equation (3.7), it can be stated as a homogeneous quadratic function of q,

$$tr(AB^{\top}) = q^{\top}Kq, \tag{3.16}$$

where K is the *symmetric traceless* matrix

$$K \equiv \begin{bmatrix} S - \mathrm{tr}(B)I & \mathbf{z} \\ \mathbf{z}^T & \mathrm{tr}(B) \end{bmatrix}, \tag{3.17}$$

where

$$S \equiv B + B^{\top} \text{and } z \equiv \begin{bmatrix} B_{23} - B_{32} \\ B_{31} - B_{13} \\ B_{12} - B_{21} \end{bmatrix} = \sum_i a_i \mathbf{b}_i \times r_i. \tag{3.18}$$

As we proved earlier, the optimal attitude can be found when $tr(AB^T)$ is maximised. Hence, $q^{\top}Kq$ is maximised subject to the constraint $|q| = 1$. The solution can be written as

$$Kq_{\mathrm{opt}} = \lambda_{\mathrm{max}} q_{\mathrm{opt}} \implies q_{\mathrm{opt}}^{\top} K q_{\mathrm{opt}} = \lambda_{\mathrm{max}}. \tag{3.19}$$

It can then be written further as

$$L(A_{\mathrm{opt}}) = \lambda_0 - \lambda_{\mathrm{max}}, \tag{3.20}$$

where the eigenvalues of the K matrix, λ_1, λ_2, λ_3 and λ_4, are related to the singular values of B,

$$\begin{bmatrix} \lambda_1 \\ \lambda_2 \\ \lambda_3 \\ \lambda_4 \end{bmatrix} = I_s \begin{bmatrix} s_1 \\ s_2 \\ s_3 \end{bmatrix},$$

where

$$I_s = \begin{bmatrix} 1 & 1 & 1 \\ 1 & -1 & -1 \\ -1 & 1 & -1 \\ -1 & -1 & 1 \end{bmatrix}. \tag{3.21}$$

The eigenvector (q_{opt}) obtained from equation (3.19) is used in

$$A(q) = (q_0^2 - |\mathbf{q}|^2)I + 2\mathbf{q}\mathbf{q}^{\top} - 2q_0[\mathbf{q}\times],$$

where

$$\mathbf{q}\times = \begin{bmatrix} 0 & -q_3 & q_2 \\ 0 & 0 & -q_1 \\ q_3 & 0 & 0 \\ -q_2 & q_1 & 0 \end{bmatrix}.$$

As K is traceless, the sum of the eigenvalues becomes zero. There is no unique solution if the two largest eigenvalues (λ_1 and λ_2) are zero, i.e. $s_2 + s_3 = 0$. This means that the data are insufficient to determine the attitudes [223].

Both the single value decomposition method and Davenport's q methods are significantly slower, even though they are robust [223].

3.3.3 Quaternion Estimation Algorithm (QUEST)

The QUaternion ESTimation (QUEST) algorithm is a popular algorithm for the single frame estimation of a quaternion that represents the attitude of a rigid body relative to a fixed coordinate system. This is widely applied in solving the Wahba problem since it was first applied in the MAGSAT mission in 1979 [223, 330]. Considering the equations (3.15) and (3.17) given under the Davenport q method, the following equivalent forms can be stated:

$$[(\lambda_{\max} + \text{tr}B)I - S]\mathbf{q} = q_0\mathbf{z},$$
$$(\lambda_{\max} - \text{tr}B)q_0 = \mathbf{q}^{\mathrm{T}}\mathbf{z}. \tag{3.22}$$

Then the optimal quaternion can be written as

$$q_{\text{opt}} = \frac{1}{\sqrt{r^2 + |\mathbf{x}|^2}}\begin{bmatrix}\mathbf{x}\\ \gamma\end{bmatrix}, r^2 + \|\mathbf{x}\|^2 \neq 0, \tag{3.23}$$

where (using the Cayley-Hamilton theorem [146])

$$\mathbf{x} = \text{adj}[(\lambda_{\max} + trB)I - S]^{-}1\mathbf{z} = [\alpha I + (\lambda_{\max} - trB)S + S^2]\mathbf{z}, \tag{3.24}$$

$$\gamma = \det[(\lambda_{\max} + \text{tr}B)I - S] = \alpha(\lambda_{\max} + \text{tr}B) - \det(S), \tag{3.25}$$

with

$$\alpha = \lambda_{\max}^2 - (\text{tr}(B))^2 + \text{tr}(\text{adj}S). \tag{3.26}$$

Substituting equation (3.23) in equation (3.22) with equation (3.24):

$$\psi(\lambda_{\max}) \equiv \gamma(\lambda_{\max} - trB) - \mathbf{z}^T[\alpha I + (\lambda_{\max} - trB)S + S^2]\mathbf{z} = 0. \tag{3.27}$$

Using the equations (3.25) and (3.26), equation (3.27) can be transformed into a fourth-order equation of $\lambda_{\max}$ which is essentially the characteristic equation of $\det(\lambda_{\max}I - K) = 0$.

The following simpler equation [330] can be written after successive application of the Cayley-Hamilton algorithm:

$$\gamma^2 + |x|^2 = \gamma(d\psi/d\lambda). \tag{3.28}$$

For the case of $r^2 + \|\mathbf{x}\|^2 = 0$, in equation (3.23), a sequential rotation-based approach was designed [330] to avoid the singularity.

Reference frame rotation in the QUEST method

Under the QUEST algorithm, the singularities problem can be solved with the attitude respect to the reference coordinate frame by 180° rotations about the x, y, z coordinate axis:

$$q^i \equiv q \otimes \begin{bmatrix}\mathbf{e}_i\\ 0\end{bmatrix} = \begin{bmatrix}\mathbf{q}\\ q_4\end{bmatrix} \otimes \begin{bmatrix}\mathbf{e}_i\\ 0\end{bmatrix} = \begin{bmatrix}q_4\mathbf{e}_i - \mathbf{q}\times\mathbf{e}_i\\ -\mathbf{q}\dot{\mathbf{e}}_i\end{bmatrix} \tag{3.29}$$

for $i = 1, 2, 3$, where $\mathbf{e}_i$ is the unit vector along ith coordinate system.

Using this equation, it is convenient to find coordinates from a reference frame. It can be proved that the inverse transformation is the same as this equation.

3.3.4 Fast optimal attitude matrix (FOAM)

This is a method that is simple and convenient to use and it is less subject to error by using a singular value decomposition of the B matrix [5]:

$$0 = \psi(\lambda_{\max}) \equiv (\lambda^2_{\max} - ||B||^2_F)^2 - 8\lambda_{\max}\mathrm{det}B - 4||\mathrm{adj}B||^2_F. \tag{3.30}$$

This equation is numerically identical to infinite-precision computations in equation (3.25) in the QUEST method.

The error covariance of the FOAM algorithm is as follows:

$$P = (K\lambda_{\max} - \det B)^{-1}(KI + BB^T). \tag{3.31}$$

3.3.5 Estimator of the optimal quarternion (ESOQ or ESOQ1) method

This method avoids a priori quaternion as in Quest and computes all diagonal elements; hence it is expensive [5]. In the ESOQ method rotation is not performed and an a priori quaternion with maximum magnitude is considered. Two methods are used in the ESOQ method, which is based on Davenport's eigenvalue equation (3.23).

1. First estimator of the optimal quaternion (ESOQ1)
   Under this method, the optimal quaternion is given by the following equations:

$$(q_{\mathrm{opt}})_k = -c[\det F^o + (\delta\lambda)tr(\mathrm{adj}F^o)] \tag{3.32}$$

   and

$$(q_{\mathrm{opt}})_{1,\dots,k-1,k+1,\dots,4} = c(\mathbf{g} + \mathbf{h}\delta\lambda). \tag{3.33}$$

2. Second estimator of the optimal quarternion (ESOQ2)
   This algorithm uses the rotation axis angle form of the optimal quarternion [5]. This method is based on the unit quaternion in Davenport's q method and two equations in the QUEST algortihm (equations 3.26 and 3.27), which are eqivalent to Davenport's eigenvalue method. Under this method, the following equation is derived:

$$q_{\mathrm{opt}} = \frac{1}{\sqrt{(\lambda_{\max} - trB)\mathbf{y}|^2 + (\mathbf{z}\cdot\mathbf{y})^2}} \begin{bmatrix} (\lambda_{\max} - trB)\mathbf{y} \\ \mathbf{z}\cdot\mathbf{y} \end{bmatrix}. \tag{3.34}$$

3.4 Quaternion Propagation

Using quaternions eliminates the need to use trigonometric functions [329]. We use a derivative of the dynamic model given in Yun and Bachmann [394]. Denoting the orientation quaternion in the earth coordinate frame as q and angular velocity ω, we state the the following equation:

$$\dot{q} = \frac{1}{2} q \otimes \omega, \tag{3.35}$$

where $\otimes$ denotes the quaternion multiplication with $\omega = [0\ \omega_1\ \omega_2\ \omega_3]^{\mathrm{T}}$ used as a *pure* quaternion. Now we use this model to deduce a state space formulation as follows.

3.5 MARG (Magnetic Angular Rates and Gravity) Sensor Arrays-based Algorithm

The algorithm proposed in Sebastian *et al.* [220] uses the quaternion representation in minimising the use of an analytically derived, optimal gradient descent algorithm on accelerometer and magnetometer data. This algorithm was introduced by Sebastian *et al.* and is applicable to inertial measurement units (IMUs).

The quaternion representation of measured angular rate is given by ω as follows:

$$\omega = \begin{bmatrix} 0 & \omega_x & \omega_y & \omega_z \end{bmatrix}^{\mathrm{T}}. \tag{3.36}$$

The quaternion derivative of the rate of change of the earth frame relative to the sensor frame is given by ${}^S_E\dot{q}$ according to equation (3.41). The orientation of the earth frame relative to the sensor frame at time t, ${}^E_S q_{\omega,t}$, was calculated by numerically intergrating ${}^S_E\dot{q}$ values:

$${}^S_E q_{\omega,t} = {}^S_E\hat{q}_{est,t-1} + {}^S_E\dot{q}_{\omega,t}\Delta t. \tag{3.37}$$

An optimisation algorithm named the gradient descent algorithm has been used to solve the optimisation problem as in equation (3.38) to identify orientation of the sensor,

$$\min_{{}^S_E\hat{q}\in R^4} f({}^S_E\hat{q}, {}^E\hat{d}, {}^S\hat{s}), \tag{3.38}$$

where ${}^S_E\hat{q}$ is the estimated orientation of the sensor, ${}^E\hat{d}$ is the predefined referenced earth direction such as gravity or the magennetic field and ${}^S\hat{s}$ is the measured field of the sensor. The orientation of the sensor ${}^S_E\hat{q}$ was found by applying the gradient descent algorithm to n iterations with a variable step size μ as follows:

$${}^S_E q_{k+1} = {}^S_E\hat{q}_k - \mu\frac{\Delta f({}^S_E\hat{q}_k, {}^E\hat{d}, {}^S\hat{s})}{||\Delta f({}^S_E\hat{q}_k, {}^E\hat{d}, {}^S\hat{s})||}, k = 0, 1, 2, 3 \ldots n. \tag{3.39}$$

The error direction of the solution surface is calculated using the objective function f and its Jacobian, J, as shown in the following equation:

$$\Delta f({}^S_E\hat{q}_k, {}^E\hat{d}, {}^S\hat{s}) = J^T({}^S_E\hat{q}_k, {}^E\hat{d}) f({}^S_E\hat{q}_k, {}^E\hat{d}, {}^S\hat{s}). \tag{3.40}$$

3.6 Model-based Estimation of Attitude with IMU Data

Superior performance in dynamic model-based estimators provides a natural choice for human pose estimation. A dynamic model that facilitates parameter estimation in a rotating and translating frame is crucial while the model can gradually be made sophisticated by incorporating full-body human biokinematic modelling. In this work we use a standard kinematic model to highlight the key contributions of this work. In the proposed algorithm, a quaternion-based approach is preferred as it eliminates the need to use trigonometric functions [329], thus avoiding singularities and Gimbal *lock*-associated complexities inherent in Euler angle-based representations.

Denoting the orientation quaternion in the reference coordinate frame as q and angular velocity ω, we state the following equation [394],

$$\dot{q} = \frac{1}{2} q \otimes \omega \tag{3.41}$$

where $\otimes$ denotes the quaternion multiplication with $\omega = [0\ \omega_1\ \omega_2\ \omega_3]^\mathsf{T}$ used as a *pure* quaternion. The gyro drift occurs due to accumulating the white noise of gyroscope readings [220]. Defining the state vector as $x = \left[x_1\ x_2\ x_3\ x_4\ x_5\ x_6\ x_7\ x_8\ x_9\ x_{10}\right]^\mathsf{T}$ where the $\left[x_1\ x_2\ x_3\right] = \left[\omega_1\ \omega_2\ \omega_3\right] = \omega$, $\left[x_4\ x_5\ x_6\ x_7\right] = \left[q_1\ q_2\ q_3\ q_4\right] = q$ and $[x_8\ x_9\ x_{10}]^\mathsf{T} = \delta$ where ω, q and δ are angular rates, quaternions and gyro drift, respectively, we can state the dynamic model as

$$\dot{x} = A(x) + Ww, \tag{3.42}$$

where

$$A(x) = \begin{bmatrix} -\dfrac{1}{\tau_x}x_1 \\ -\dfrac{1}{\tau_y}x_2 \\ -\dfrac{1}{\tau_z}x_3 \\ \dfrac{x_3x_5 - x_2x_6 + x_1x_7}{2\sqrt{x_4^2 + x_5^2 + x_6^2 + x_7^2}} \\ \dfrac{-x_3x_4 + x_1x_6 + x_2x_7}{2\sqrt{x_4^2 + x_5^2 + x_6^2 + x_7^2}} \\ \dfrac{x_2x_4 - x_1x_5 + x_3x_7}{2\sqrt{x_4^2 + x_5^2 + x_6^2 + x_7^2}} \\ \dfrac{-x_1x_4 - x_2x_5 - x_3x_6}{2\sqrt{x_4^2 + x_5^2 + x_6^2 + x_7^2}} \\ -\dfrac{1}{d_x}x_8 \\ -\dfrac{1}{d_y}x_9 \\ -\dfrac{1}{d_z}x_{10} \end{bmatrix}, \quad W = \begin{bmatrix} I_3 & O_{3\times3} \\ O_{4\times3} & O_{3\times3} \\ O_{3\times3} & I_3 \end{bmatrix}$$

$$w = \left[T_x, T_y, T_z, B_1, B_2, B_3\right]^\mathsf{T} \text{ with } A(x) \in \mathbf{R}^{10\times1} \text{ and } W \in \mathbf{R}^{10\times6}.$$

Here, T_x, T_y, T_z indicate the torque due to uncertain human movements and B_1, B_2, B_3 indicate the uncertainty in the bias responsible for the gyroscopic drift. I_m and $O_{m\times n}$ denote *identity* and *zero* matrix of appropriate dimensions. The measurement model can be stated as follows:

$$y = \hat{C}(x) + v, \tag{3.43}$$

where $y = \left[y_1 \cdots y_9\right]^\mathsf{T} = \left[\hat{\omega}_1\ \hat{\omega}_2\ \hat{\omega}_3\ \hat{a}_1\ \hat{a}_2\ \hat{a}_3\ \hat{h}_1\ \hat{h}_2\ \hat{h}_3\right]^\mathsf{T}$ is the IMU measurement vector with angular rate from gyroscopes, acceleration from accelerometers and orientation of the earth's magnetic field from magnetometers. Here, $v = \left[v_1\ v_2\ v_3\ 0\ 0\ 0\ 0\ 0\ 0\right]^\mathsf{T}$ is the measurement noise. Further, the time constants for the motion and variance of continuous white noise are denoted, respectively, by $\tau = \begin{bmatrix}\tau_x & \tau_y & \tau_z\end{bmatrix}^T$ and $d = \begin{bmatrix}d_x & d_y & d_z\end{bmatrix}^T$ [173]. Here

$$\hat{C}(x) = \begin{bmatrix} x_1 + x_8 \\ x_2 + x_9 \\ x_3 + x_{10} \\ -2\|\hat{g}\|\left(x_5x_7 - x_4x_6\right) \\ -2\|\hat{g}\|\left(x_4x_5 + x_6x_7\right) \\ -\|\hat{g}\|\left(x_4^2 - x_5^2 - x_6^2 + x_7^2\right) \\ 2\hat{h}_2^e\left(x_5x_6 + x_4x_7\right) + 2\hat{h}_3^e\left(x_5x_7 - x_4x_6\right) \\ \hat{h}_2^e\left(x_4^2 - x_5^2 + x_6^2 - x_7^2\right) + 2\hat{h}_3^e\left(x_6x_7 + x_4x_5\right) \\ 2\hat{h}_2^e\left(x_6x_7 - x_4x_5\right) + \hat{h}_3^e\left(x_4^2 - x_5^2 - x_6^2 + x_7^2\right) \end{bmatrix}.$$

Let the measurement in the reference frame have $\hat{g} = \begin{bmatrix}0 & 0 & -\|\hat{g}\|\end{bmatrix}^T$ and $\hat{h}^e = \begin{bmatrix}0 & \hat{h}_2^e & \hat{h}_3^e\end{bmatrix}^T$ acceleration and magnetometer readings, respectively.

Remark 1. *Without the loss of generality we have aligned the X axis of the stationary reference coordinate frame in a perpendicular direction to the magnetic direction to simplify the resulting expressions.*

Referring to equation (3.43), we have a unique solution given by

$$\begin{aligned} \hat{x}_m &= \sqrt{\frac{K_m + \sqrt{K_m^2 + 4L_m^2}}{2}} \quad m \in \{4, 6\} \\ \hat{x}_n &= \sqrt{\frac{-K_n + \sqrt{K_n^2 + 4L_n^2}}{2}} \quad n \in \{5, 7\} \\ \text{where } K_4 &= K_5 = \frac{p_2 + q_3}{2},\ L_4 = L_5 = \frac{q_2 - p_3}{2}, \\ K_6 &= K_7 = \frac{p_2 - q_3}{2},\ L_6 = L_7 = \frac{q_2 + p_3}{2} \\ \text{with } p_2 &= \frac{\hat{h}_2\|\hat{g}\| + \hat{a}_2\hat{h}_3^e}{\hat{h}_2^e\|\hat{g}\|},\quad p_3 = \frac{\hat{h}_3\|\hat{g}\| + \hat{a}_3\hat{h}_3^e}{2\hat{h}_2^e\|\hat{g}\|} \\ q_2 &= \frac{-\hat{a}_2}{2\|\hat{g}\|},\quad q_3 = \frac{-\hat{a}_3}{\|\hat{g}\|}. \end{aligned} \tag{3.44}$$

Considering the measurement uncertainty, let the measurement model for the converted measurements with respect to magnetometer and accelerometer measurements be:

$$\begin{aligned} &\hat{a}_i = a_i + v_i^a, \hat{h}_j^e = h_j^e + v_j^e, \hat{h}_j = h_j + v_j^h \\ &\forall i \in \{1, 2, 3\} \text{ and } j \in \{2, 3\}, \end{aligned} \tag{3.45}$$

where $\hat{a}_i$, $\hat{h}_j^e$ and h_j indicate the accelerometer readings subjected to measurement noise (v_i^a) and magnetometer measurement subjected to measurement noise (v_j^e and v_j^h) in the reference frame and the mobile frame, respectively. The error bounds are described in the following form.

Assumption 1. *The following holds.*

For given constants α and β, let $0 \le \upsilon_i^a \le \beta a_i$, $0 \le \upsilon_j^e \le \alpha h_j^e$ and $0 \le \upsilon_j^h \le \alpha h_j$ $\forall i \in \{1, 2, 3\}$ and $j \in \{2, 3\}$.

In the case of converted measurements, let us define the following:

$$\mu = \frac{(1+\alpha)(1+\beta)}{(1-\alpha)(1-\beta)}, \sigma = \frac{(1-\alpha)(1-\beta)}{(1+\alpha)(1+\beta)} \tag{3.46}$$

$$\lambda = \sqrt{\frac{\mu+\sigma}{2}}, \phi = \sqrt{\frac{\mu-\sigma}{2}}. \tag{3.47}$$

Now we can state the converted measurement as

$$\hat{x}_i = \lambda x_i + n_i \tag{3.48}$$

$$\text{where} \quad \|n_i(t)\| \le \|\phi x_i\| \quad \forall i \in [4, 5, 6, 7]. \tag{3.49}$$

Denoting

$$C = \left[\begin{array}{c:c:c} I_3 & O_{3\times4} & I_3 \\ \hdashline O_{4\times3} & \lambda I_4 & O_{4\times3} \end{array}\right] \text{and}$$

$$K = \left[\begin{array}{c:c:c} I_3 & O_{3\times4} & I_3 \\ \hdashline O_{4\times3} & \phi I_4 & O_{4\times3} \end{array}\right],$$

the converted measurement model corresponding to the non-linear measurement model in equation (3.43) can be stated in the following linear form:

$$y_c(t) = Cx(t) + n(t). \tag{3.50}$$

Here $y_c = \left[\omega_1\ \omega_2\ \omega_3\ \hat{x}_4\ \hat{x}_5\ \hat{x}_6\ \hat{x}_7\right]^{\mathrm{T}}$, $n(t) \triangleq \left[n_1(t)n_2(t)n_3(t)n_4(t)n_5(t)n_6(t)n_7(t)\right]$.

3.7 Robust Optimisation-based Approach for Orientation Estimation

The x_4, x_5, x_6 and x_7 of the state vector denotes the orientation quaternion. With $\mathbb{R}_+$ denoting the set of non-negative real numbers, define

$$\begin{aligned}
F(x) &= (x_4 - P)^2 + (x_5 - Q)^2 + (x_6 - R)^2 + (x_7 - S)^2, \\
G(x) &= -2Px_4 - 2Qx_5 - 2Rx_6 - 2Sx_7, \\
x &= [x_4\ x_5\ x_6\ x_7]^T, A_1 = [1\ 0\ 0\ 0]^T, \\
A_2 &= [0\ 1\ 0\ 0]^T, A_3 = [0\ 0\ 1\ 0]^T, A_4 = [0\ 0\ 0\ 1]^T, \\
\Gamma &= \sqrt{P^2 + Q^2 + R^2 + S^2}, p_1 = \frac{1}{\Gamma}[P\ Q\ R\ S]^{\mathrm{T}}, \\
p_2 &= \frac{-1}{\Gamma}[P\ Q\ R\ S]^{\mathrm{T}}, \\
\Omega &= \{x \in \mathbb{R}_+^4 : x_4^2 + x_5^2 + x_6^2 + x_7^2 = 1\}, \\
\Lambda &= \left\{ x \in \mathbb{R}_+^4 : \begin{array}{c} x_4^2 + x_5^2 + x_6^2 + x_7^2 \le 1 \\ \text{and} \\ x_4 + x_5 + x_6 + x_7 \ge 1 \end{array} \right\},
\end{aligned}$$

$\partial\Lambda$ is a boundary of Λ;

$h_4 = \{x \in \mathbb{R}^4_+ | \; x_4 = 0\}, h_5 = \{x \in \mathbb{R}^5_+ | \; x_5 = 0\},$

$h_6 = \{x \in \mathbb{R}^4_+ | \; x_6 = 0\}, h_7 = \{x \in \mathbb{R}^5_+ | \; x_7 = 0\},$

$h_8 = \{x \in \mathbb{R}^4_+ | \; x_4 + x_5 + x_6 + x_7 = 1\},$

$\Lambda_4 = (\partial\Lambda \backslash \Omega) \bigcap h_4, \Lambda_5 = (\partial\Lambda \backslash \Omega) \bigcap h_5,$

$\Lambda_6 = (\partial\Lambda \backslash \Omega) \bigcap h_6,$

$\Lambda_7 = (\partial\Lambda \backslash \Omega) \bigcap h_7, \Lambda_8 = (\partial\Lambda \backslash \Omega) \bigcap h_8.$

Now we can state the following lemma.

Lemma 1. *The solution to the following problem of*

min $F(x)$ *subjected to* $x \in \Omega$

can be stated as follows:

L1.1 If $P = Q = R = 0$ *then* $\begin{bmatrix} \frac{1}{2} & \frac{1}{2} & \frac{1}{2} & \frac{1}{2} \end{bmatrix}^{\mathrm{T}}$ *is the optimal solution.*

L1.2 If $P \geq 0, Q \geq 0, R \geq 0, S \geq 0$ *then the optimal value of* $(OP)_1$ *is:* $\min\{F(A_1), F(A_2), F(A_3), F(A_4), F(p_1)\}$

L1.3 If $P \leq 0, Q \leq 0, R \leq 0, S \leq 0$ *then the optimal value of* $(OP)_1$ *is:* $\min\{F(A_1), F(A_2), F(A_3), F(A_4), F(p_2)\}$

L1.4 Else the optimal value of $(OP)_1$ *is:* $\min\{F(A_1), F(A_2), F(A_3), F(A_4)\}$.

Proof: From L1.3 and L1.4, we see that if $x^* \in \Omega$ is an optimal point of problem $(OP)_3$, then it also is an optimal point of problem $(OP)_1$. Therefore, to solve problem $(OP)_3$, we only need to find an optimal point $x^* \in \Omega$ for problem $(OP)_3$. For $\gamma \in \mathbb{R}$, we denote the γ-level set for linear functional $G(x)$ as follows:

$$G_\gamma = \{x \in \mathbb{R}^4 | \; G(x) = \gamma\}.$$

Clearly, $G_\gamma, \gamma \in \mathbb{R}$ are parallel hyperplanes. Therefore, if G_{γ_0} is a supporting hyperplane of the convex set Λ at $x^0 \in \partial\Lambda$ then x^0 is an optimal point and $G(x^0) = \gamma_0$ is the optimal value of problem $(OP)_3$. Similar to the proof of L1.3, if x^0 belongs to one of five sets $\Lambda_i, i = 4, 5, \ldots, 8$ then one of four points A_1, A_2, A_3, A_4 is an optimal point of problem $(OP)_3$. On the other hand, G_{γ_0} is a supporting hyperplane of the convex set Λ at $x^0 = [x_4^0 \; x_5^0 \; x_6^0 \; x_7^0]^T \in \Omega$ if

$$\frac{x_4^0}{P} = \frac{x_5^0}{Q} = \frac{x_6^0}{R} = \frac{x_7^0}{R} \tag{3.51}$$

(for case $P \neq 0, Q \neq 0, R \neq 0$). In this case, equation (3.51) implies that

$$\frac{(x_4^0)^2}{P^2} = \frac{(x_5^0)^2}{Q^2} = \frac{(x_6^0)^2}{R^2} = \frac{(x_7^0)^2}{R^2}$$
$$= \frac{(x_4^0)^2 + (x_5^0)^2 + (x_6^0)^2 + (x_7^0)^2}{P^2 + Q^2 + 2R^2}. \tag{3.52}$$

If $P > 0, Q > 0, R > 0$, then by using L1.2 we have a unique solution that belongs to Ω of equation (3.51), which is p_1. If $P < 0, Q < 0, R < 0$ then by using L1.2 we have a unique solution that belongs to Ω of equation (3.51), which is p_2. Note that if $P = 0$, $Q = 0$, $P = 0$, then we conclude that $x_4^0 = 0$, $x_5^0 = 0$, $x_6^0 = x_7^0 = 0$, respectively. Otherwise equation (3.51) has no solution belonging to Ω. □

3.8 Implementation of the Orientation Estimation

The process of prefiltering is to ensure that the frequency-bounded noise is filtered out via simple low-pass filtering. Using the empirical knowledge, we set the bandwidth of the low-pass filters. We use the converted measurement as raw estimates for the linear robust Kalman filter while standard extended Kalman filtering and also robust extended Kalman filtering use the raw measurements when evaluating the performance of the estimators. Indeed, all these use the optimisation framework we mathematically justified to ensure that the standard quarternion constraints are met. As depicted in Figure 3.1, in the first step, the converted measurement approach is used to compute the quaternion with the magnetometer (h) and the accelerometer readings (a). The magnetometer readings suffer from scaling errors and offset biases. The errors are indeed device-specific and hence the normalised readings were used to calculate the quaternions.

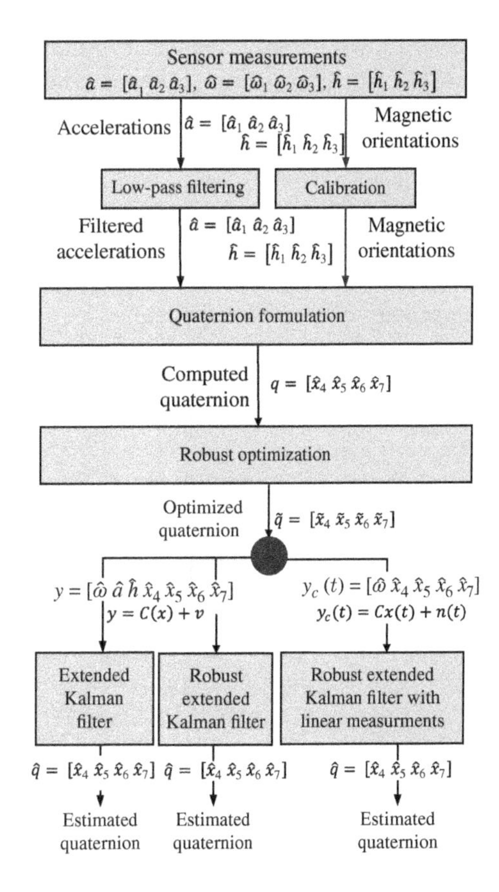

Figure 3.1 Block diagram of the algorithm.

3.8.1 Extended Kalman filter-based approach

The non-linear dynamic and measurement model described in equations (3.42) and (3.43), respectively, are used in the standard extended Kalman filter implementation

$$E(w^{\mathrm{T}}w) = \left[\begin{array}{c|c} Q_1 I_3 & O_3 \\ \hline O_3 & Q_2 I_3 \end{array}\right]. \tag{3.53}$$

The numerical values for Q_1 and Q_2 are evaluated as given in Yun and Bachmann [394].

3.8.2 Robust extended Kalman filter implementation

The non-linear dynamic and measurement model described in equations (3.42) and (3.43), respectively, are used under the normal bounded uncertainty assumption given in inequality (6.26).

3.8.3 Robust extended Kalman filter with linear measurements

The non-linear dynamic and measurement model described in equations (3.42) and (3.50), respectively, are used under the norm bounded uncertainty assumption given in inequality (6.26). The non-linear measurement model given in equation (3.43) is converted to the underlying linear form with the measurement assumptions in equation (3.45) resulting in equation (3.50). The quaternions obtained in equation (3.44) as converted measurements are in fact considered as a timewise observation in the linear measurement model in equation (3.50). Hence the measurement vector can be updated as $y = [y_1 \cdots y_{13}]^{\mathrm{T}} = [\omega_1\ \omega_2\ \omega_3\ a_1\ a_2\ a_3\ h_1\ h_2\ h_3\ \tilde{x}_4\ \tilde{x}_5\ \tilde{x}_6\ \tilde{x}_7]^{\mathrm{T}}$ with the angular rates from gyroscopes, accelerations from accelerometers, orientation of the earth's magnetic field from magnetometers and the measurement converted quaternions from equation (3.44). Furthermore, the time constants for the motion and variance of continuous white noise are denoted by τ and d, respectively.

3.9 Computer Simulations

Two hypothetical scenarios were considered to validate the underlying assertions by employing torques T_x, T_y, T_z and time constants τ_x, τ_y, τ_z in the respective Cartesian axes to emulate the relevant kinematics of the human arm. The torque gradually increases while the arm is being lifted and then is kept constant prior to reducing to the resting state, which corresponds to the upright position. Gyroscope, accelerometer and magnetometer readings were captured as the simulated kinematics using equations (3.42), (3.43) and (3.44). The resulting measurements were used with different estimators: extended Kalman filter (EKF), robust extended Kalman filter (REKF) and robust extended Kalman filter with linear measurements (REKFLM) for real-time estimation of the arm orientation. Figure 3.2 shows the actual angle variation with time and the estimated angle variation from each of the algorithms simultaneously for this hypothetical scenario. Notably, the shoulder pitch, yaw and roll angles deduced from the estimated state are the same for

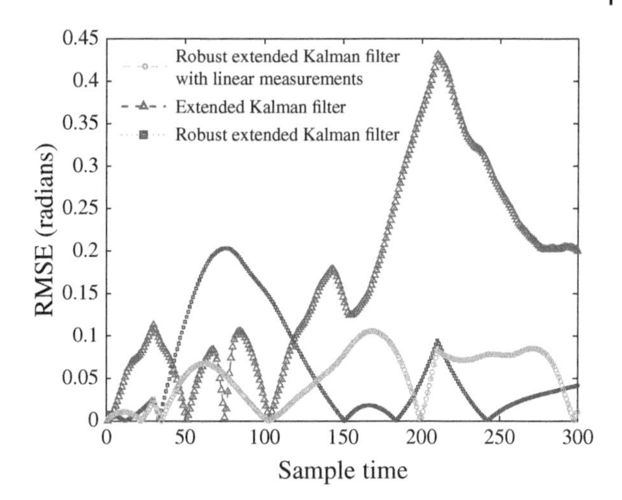

Figure 3.2 RMSE of the estimated angle.

each algorithm compared to the simulated actual angles when the uncertainty is low. However, the gyroscopic bias uncertainty ($\sqrt{B^T B}$), where $B \triangleq [B_1\ B_2\ B_3]^\top$ and B_1, B_2 and B_3, is taken as identical with 0.000 05 increments from 0.000 01. The estimation error is increased significantly, as depicted in Figure 3.3. Further, Gaussian noise was introduced to the generated measurements to validate the robustness of each algorithm under measurement noise uncertainty. The signal-to-noise ratio between 60 and 20 dB was introduced to the simulated accelerometer, magnetometer and gyroscope readings with the kinematic model parameters of τ_x,τ_y and τ_z set to 0.25 s^{-1} and $[B_1\ \ B_2\ \ B_3]$ set to $[0.0001\ \ 0.0001\ \ 0.0001]$. The second simulation is designed to investigate the optimisation algorithm discussed in Section 3.7. Unlike the previous case, the estimated quaternion ($[\hat{X}_4\ \ \hat{X}_5\ \ \hat{X}_6\ \ \hat{X}_7]$), prior to using as input to the estimator, is optimised using the proposed algorithm. Indeed, it is standard practice to normalise the quaternion and here we establish a mathematical justification for this process. The model parameters such as the time constant and the uncertainty constant are the same as they were for the first simulation. Gaussian noise (60 dB to 20 dB signal-to-noise ratios) was introduced to gyroscope, magnetometer and accelerometer readings in the first simulation.

3.10 Experimental Setup

An inertial measurement sensor in an integrated system with wireless communication was positioned on the wrist of the subject in order to capture the movement of the shoulder joint. Only one wearable sensor was used for the experiment to highlight the crucial requirement of engaging the smallest number of sensors. People with disabilities are usually reluctant to wear a number of sensors because it is inconvenient to engage in their day-to-day activities wearing several medical accessories [221]. Hence, we maintained the minimal sensor usage to ensure the comfort and facilitate the uptake.

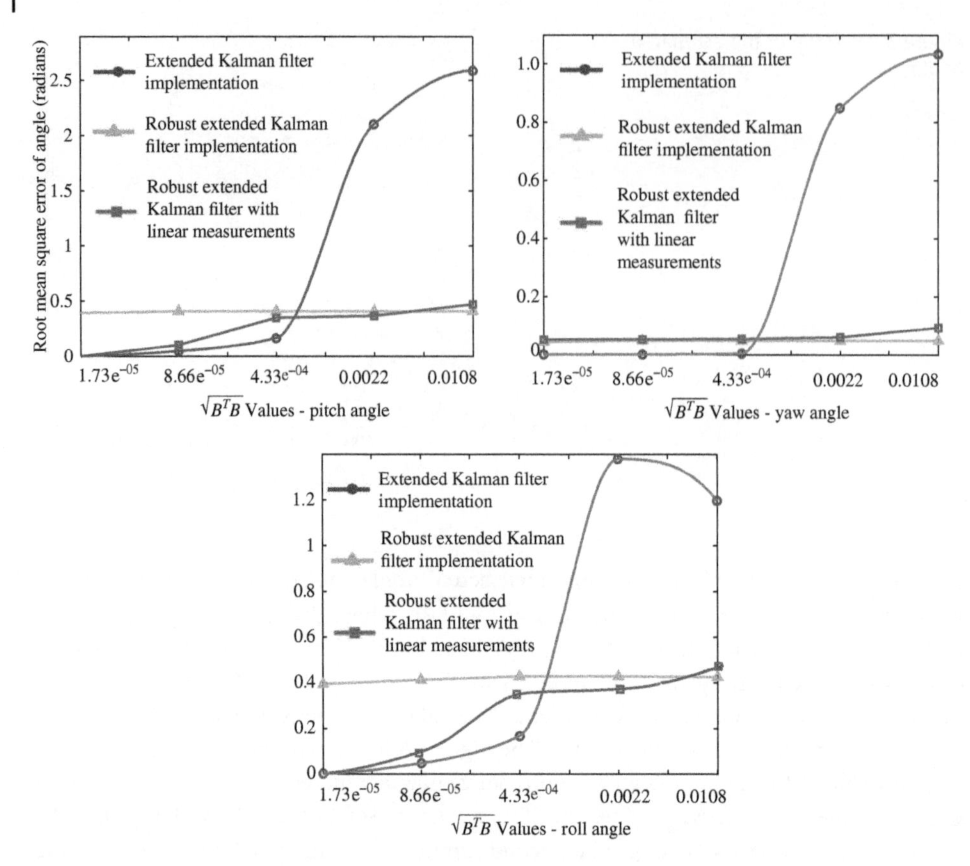

Figure 3.3 The error in estimated angle against the uncertainty bias ($\sqrt{B^T B}$).

The scapular movements for identifying normal and abnormal movements based on three-dimensional measurements have been described [233, 360]. Rotational motion of the scapula with respect to the thorax was described on the basis of a Euler angle sequence of external/internal rotation (ZS axis), upward/downward rotation (YS axis) and posterior/anterior tilting (XS axis). In our experiment, we have recorded and investigated shoulder movements in these three common movement scenarios present in a multitude of shoulder movement-related disabilities: upward/downward rotation (YS axis) in the sagittal plane, posterior/anterior tilting (XS axis) in the coronal plane and horizontal rotation in the transverse plane (ZS axis). These exercises essentially represent flexion–extension, abduction–adduction and internal–external rotation and are commonly used in the examination of shoulder and arm motor functionality using inertial sensors [288, 394]. Typically they are conducted in the biokinematic laboratories or under clinical scenarios to assess motor functionality of the shoulder as a subjective assessment for performing day-to-day activities such as lifting a bottle of water or placing a book on a shelf [188, 359], etc.

The validation of the underlying algorithms was conducted through data captured from four healthy subjects (two males and two females) using a Kinect© optical system and

10 healthy subjects (eight males and two females) using a Vicon optical system without any history of joint or muscle impairments. Each subject was asked to do three simple exercises:

1. Lifting the arm in front of the body by 90° (forward flexion–extension as Figure 3.6a,b).
2. Lifting the arm along the side of the body (abduction–adduction as Figure 3.6a,c).
3. Lifting the arm to the back of the body (backward flexion–extension as Figure 3.6a,d).

Each exercise was repeated three times over approximately 10 minutes with the inertial sensor worn on the distal left arm. The experiment setup is shown in Figure 3.6. The exercise routines were recorded using the Vicon optical motion capture system (Vicon T40S System) and a Microsoft Kinect© system separately.

Despite these exercises, the five subjects were employed to conduct horizontal flexion and extension in front of a Vicon optical motion capture system. The subjects were asked to swing the whole arm by the shoulder. The exercises were repeated three times over approximately 10 minutes with the inertial sensor worn on the distal left arm. The subject was in the orthostatic position with the sensor frames and reference frames approximately aligned initially. In the underlying formulation, the torques are considered to be uncertainty inputs and the time constants are determined in line with the prior computer simulations discussed in Section 3.9.

3.11 Results and Discussion

3.11.1 Computer simulations

The root mean squared error (RMSE) is plotted in Figure 3.4 for the three estimators considered, EKF, REKF and REKFLM, with the subjected (60 dB to 20 dB) noise levels. Irrespective of engaging the optimised quaternion (Section 3.7), the RMSE was less for REKFLM. This is particularly observable when the uncertainties are significant. Indeed, the filter accuracy in estimating the rotation angle improved when the noise level was reduced from 20 dB to 60 dB. The error in EKF increased markedly and the error in REKF was exaggerated compared to REKFLM. In all the estimation algorithms considered, quaternion optimisation had a positive yet reduced impact on lower noise levels (50 dB to 60 dB) on the angle estimation accuracy, unlike that for larger noise levels (20 dB to 30 dB). Indeed, the superior estimation accuracy in the robust extended Kalman filter with linear measurements (REKFLM) is further enhanced with the use of quaternion optimisation, as depicted in Figure 3.4. As shown in Figure 3.5, quaternion optimisation resulted in an approximately 30% RMSE improvement in the EKF implementation when the SNR was 20 dB, in addition to a more prominent improvement when the SNR was between 28 dB and 20 dB. In contrast, RMSE improvement in the REKF implantation was 42% when the SNR was 20 dB, with noticeable improvements in the 20 dB to 30 dB noise range. The RMSE improvement in REKFLM due to quaternion optimisation was relatively less in comparison to the other two algorithms, with an approximately 9% improvement when the SNR was 20 dB. REKFLM

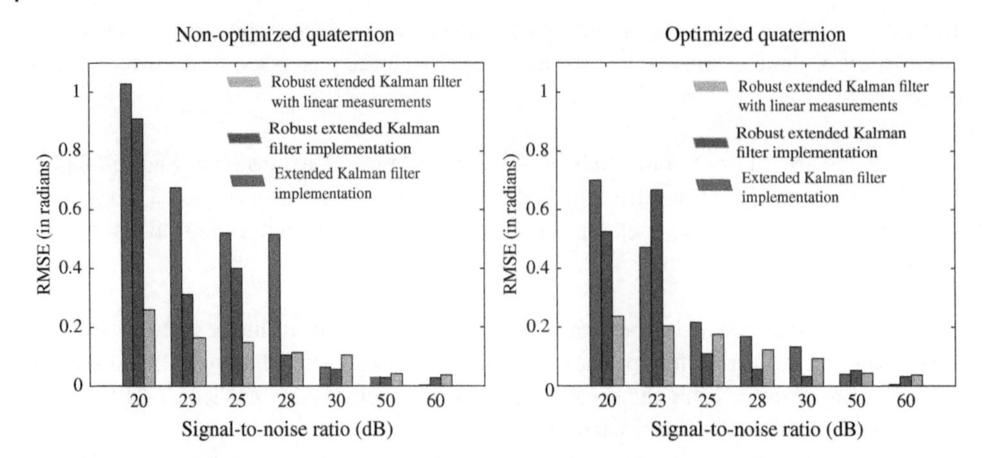

Figure 3.4 The RMSE subjected to introduced noise.

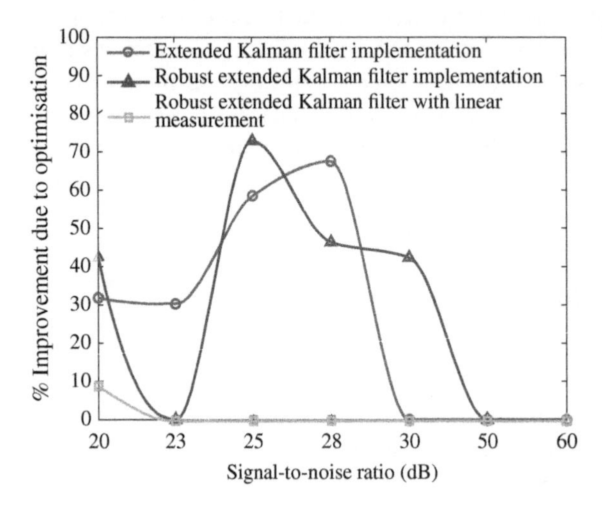

Figure 3.5 Percentage improvement due to quaternion optimisation.

outperforms the other estimators, albeit all approaches proclaim the benefit of quaternion optimisation to varying degrees.

3.11.2 Experiment

Figure 3.2 shows the RMSE in the estimated shoulder movement angles for the simple exercise of forward extension, when the movement replicated the execution in a simulated environment. Here the physical movements were carried out as close as possible to the simulated movements and the IMU measurements were then used to estimate the actual angle turned. The arm motion was along a planar trajectory in order to ensure minimal system complexity. This allowed the primary focus to be the assessment of the underlying filtering algorithms. This indeed avoided more complex torques necessary to generate arbitrary trajectories generally experienced in reality. Figure 3.7 shows the estimated angle (roll angle) difference compared to the Vicon optical system for the same exercise.

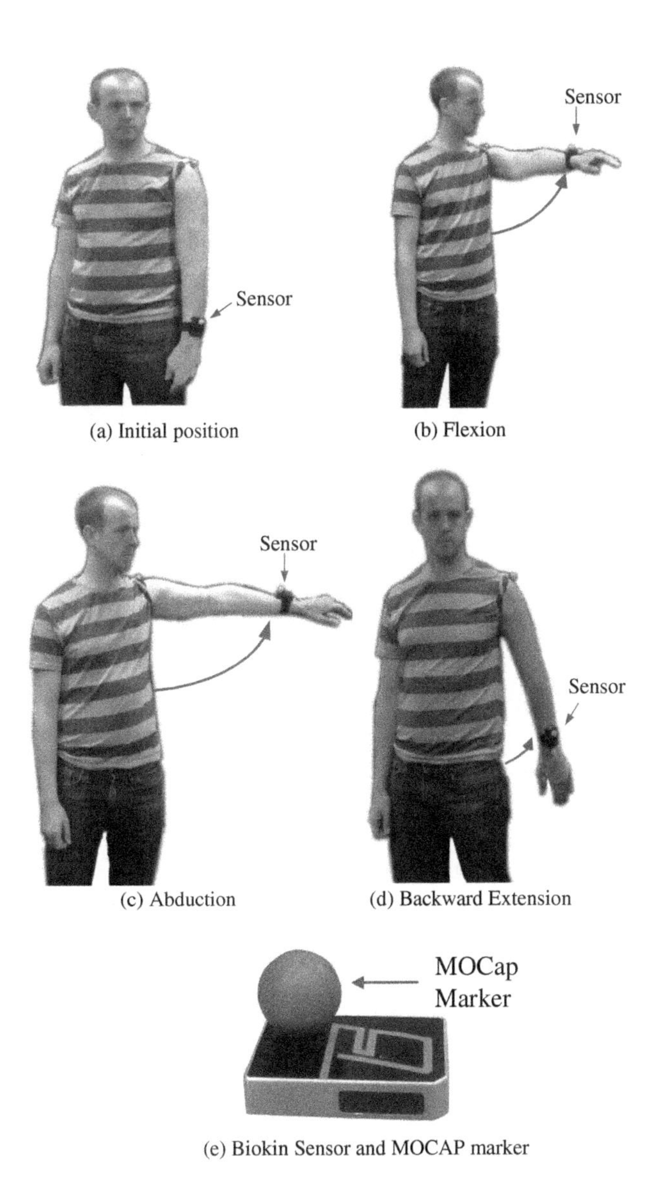

Figure 3.6 Experiment setup and procedure: the sensor and marker are worn on the wrist. Images taken with consent.

Here, Figure 3.7(a) and (b) show the RMSE in the corresponding angle differences optimised and non-optimised quaternions respectively. Angles derived from REKFLM were similar to the angles measured from the Vicon system irrespective of the engagement of quaternion optimisations (see Figure 3.7). Quaternion optimisation markedly improved each estimation algorithm, reducing the angle estimation error significantly. Table 3.1 lists the average RMSE for three exercises (forward flexion–extension, abduction–adduction and backward flexion–extension) when IMU measurements were compared to both Kinect© and Vicon systems. The graph in Figure 3.8 shows the performance, in terms of RMSE,

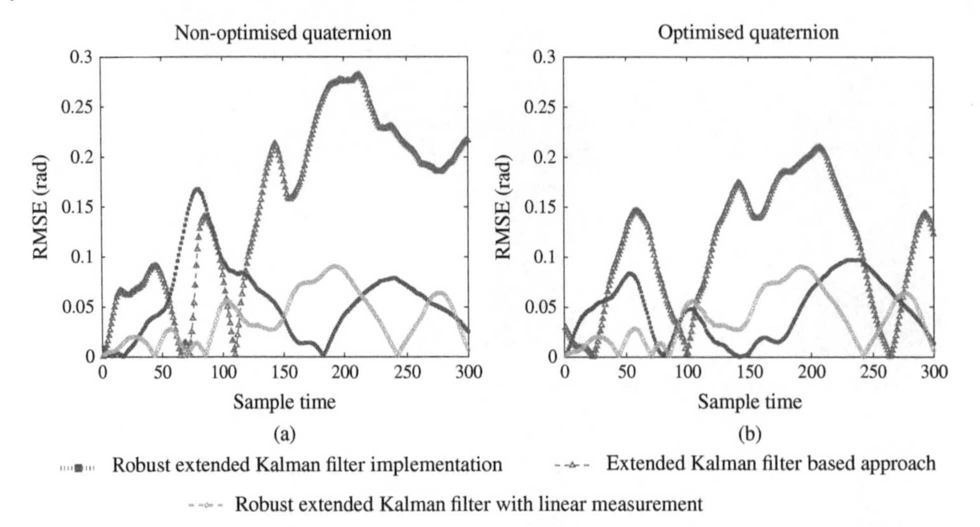

Figure 3.7 RMSE in angle estimation for forward extension exercise in comparison to VICON optical system.

of each algorithm. Figure 3.8(a) and (b) shows the RMSE over four healthy subjects with respect to the Kinect© optical system. Figure 3.8(c) and (d) shows the RMSE over 10 healthy subjects with respect to Vicon measurements. Similar to computer simulations, the EKF and REKFLM were the least and most accurate algorithms, respectively. Figure 3.8(b) and (d) depict improvement, in terms of RMSE and with respect to Kinect and Vicon measurements, due to the engagement of quaternion optimisation for each filter and subject respectively. As depicted in Table 3.1, the averaged RMSE with respect to Kinect© when using EKF was reduced by 43%, 34% and 10% for the three exercises, respectively, due to quaternion optimisation, while the averaged RMSE in the EKF algorithm was reduced by 36%, 21% and 19% with respect to the Vicon optical system. For the case of REKF, accuracy improved by 47%,49% and 14% with respect to the Kinect© system and 30%, 38% and 24% with respect to the Vicon optical system across the aforementioned exercises. This result implies that the accuracy in EKF and REKF methods improved significantly due to quaternion optimisation, yet the accuracy of the REKFLM algorithm improved by about 14%, 23% and 21%, respectively, for the above exercises with the Kinect© system. Accuracy of the REKFLM approach was improved by 20%, 21% and 7% for the forward flexion–extension exercise, abduction–adduction exercise and backward flexion–extension exercise, respectively, compared to the Vicon optical system when engaged with quaternion optimisation. Further, the accuracies in EKF, REKF and REKFLM were improved by 60%, 56% and 43% for the horizontal flexion–extension due to quaternion optimisation.

We notice that, generally, the REKFLM algorithm outperforms EKF and REKF. Furthermore, quaternion optimisation significantly improves the state estimation irrespective of the estimator. Further, our novel estimator was compared with two commonly used algorithms: the extended Kalman filter proposed by Bachman *et al.* and the Madgwick algorithm proposed by Madgwick *et al.* According to [394], the maximum error at the dynamic state was 9° in an extended Kalman filter against optical systems. This error was presented while

Table 3.1 Averaged RMSE error in the angle estimation for arm exercises in comparison to Kinect© and Vicon system-based measurements.

	Averaged RMSE of Non- Optimization Quaternion						Averaged RMSE of Optimised Quaternion					
	Compared to Kinect©		Optical System	Compared to Vicon		Optical System	Compared to Kinect©		Optical System	Compared to Vicon		Optical System
	EKF	REKF	REKFLM	EKF	REKF	REKFLM	EKF	REKF	REKFLM	EKF	REKF	REKFLM
Forward flexion–extension	0.2576	0.1631	0.0712	0.1469	0.0833	0.0491	0.1476	0.0874	0.0613	0.0911	0.0574	0.0393
Abduction–adduction	0.206	0.1242	0.0698	0.1181	0.0933	0.0531	0.1352	0.063	0.0537	0.0924	0.0569	0.0415
Backward flexion–extension	0.1622	0.117	0.0527	0.0904	0.058	0.0376	0.1475	0.1004	0.0417	0.0723	0.0438	0.0351
Horizontal flexion–extension	—	—	—	0.2674	0.1452	0.0957	—	—	—	0.1061	0.0641	0.0547

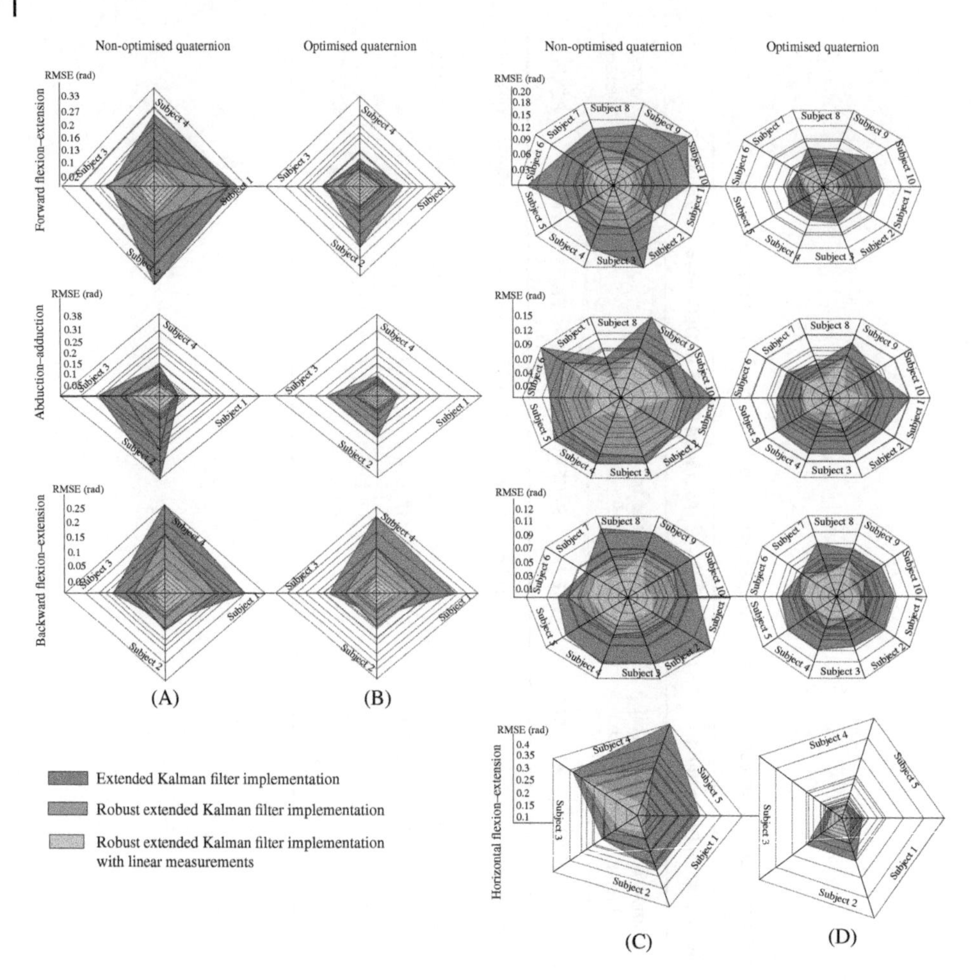

Figure 3.8 Filter performance comparison: RMSE in angle estimation for the upper arm exercises with respect to the Kinect© optical system (a,b), and with respect to the Vicon optical system (c,d).

conducting arm exercises such as extension–flexion. According to Madgwick *et al.* [220], the error of the Madgwick algorithm at the dynamic state was less than 7° compared to the optical system. The error of our novel robust extended Kalman filter with liner measurements (REKFLM) was less than 3°. According to the maximum errors presented for each of these three algorithms, REKFLM was outperforming the Madgwick algorithm and extended Kalman filter.

3.12 Conclusion

It has been demonstrated that adopting a linear formulation in the measurement scheme provides improved results for a real-time human kinematic movement estimation as opposed to the standard approach involving extended Kalman filtering or even a robust version of extended Kalman filtering. The measurement conversion-based linear approach does, in fact, result in improved estimation accuracy. Indeed, the quaternion normalisation improved the estimation accuracy of all estimators in general and the mathematical verification of the process completes the justification of the current practice in place. Although there is relatively less improvement due to quaternion estimation for the converted measurement Kalman filtering, the proposed approach still outperforms traditional approaches. These assertions have been verified by computer simulations as well as hardware experimentation.

4

Capturing Finger Movements

4.1 Introduction

Our hands play a pivotal role in performing daily activities and interacting with the surrounding world. Concerning the kinematics of the hand, we look at four main bones of each finger with the anatomical terminology given as metacarpal, proximal phalanx, middle phalanx, distal phalanx. The joint between the metacarpal and proximal phalanx is named the metacarpophalangeal joint (MCP). Similarly, the joint between the proximal phalanx and middle phalanx is called the proximal interphalangeal joint (PIP) while the joint between the middle phalanx and distal phalanx is called the distal interphalangeal joint (DIP). The joints of the human hand are classified into three types: flexion, directive or spherical joints, which consist of one degree of freedom (DOF); extension/flexion, two DOFs: one for extension/flexion and one for adduction/abduction; and three DOFs: rotation. For each finger and thumb, there are four DOFs. Considering the three DOFs for the rotation of the wrist, the hand model has 23 DOFs [183]. The position of each joint and the angles of interest of the finger joints are depicted in Figure 4.1.

Understanding the way our hands move provides an insight into how daily activities are performed, an integral part in rehabilitation following hand injuries. Measuring the phalangeal range of motion (ROM) or the flexibility of the hand is an essential part of clinical practice, particularly in exercise-based rehabilitation. An accurate and standardised tool to measure the active ROM of the hand is essential to any progressive assessment scenario in phalangeal rehabilitation. Historically, subjective visual examinations have been used to examine the ROM and, subsequently, declination joint angles were measured with universal goniometers to evaluate flexibility [73, 106, 296, 305]. With the development of technology, new goniometer models were gradually introduced and improved to assist clinicians [18, 32, 75, 98, 195, 201, 236, 264, 285, 380]. In measuring rotary motions of the forearm and shoulder, Laupattarakasem *et al.* [195] introduced an axial rotation gravity goniometer to improve reliability. In another study, one of the first two-element fibre-optic goniometers was built using graded-index microlens receivers [236]. The fibre-optic goniometer was improved in a later study by Donno *et al.* [98]. When personal computers became popular and were capable of effortlessly communicating with a variety of hardware, Barreiro *et al.* built a computer-based goniometer that can directly record declination angles into a personal computer [32]. Researchers also wanted to reduce the production cost of goniometers such as in Coburn *et al.*'s study [75], wherein they used

Human Motion Capture and Identification for Assistive Systems Design in Rehabilitation, First Edition.
Pubudu N. Pathirana, Saiyi Li, Yee Siong Lee and Trieu Pham.

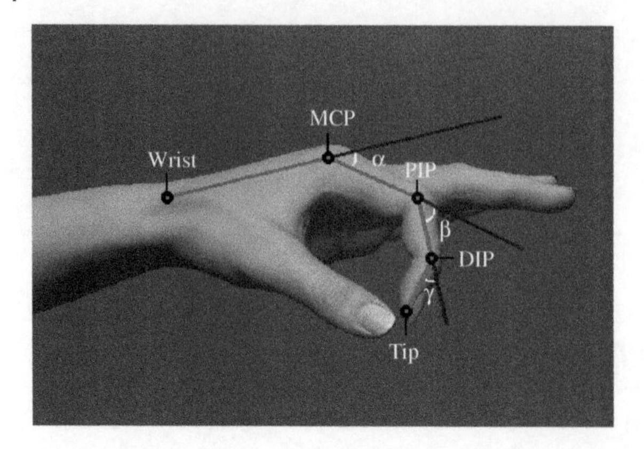

Figure 4.1 Position of phalangeal joints. Source: Trieu Pham.

remote sensors to build a goniometer. In more recent research, the development of MoCap systems provided a convenient and accurate approach to evaluate ROM such as the use of the Vicon system in Windolf *et al.*'s study [380] and the Kinect-based system in Pfister *et al.*'s study [285]. More interestingly, smartphones with integrated accelerometer sensors have also been considered [264].

With the sensor physically in contact with the finger, the joint measurements can be affected [52]. Another challenge in the current practice is that the assessment tools, including the universal goniometer, electro-goniometer, fibre-optic goniometer, Vicon and accelerometer integrated smartphones, require physical contact with the finger and therefore injuries such as burns, wounds, lacerations or even dermatological conditions can cause difficulties with the assessment tool. To avoid physical contact with the finger, users align goniometers along phalangeal bones while maintaining a small gap with the hand. However, this technique is inconvenient and tends to be subjective and prone to error. Another significant challenge is intra- and inter-rater reliability [105]. Studies into the reliability of a universal goniometer report a variance of 7–9° between therapists [105, 228] when measuring joint angles, leading to a 27° difference over the three joints of a finger. Rome *et al.* [305] compared the reliability of three types of goniometers: universal goniometer, fluid goniometer and electro-goniometer. The study suggested that each device cannot be used interchangeably. Due to the subjectiveness of the observers, the study also suggested that measurements of ankle dorsiflexion should be restricted to a single observer using only one device. These limitations of the current tools can influence assessments, treatments and the overall management of the rehabilitation process. Research has been conducted on the reliability of universal goniometers and proposed devices [18, 64, 161, 179, 184, 246, 252, 264], as reliability is an important aspect in clinical practice.

Adapting optical measurement systems, or computer vision-based approaches provides a non-contact form of measurements that can address the current challenges. In recent years, captures of hand movements have attracted attention, particularly with the development of a number of pervasive devices such as the Microsoft Kinect Sensor and Leap Motion Controller, as they offer better solutions in measuring both body and finger movements [377]. More recent research in this area used a Microsoft Kinect© to build a 3D skeletal hand tracking system [235, 266]. Metcalf *et al.* [237] recently proposed a Kinect-based system to capture motion and measure hand kinematics. However, Kinect© is primarily aimed

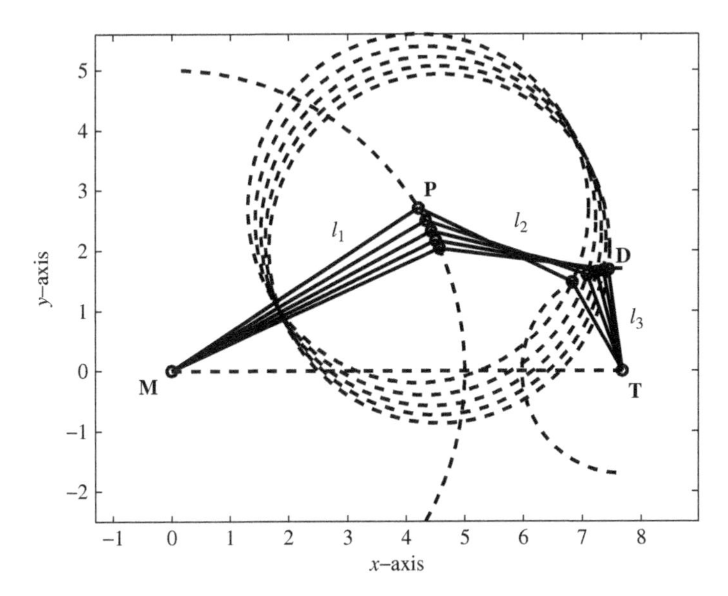

Figure 4.2 For a fixed base position of the metacarpal phalangeal joint **M**, there are infinitely many solutions of θ_{MCP}, θ_{PIP} and θ_{DIP}, or in other words **P** and **D**, to place the fingertip **T** into a particular position, in addition to the knowledge of the lengths of the phalangeal bones (l_1, l_2 and l_3).

at full-body movements and has limitations in respect of the accuracy required for finer finger movements. Although the study claims an accuracy of less than 15°, the system is not reliable to use in medical and rehabilitation applications due to a lack of an appropriate level of confidence in the measurements. While developing the system, Metcalf introduced a hypothesis that for a fixed base position of the metacarpal phalanx joint (MCP), there will be only one combination of θ_{MCP}, θ_{PIP} and θ_{DIP} to place the fingertip into a particular position. While this hypothesis was experimentally verified in their paper, the trial was limited to only healthy subjects. There is no guarantee that all the joints and muscles in an injured hand can follow regular movements and constraints. Moreover, it is clear that the hypothesis failed from its mathematical standpoint (Figure 4.2). The Leap Motion Controller is a promising device for hand gesture recognition and fingertip tracking but lacks the interphalangeal joint detection capability necessary for the underlying medical application. In its hand-tracking model, the proximal interphalangeal joint and the distal interphalangeal joint are estimated based on the constraint $\theta_{\text{DIP}} = \frac{2}{3}\theta_{\text{MCP}}$, which is defined only for healthy hands. As reported by ElKoura *et al.* [104], this constraint is a rough approximation for the intricate control of the hand. In his experiments with 20,000 samples, the difference of the value $\theta_{\text{DIP}} - \frac{2}{3}\theta_{\text{MCP}}$ ranges from −36° to 42° with a 95% confidence level. In therapeutic rehabilitation, where impaired hands are the major objective, the hand tracking model of the Leap Motion Controller cannot be applied to the finger joint position of injured hands.

A non-contact measurement scheme using a Creative Senz3D Camera (Figure 4.3) can be used to improve the accuracy in measuring the finger joint angles. While Kinect© is designed for full body capture, the Creative Senz3D is optimized for short-range gesture

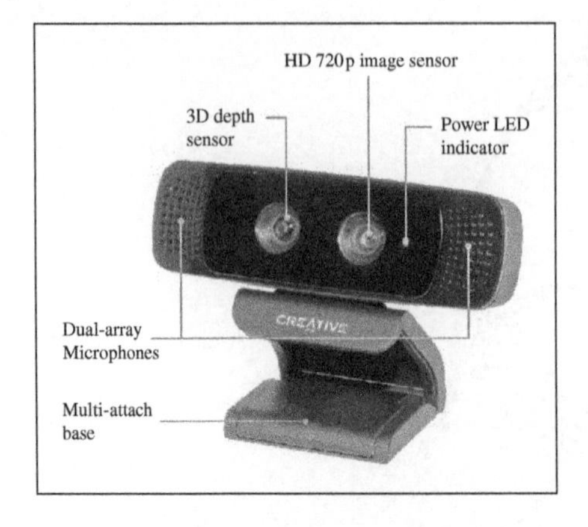

Figure 4.3 The Creative Senz3D Camera. Source: Trieu Pham.

interaction and, hence, it is a better alternative for hand measurements. Albeit optical measurement schemes [86, 362] providing a non-contact form of measurements eliminating such issues in addition to improved hygiene, one limitation of such optical devices is that it may not be possible to measure all the joint angles due to occlusions and may require the use of multiple sensors. The system proposed in Pham *et al.* [286] is the first non-contact measurement system utilising Intel Perceptual Technology and a Senz3D Camera for measuring phalangeal joint angles. A new approach to achieve the total active movement without measuring three joint angles individually in order to enhance the accuracy of the system is proposed. An equation between the actual spatial position and the measurement value of the proximal interphalangeal joint was established through the measurement values of the total active movement so that its actual position can be inferred. Verified by computer simulations, experimental results demonstrated a significant improvement in the calculation of the total active movement and successfully recovered the actual position of the proximal interphalangeal joint angles. A trial that was conducted to examine the clinical applicability of the system involving 40 subjects confirmed the practicability and consistency in the proposed system. The time efficiency conveyed a stronger argument for this system to replace the current practice of using goniometers.

4.2 System Overview

The measurement system consists of an affordable optical sensor Creative Senz3D and a computer for data acquisition from the optical sensor (Figure 4.4) and processing. The depth image of the Creative Senz3D has a 640 × 480 pixel resolution and refresh frequency of 30 Hz. The base tracking algorithm behind the system was first introduced by Melax [235] and later integrated into the Intel® Perceptual Computing Software development kit. It uses a convex rigid model to approximate the hand. Each phalangeal bone of the hand is approximated by a convex rigid body, with one rigid body to describe the palm. The rigid hand model is 20 cm long and adjustable to fit most hand dimensions in order to improve accuracy. Tracking information of the hand is optimised with our proposed method represented in Section 4.3. Our proposed approach improves the accuracy specifically for joint angle measurements as it is based on the anatomical structure of the hand.

Figure 4.4 Setup of the measurement system.

4.3 Accuracy Improvement of Total Active Movement and Proximal Interphalangeal Joint Angles

To mathematically describe physiological structure (Figure 4.5), let the wrist joint, metacarpophalangeal joint, proximal interphalangeal joint, distal interphalangeal joint and tip joint in the finger plane be denoted by the letters W, M, P, D and T, respectively. Let α, β, γ be the declination angles for the metacarpophalangeal joint, the proximal interphalangeal joint and the distal interphalangeal joint, respectively. Then,

$$\alpha = \arccos \frac{\overrightarrow{WM}.\overrightarrow{MP}}{\left|\overrightarrow{WM}\right|.\left|\overrightarrow{MP}\right|}, \tag{4.1}$$

$$\beta = \arccos \frac{\overrightarrow{MP}.\overrightarrow{PD}}{\left|\overrightarrow{MP}\right|.\left|\overrightarrow{PD}\right|}, \tag{4.2}$$

$$\gamma = \arccos \frac{\overrightarrow{PD}.\overrightarrow{DT}}{\left|\overrightarrow{PD}\right|.\left|\overrightarrow{DT}\right|}. \tag{4.3}$$

Also, as defined in a global coordinate frame, let $e_i, i \in [1, \ldots, 4]$ denote the measurement noise of directions of $\overrightarrow{WM}, \overrightarrow{MP}, \overrightarrow{PD}, \overrightarrow{DT}$ respectively. Assuming a normal probability distribution, $p\left(e_i\right) \sim N\left(0, Q_i\right)$ for $i \in [1, \cdots, 4]$, where Q_is indicate the noise power spectral densities. Using ^ to indicate the measured value and $_0$ to indicate the actual value, the

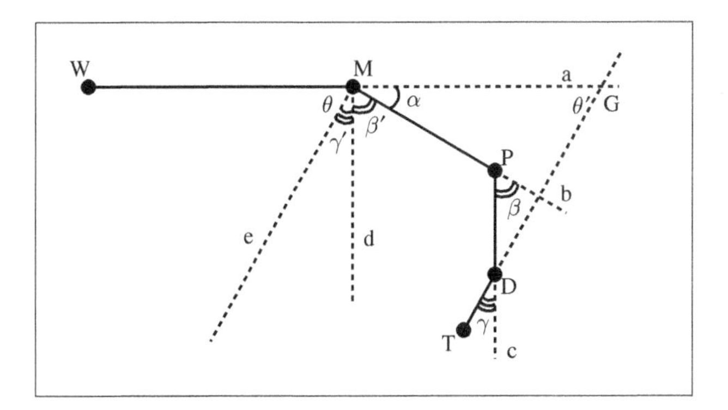

Figure 4.5 Geometry model of the finger.

measured declination angles can be denoted as follows:

$$\hat{\alpha} = \alpha_0 + e_1 + e_2,$$
$$\hat{\beta} = \beta_0 + e_2 + e_3, \tag{4.4}$$
$$\hat{\gamma} = \gamma_0 + e_3 + e_4.$$

The concept of total active movement (TAM) has traditionally been used as a means of determining the ROM at the MCP, PIP and DIP joints, and it has been used in the evaluation of hand performance in many studies [111, 135, 174, 306]. The TAM of a finger is defined as

$$\mathbf{S} = \alpha + \beta + \gamma.$$

Consequently, the TAM denoted interim of declination angles can be stated as

$$\hat{\mathbf{S}}_1 = \hat{\alpha} + \hat{\beta} + \hat{\gamma} = e_1 + 2e_2 + 2e_3 + e_4 + \alpha_0 + \beta_0 + \gamma_0. \tag{4.5}$$

Measuring three phalangeal joint angles to calculate the TAM is invariably subjected to the four relevant measurement noise inputs (e_1, e_2, e_3 and e_4). Noticing that the TAM can be obtained with lesser measurements ($\overrightarrow{WM}$ and $\overrightarrow{DT}$), and so lesser noise (e_1 and e_4), we now look at the geometry model of the finger. Through point M, we construct line Md and Me parallel to $\overrightarrow{PD}$ and $\overrightarrow{DT}$, respectively. Consequently, $\hat{\beta}$ and $\hat{\beta}'$ are a pair of corresponding angles and $\hat{\gamma}$ and $\hat{\gamma}'$ are a pair of corresponding angles:

$$\hat{\beta} = \hat{\beta}',$$
$$\hat{\gamma} = \hat{\gamma}'.$$

Let θ be the angle between $\overrightarrow{MW}$ and Me; θ can be negative if Me is in the opposite half-plane to the half-plane consisting of P and D. For the purpose of this section as well as the human finger joint limitations, we restrict angles α, β and γ to the range of 0° to 90° each. Let G be the intersection of the lines DT and WM. As $\overrightarrow{DT}$ and Me are parallel, $\widehat{WGT}$ is equal to θ. Then, it is possible to achieve the TAM through the measurement of $\overrightarrow{WM}$ and $\overrightarrow{DT}$ as follows:

$$\mathbf{S}_2 = \alpha + \beta + \gamma = \pi - \theta = \pi - \widehat{WGT}. \tag{4.6}$$

Angle $\widehat{WGT}$ is dependent only on two measurement noise processes:

$$\widehat{WGT} = \hat{\theta} = \theta_0 - e_1 - e_4.$$

The advantage of this approach is that it eliminates noise from measurements of $\overrightarrow{MP}$ and $\overrightarrow{PD}$ compared to equation (4.5):

$$\hat{\mathbf{S}}_2 = \pi - \hat{\theta} = \pi - \theta_0 + e_1 + e_4. \tag{4.7}$$

Exploiting this result, we can improve the accuracy of the proximal interphalangeal joint angle. From equations (4.5) and (4.6), we represent the first approach for calculating TAM:

$$\hat{\mathbf{S}}_1 = e_1 + 2e_2 + 2e_3 + e_4 + \pi - \theta_0. \tag{4.8}$$

From equations (4.7) and (4.8),

$$\hat{\mathbf{S}}_1 - \hat{\mathbf{S}}_2 = 2\left(e_2 + e_3\right). \tag{4.9}$$

From equations (4.4) and (4.9),

$$\beta_0 = \hat{\beta} - \frac{\hat{\mathbf{S}}_1 - \hat{\mathbf{S}}_2}{2}. \tag{4.10}$$

Equation (4.10) shows that we can find the real declination angle for PIP. We choose the estimated value $\tilde{\beta}$ of β as described in equation (4.10). This is important for the development of a monocular measurement system. There is a high chance of the proximal interphalangeal joint being occluded by the distal phalangeal but our contribution helps to overcome this problem.

In practice, the vectors $\overrightarrow{MP}$, $\overrightarrow{PD}$ and $\overrightarrow{DT}$ are calculated from the point cloud using orthogonal regression for a linear fitting. Tracking information of the SDK is only used for segmentation purposes since the hand tracking algorithm is optimised for speed and recognition rate rather than the precise bone positions. As the application is focused on the phalangeal joint angle, finding the directions of vectors $\overrightarrow{MP}$, $\overrightarrow{PD}$ and $\overrightarrow{DT}$ is crucial. Consider the line $\mathbf{L}$ so that the vector $\overrightarrow{MP}$ is in the direction vector of $\mathbf{L}$:

$$\mathbf{L}(t) = t\vec{\mathbf{D}} + \mathbf{A},$$

where $\vec{\mathbf{D}}$ is along the line $\mathbf{L}$. Define $\mathbf{X}_i$ to be the cloud point of the proximal phalangeal bone; then

$$\mathbf{X}_i = \mathbf{A} + d_i\vec{\mathbf{D}} + p_i\vec{\mathbf{D}}_i^{\top},$$

where $d_i = \vec{\mathbf{D}} \cdot (\mathbf{X}_i - A)$ and $\vec{\mathbf{D}}_i^{\top}$ is some unit vector perpendicular to $\vec{\mathbf{D}}$ with the appropriate coefficient p_i. Define $\mathbf{Y}_i = \mathbf{X}_i - \mathbf{A}$. The vector $\vec{\mathbf{N}}^{\mathbf{A}}$ from $\mathbf{X}_i$ to its projection on to the line is

$$\mathbf{Y}_i - d_i\vec{\mathbf{D}} = p_i\vec{\mathbf{D}}_i^{\top}.$$

The cost function for the least squares minimisation is $C(\mathbf{A}, \mathbf{D}) = \sum_{i=1}^{m} \| p_i \|^2$. Using equation (1.12), the cost function can be rewritten as

$$C(\mathbf{A}, \mathbf{D}) = \sum_{i=1}^{m} \left(\mathbf{Y}_i^{\top} \left[I - \vec{\mathbf{D}}\vec{\mathbf{D}}^{\top} \right] \mathbf{Y}_i \right).$$

Taking the derivative with respect to A,

$$\frac{\partial C}{\partial A} = -2 \left[I - \vec{\mathbf{D}}\vec{\mathbf{D}}^{\top} \right] \sum_{i=1}^{m} \mathbf{Y}_i.$$

The cost function is minimized when the derivative is zero. A similar approach is used to approximate vectors $\overrightarrow{PD}$, $\overrightarrow{DT}$.

The direction of the vector $\overrightarrow{WM}$ is along the intersection of the finger plane and the palm plane. The palm plane is also approximated using orthogonal regression. The palm plane p has a unit normal vector $\vec{\mathbf{N}}$ and a central point of the palm $\mathbf{A}$. Define $\mathbf{X}_i$ to be cloud points of the palm; then

$$\mathbf{X}_i = \mathbf{A} + \lambda_i\vec{\mathbf{N}} + p_i\vec{\mathbf{N}}_i^{\perp},$$

where $\lambda_i = \vec{\mathbf{N}} \cdot (\mathbf{X}_i - \mathbf{A})$ and $\vec{\mathbf{N}}_i^{\perp}$ is some unit vector perpendicular to $\vec{\mathbf{N}}$ with an appropriate coefficient p_i. Define $\mathbf{Y}_i = \mathbf{X}_i - \mathbf{A}$. The vector $\vec{\mathbf{N}}^{\mathbf{A}}$ from $\mathbf{X}_i$ to its projection on to the palmar plane is $\lambda_i \vec{\mathbf{N}}$. Then,

$$\| \vec{\mathbf{N}}^{\mathbf{A}} \|^2 = \lambda_i^2 = (\mathbf{N} \cdot \mathbf{Y}_i)^2.$$

The cost function for the least squares minimisation is $C(\mathbf{A}, \vec{\mathbf{N}}) = \sum_{i=1}^{m} \lambda_i^2$. The cost function can be rewritten as

$$C\left(\mathbf{A}, \vec{\mathbf{N}}\right) = \mathbf{N}^{\mathrm{T}} \left(\sum_{i=1}^{m} \mathbf{Y}_i \mathbf{Y}_i^{\mathrm{T}} \right) \mathbf{N} = \mathbf{N}^{\mathrm{T}} M(A) \mathbf{N},$$

where $M(A)$ is given by

$$M(A) = \begin{bmatrix} \sum_{i=1}^{m} (x_i - x_{\mathbf{A}})^2 & \sum_{i=1}^{m} (x_i - x_{\mathbf{A}})(y_i - y_{\mathbf{A}}) & \sum_{i=1}^{m} (x_i - x_{\mathbf{A}})(z_i - z_{\mathbf{A}}) \\ \sum_{i=1}^{m} (x_i - x_{\mathbf{A}})(y_i - y_{\mathbf{A}}) & \sum_{i=1}^{m} (y_i - y_{\mathbf{A}})^2 & \sum_{i=1}^{m} (y_i - y_{\mathbf{A}})(z_i - z_{\mathbf{A}}) \\ \sum_{i=1}^{m} (x_i - x_{\mathbf{A}})(z_i - z_{\mathbf{A}}) & \sum_{i=1}^{m} (y_i - y_{\mathbf{A}})(z_i - z_{\mathbf{A}}) & \sum_{i=1}^{m} (z_i - z_{\mathbf{A}})^2 \end{bmatrix}.$$

The cost function is in the quadratic form and the minimum is the smallest eigenvalue of $M(A)$. The corresponding eigenvector N is the normal to the palm plane that we need to find.

4.4 Simulation

The inference in Section 4.3 is ascertained through a simulation exercise presented in this section. A configuration of a finger is a set of three declination joint angles of the finger. For each joint, 19 joint angle values of 0–90° were evenly generated with the resolution of 5°; hence there were a total of $19^3 = 6851$ configurations generated for the hand. Random Gaussian noise was added to each joint angle value. The original value of each joint is assumed to be the actual value of the joint angle, and the later value, which was added Gaussian noise, is assumed to be the measured value from the tracking algorithm of the camera. These configurations were classed into six groups based on the total active movement value: 0–45°, 45–90°, 90–135°, 135–180°, 180–225° and 225–270°. The number of configurations for each group is presented in Table 4.1. After generating data using the above protocol, we had data that covered almost all possible configurations of a human finger. In the next step, we computed the TAM value of each configuration using two methods and compared those with the actual value of TAM. In Table 4.1, the difference between the actual TAM value and the measurements using two methods ($\mathbf{S}_1$ and $\mathbf{S}_2$) mentioned above is denoted by

$$\Delta\mathbf{S}_1 = \left| \hat{\mathbf{S}}_1 - \mathbf{S}_0 \right|,$$

$$\Delta\mathbf{S}_2 = \left| \hat{\mathbf{S}}_2 - \mathbf{S}_0 \right|.$$

Overall, the values of $\Delta\mathbf{S}_1$ are larger than the values of $\Delta\mathbf{S}_2$ in the same group. The mean values of 6851 configurations for $\Delta\mathbf{S}_1$ and $\Delta\mathbf{S}_2$ are 12.47° and 5.62°, respectively. From these figures, implementing the proposed approach to find the TAM value can generally reduce

Table 4.1 Number of configurations for six groups of TAM.

Group	Quantity	ΔS_1	ΔS_2	$\Delta\beta_1$	$\Delta\beta_2$
0–45°	165	13.7	5.9	6.1	0
45–90°	975	12.3	5.7	5.5	0
90–135°	2154	12.5	5.6	5.6	0
135–180°	2235	12.1	5.5	5.6	0
180–225°	1110	13	5.8	5.8	0
225–270°	220	12.7	5.4	5.6	0

the error of measurements of TAM from 12.47° to 5.62°, which is significant. Similarly, the differences between the actual PIP joint angle β_0 and the initial measurement $\hat{\beta}$ is denoted as $\Delta\beta_1$; the difference between the actual PIP joint angle β_0 and the estimated value $\tilde{\beta}$ is denoted by $\Delta\beta_2$, i.e.

$$\Delta\beta_1 = \hat{\beta} - \beta_0,$$
$$\Delta\beta_2 = \tilde{\beta} - \beta_0.$$

The mean value of 6851 configurations for $\Delta\beta_1$ is 5.64° while the one for $\Delta\beta_2$ is 0°. This result infers that the mean error of the initial measurement is 5.64°, and the value of approximately 0° is the difference between the actual β_0 and the estimated $\tilde{\beta}$ obtained by equation (4.10), thus confirming our assertions. Figure 4.6 shows further details of statistical characteristics including median, quartiles and whiskers. In summary, this simulation shows the reductions in error if the proposed approach is employed on top of the camera measurements.

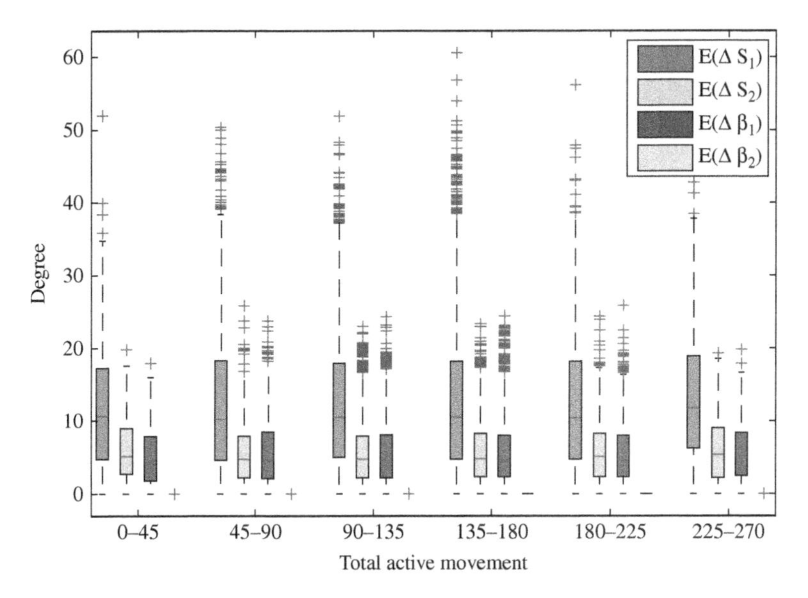

Figure 4.6 Simulation results for accuracy improvement of the TAM and PIP.

4.5 Trial Procedure

In this section, we present a validation procedure to ascertain the reliability of the system. The final target of this procedure was to examine whether the system can be used in clinics to replace the traditional universal goniometer. The experimental procedure was approved by Deakin University Human Ethics Advisory Group (HEAG) and all participants provided their written informed consent to participate. Two professional hand therapists conducted this trial using a universal goniometer. Although human ratings can be subjective and may not possess a degree of accuracy, it is the current practice. The procedure was conducted following the clinical recommendations of the American Society of Hand Therapists. The dorsal measurement technique, which was proven to be as reliable as a lateral placement [260], was used during the procedure.

Forty participants were recruited to participate in the trial. There were 35 females and five males in the study, with the mean age of 29.3 years ($\sigma = 11.5$); 35 subjects were right hand dominant and five subjects were left hand dominant. Both hands of the participants were measured in the trial. Participants were seated at a table with their arm on the hand elevator and elbow supported on the table. This position put the shoulder in an approximately 45–80° flexion, the elbow in approximately 40–60° flexion and the wrist in a neutral, pronated position. The Creative Senz3D camera was mounted on a tripod and placed below, in front of the participant's hand, as depicted in Figure 4.7. A training trial was conducted

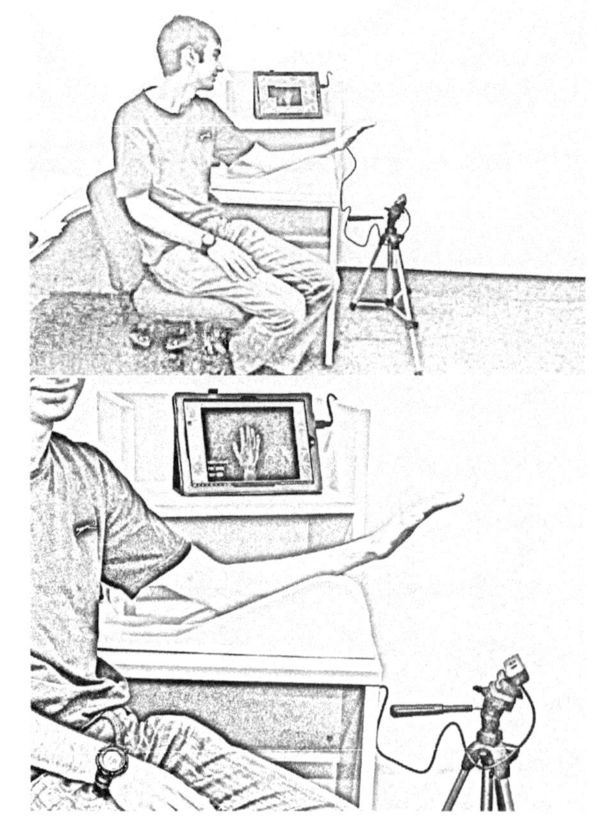

Figure 4.7 The left hand of the man was measured by our system. Source: Trieu Pham.

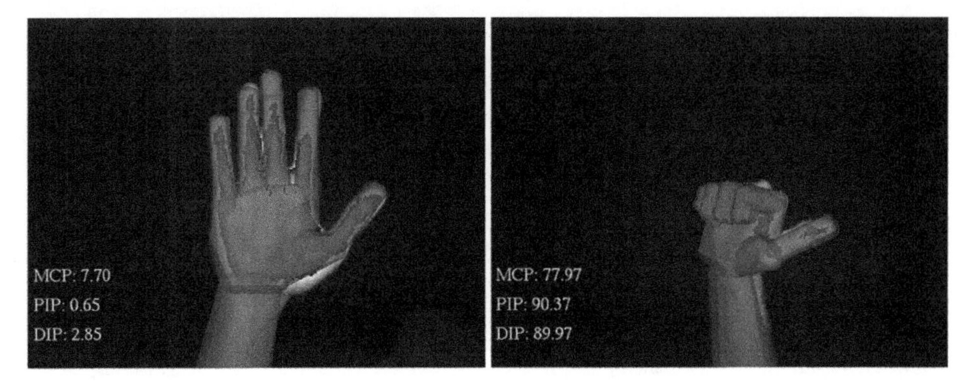

Figure 4.8 Extension and flexion positions of the hand in the tracking application. Source: Trieu Pham.

prior to the commencement of the research to practise protocols and logistics of the study. Before measuring, the hand model was manually scaled to fit with the size of the participant's hands. Following the real treatment scenario, the hands were measured in two standardised positions: extension and flexion.

The Creative Senz3D camera was connected to a computer with Intel Core i7-3740QM 2.7Ghz CPU, 8GB RAM, Nvidia NVS 5400M VGA, and installed Windows 7 64-bit operating system. The Intel® Perceptual Computing SDK was installed for aiding the hand tracking stage. The declination joint angles of the hand were computed from the hand model when the model precisely matched the subject's hand (Figure 4.8). Then these values were employed to compute the estimation of the TAM and PIP angle values utilising our proposed approaches. After the position of the hand was captured by the camera, the subject kept their hand still for therapists to measure joint angles using a dorsal measurement technique. At the end, the final results from the camera and the therapists were compared to evaluate the system.

4.6 Results

4.6.1 Concurrence validity

We established the validity of the proposed approach by comparing with the scores obtained from a well-established measurement procedure for the same cohort of subjects; i.e. there exists a consistent correlation between the scores from the two contrasting and independent measurement procedures [337]. Concurrent validity is often determined using the Pearson product-moment correlation. Following the trial procedure, there were a total of 480 measurements taken from 40 participants of both hands, three joints and two positions (extension and flexion). The correlation between the proposed system and the universal goniometer was $r = 0.95$. At the joint level, the correlation between the proposed system and the universal goniometer was $r_{\text{MCP}} = 0.96$, $r_{\text{PIP}} = 0.98$ and $r_{\text{DIP}} = 0.87$. The relationship between the proposed approach and the manual goniometer-based measurements indicates that all the variables demonstrate a higher degree of correlation ($r > 0.85$).

Additional analysis was completed to determine the clinical significance of the proposed system. Five degrees of accuracy are required between two measuring devices on any

Table 4.2 Clinical significance between the proposed system and universal goniometer.

Variable	Mean difference	Standard deviation
Extension, left MCP	8.25	5.81
Extension, left PIP	5.23	4.41
Extension, left DIP	3.83	2.92
Extension, right MCP	8.57	6.08
Extension, right PIP	4.76	3.79
Extension, right DIP	5.32	5.97
Flexion, left MCP	11.05	10.73
Flexion, left PIP	13.00	15.98
Flexion, left DIP	18.45	13.47
Flexion, right MCP	10.49	11.92
Flexion, right PIP	7.37	5.83
Flexion, right DIP	18.09	14.62

particular joint to be clinically significant [166]. In the trial with 480 measurements, the mean of the difference between the proposed system and the universal goniometer was 9.53° ($\sigma = 10.54$). More specifically, the difference between 240 extension measurements was 5.99° ($\sigma = 5.24$), and the difference between 240 flexion measurements was 13.07° ($\sigma = 13.04$). Further details of the analysis at joint levels are depicted in Table 4.2.

4.6.2 Internal reliability

The results of two measurements taken by the same tool, either the proposed system or the universal goniometer, were compared to infer the intraclass correlation coefficient (ICC) with a 95% confidence interval. According to Portney *et al.* [292], the ICC value of above 0.75 is considered as good reliability, while a good clinical measurement should have a value above 0.9. The overall ICC of the proposed system was 0.998 at the 95% confidence level. In comparison, the overall ICC of the universal goniometer is 0.994 at the 95% confidence level. The results of the analysis on extension and flexion, and at the joint level are shown in Table 4.3. The results show that the flexion has better ICC values than extension in both the proposed system and the universal goniometer. At the joint level, the DIP demonstrated the lowest ICC values in both the proposed system and the universal goniometer, but these values are still considered to have clinical good reliability (ICC > 0.9).

To determine the clinical significance of the repeated measurements, the results of two measurements taken by the same tool, either the proposed system or the universal goniometer, were compared. The mean of the difference of the proposed system was 1.94° ($\sigma = 2.95$) while the mean of the difference of the universal goniometer was 4.45° ($\sigma = 5.15$). For extension and flexion, the means of the difference of the proposed system was 1.19° ($\sigma = 1.53$) and 2.70° ($\sigma = 3.75$), respectively. The means of the difference of the

Table 4.3 The internal reliability of the proposed system.

Variable	Proposed system		Universal goniometer	
	ICC	Confidence interval (95%)	ICC	Confidence interval (95%)
Extension	0.974	0.963-0.981	0.908	0.882-0.929
Flexion	0.990	0.985-0.993	0.959	0.947-0.968
MCP	0.999	0.999–1	0.995	0.993-0.996
PIP	0.999	0.999–1	0.998	0.997-0.998
DIP	0.993	0.990–0.996	0.986	0.980-0.990

Table 4.4 Internal clinical significance of the proposed system and the universal goniometer.

Variable	Proposed system		Universal goniometer	
	Mean difference	Standard deviation	Mean difference	Standard deviation
Extension, left MCP	1.04	1.5	2.75	3.39
Extension, left PIP	1.31	1.43	2.89	3.74
Extension, left DIP	1.77	2.10	1.38	2.99
Extension, right MCP	0.35	0.44	4.5	3.16
Extension, right PIP	1.08	1.11	2	3.16
Extension, right DIP	1.58	1.77	1.25	3.54
Flexion, left MCP	2.65	2.50	6.25	5.51
Flexion, left PIP	1.76	2.67	5.50	4.36
Flexion, left DIP	5.81	5.41	7.63	6.79
Flexion, right MCP	1.09	1.11	5.50	5.97
Flexion, right PIP	0.99	1.92	4.75	4.38
Flexion, right DIP	3.90	4.64	9.0	5.15

universal goniometer for extension and flexion were larger than the proposed system, being 2.46° ($\sigma = 3.48$) and 6.44° ($\sigma = 5.76$), respectively. Further details of the analysis at the joint level are shown in Table 4.4. In three joints, the DIP often had a greater difference and greater variance between repeated measurements than the two remaining joints. The resulting analysis of repeated measurements clinically and statistically demonstrates the reliability of the proposed system.

4.6.3 Time efficiency

The subjects were asked to keep their hands still in different positions for measuring and a timer was used to record the measurement duration. For each position of the hand, the proposed system was able to capture the joint angles of 15 joints of the hand, whereas, for

Table 4.5 Time of measurement per joint.

Item	Mean (seconds)	SD (seconds)
Universal goniometer		
Extension of left hand	2.77	0.78
Extension of right hand	5	0.83
Flexion of left hand	5.9	1.72
Flexion of right hand	5.96	1.59
Proposed system		
Extension of left hand	0.54	0.29
Extension of right hand	0.62	0.29
Flexion of left hand	1.13	0.61
Flexion of right hand	1.38	0.96

a manual assessor, only three joints were measured. Therefore, the proposed system needs only one measurement for each extension or flexion position while the manual assessor needs five measurements for each extension or flexion position. The time of measurement was divided by the number of joints measured. Consequently, the measurement time using the proposed system was divided by 16 and the measurement time using a manual assessor was divided by 3. Details of measurement times are reported in Table 4.5. As depicted in Table 4.5, the proposed system completed the measurements 2–4 seconds faster than the therapists with a universal goniometer.

4.7 Discussions

The correlation between the proposed system and the universal goniometer is strong enough to confirm that the proposed system provides similar measurements of finger ROM to the universal goniometer. The mean difference between the proposed system and the universal goniometer of 9.53° ($\sigma = 10.54$) is close to previous studies in which the mean difference between two therapists using a universal goniometer is 7–9°. From these analyses, it can be inferred that the proposed system is a valid tool to use in finger ROM measurements although the clinical differences are greater than the 5° required for clinical significance.

The ICC of the proposed system was found to be very strong ($ICC = 0.998$, $ICC_{ext} = 0.974$, $ICC_{flx} = 0.99$). This implies that the proposed system is capable of reproducing almost the same results for the same joint angle. The consistency of the system enables it to be more accepted as a standard measurement tool. The analysis also demonstrates that the mean difference of the proposed system in two repeated measurements is 1.94° ($\sigma = 2.95$). This number is less than 5° of clinical significance, thus inferring that the proposed system is more reliable when being used in a test-retest scenario, which is common in clinical practice.

The time efficiency of the proposed system significantly improves current practices in hand therapy. Measurement of the phalangeal ROM is a regular occurrence in hand therapy clinics and shortening the duration of measurements enhances patients' comfort. This also enhances the efficiency of the provided service by clinicians and shortens therapy sessions so that waiting lists can be reduced. Experimental results demonstrated that our system is much more effective in terms of operating time than the manual process, saving up to 4 seconds per joint. In practice, when measuring all joints of the hand, the system is capable of saving up to several minutes due to the additional time required for data entry.

The findings of this study may not be generalised to a clinical setting for a number of reasons. As the target of this study was to introduce a non-contact ROM measurement for the hand, we only recruited healthy subjects, without oedema or wound, to simplify the procedure. In addition, two therapists had to memorise the results of three joint measurements before entering the numbers into the report, thus increasing the chance of observer error. Although the universal goniometer is most popularly used in clinical practice, it is not as accurate as other methods such as the X-ray. Lastly, despite endeavouring to ensure the subject's hand remained in the same position, it was impossible to maintain the position of the hand for 30 seconds without any changes. Because of the short length of the phalangeal bone, even a small movement of the hand can largely impact the reading of both the therapists and the proposed system.

The improvement of ROM is considered a key aspect in exercise-based rehabilitation. Busy clinics often have their clients treated by different therapists, which raises the issue of accuracy and consistency in measuring ROM among therapists. An automated, non-contact system replacing humans in measurement tasks ensures the consistency of measurements and effectiveness of treatments. In order to be used as an effective ROM tool, the measurements need to be reliable and valid [55]. Current devices such as manual goniometers or electro-goniometers are often inconsistent in terms of reliability and accuracy of measurement experiences due to a higher degree of subjectivity based on the assessor. In certain situations, it can be uncomfortable for the patient and may even cause hygiene and health issues upon contact. The development of the proposed system was motivated by the demand for a standardised, reliable, objective and time-efficient system for measuring phalangeal ROM.

Using advanced technology reduces the cost of healthcare and shortens appointment sessions [206] without compromising the quality of care. In recent decades, computer-based evaluation systems for hand therapy have evolved and their use in hand therapy clinics is becoming common practice. Computer-based evaluation systems are in demand and are destined to be integrated into existing healthcare systems and hospitals to enhance rehabilitation processes. The portability and convenience of this system allow it to be used in different locations, enabling it to have a significant impact on telerehabilitation.

4.8 Approaching Finger Movement with a New Perspective

Technology use in healthcare provision reduces cost and enhances the quality of service and patient comfort. Our proposed system provides clinicians with an innovative and effective solution to evaluate hand flexibility. More importantly, the non-contact and faster measurements help improve patient comfort as the time of measurement is shortened significantly. In addition, the proposed approach can be used in many other potential applications involving quantitative assessment of the finger functionality. The system has the potential to become a standard facility to measure hand functionality and replace current manual goniometers. An expansion of this system to telehealth and other e-health applications is a necessary step to integrate the system into hospitals and to enhance client satisfaction.

Historically, subjective visual examinations have been used to examine the range of finger movements and, subsequently, declination joint angles were measured with goniometers [73] to evaluate flexibility in a more quantitative form. Despite the wide use by clinicians and other advantages, goniometer-based declination angle measurements fall short of obtaining a complete dextral profile. According to Kendall *et al.* [169], "For muscles that pass over two or more joints, the normal range of muscle length will be less than the total range of motion of the joints over which the muscle passes". Therefore, in measuring joint movement in which two joints are involved, the second joint should be placed in a shortened position – demonstrating that hand movement is not just simply summing measured declination angles or even separately considering each angle as per the current practice. This is particularly the case when flexing the fingers into a fist where joint angles are dependent on the position of the proximal joints. A more effective and insightful form of assessing hand function is indeed a necessity. Biggs *et al.* [38] confirmed that extrinsic muscles of the hand signal fingertip locations more accurately than they signal the angles of individual finger joints. This work found that even all three extrinsic muscles were not adequate to accurately estimate the flexion angles of all the joints whilst just two extrinsic muscles taken together could always provide information adequate to estimate fingertip location.

Many dexterous daily activities are determined by fingertip trajectories rather than finger joint angles and therefore an approach based on the fingertip locations and trajectory can provide a more effective representation of the flexibility of the hand. In fact, when controlling the human end effector system comprising a higher degree of freedom, spatial aspects of the requisite movements are controlled by the higher levels of the nervous system rather than the specific joints or muscles [36]. The coordinates of fingers have captured the attention of researchers over the last decade [46, 87, 203, 212, 402]. However, most of these studies have proposed a reflective marker-based tracking system [203, 212] or a glove-based tracking system [46, 402] and the therapist is required to position the markers or to wear gloves, making it time-consuming and not viable in busy clinical environments.

In recent years, the measurement of fingertip trajectories has become quite popular, particularly with the development of a number of pervasive devices such as the Microsoft Kinect Sensor, the Leap Motion Controller and the Creative Senz3D Camera, as they offer a better solution in measuring a range of joint movements. More recent researchers in this arena used a Microsft Kinect© to build a 3D skeletal hand tracking system [235, 237, 266]. However, Kinect© is aimed at full-body movements and has limitations concerning the accuracy required for finer finger movements.

The set of all reachable fingertip positions is denoted as the *"reachable space"*. In comparison to a subjective visual assessment of reachable spaces, optical positioning devices can be used to obtain a descriptive visualisation of finger movements accurately and in a non-contact form. Accordingly, the reachable space of fingertips warrants closer attention, particularly in areas involving clinical rehabilitation. In studying the characteristics of planar fingertip movements, Cruz *et al.* [82] used the Optotrak system to measure the fingertip location. Using 10 subjects (six males, four females), a generic approximation to the reachable space was obtained. In an earlier study, Venema *et al.* [363] deduced a *workspace* by recursively capturing a range of hand movements, although a specific mathematical postulation was not provided.

A reachable space concept has been discussed in a multitude of areas including robotics [190, 191], biomechanics and also certain natural sciences [38, 138, 164, 186, 399]. Leitkam *et al.* [203] compared the experimental data from a reflective marker-based system to a biomechanical model by forming a reachable space using a computationally challenging approach of point by point volumetric calculation. Though Leitkam and other researchers [82, 133, 164, 274] obtained the motion path of the fingers and hence the reachable space, these underlying techniques are not suitable for real-time applications destined to be used within prescribed therapeutic sessions. Zheng *et al.* [399] analysed the reachable space of the fingertip and introduced an exoskeleton-type hand rehabilitation assistive device. Kuo *et al.* [186] presented a quantitative method for measuring the functional workspace of a human hand through the reachable space. However, all of the above studies have used numerical feed-forward kinematics and not closed-form solutions to obtain the reachable space. Although Alciatore *et al.* [23] presented a Monte Carlo-based method to effectively determine the boundary of the reachable space, the process of determining the boundary and implementation is not simple due to random sampling, and, moreover, this is not a closed-form solution. In the general literature, the problem of determining the boundary of the workspace with a closed-form solution was addressed in mathematics and robotics for N-manipulators [65] in 1996, but the approach did not consider the limitations in the range of motion of the human hand, and it is complex for this simpler case. The present study intends to address this lacuna to provide a better means of capturing the reachable space to analyse and assess human hand functionality.

The finger flexibility assessment by means of reachable space is currently considered an effective tool to describe the range of motion of the hand. Existing approaches numerically compute the reachable space using forward kinematics such as exhaustive scanning or Monte Carlo methods. In the following section, we present a mechanism to compute, quantify, compare and determine the size of the reachable space via a computationally efficient boundary determining approach. Here we provide explicit formulas mathematically, determining the reachable space boundary and its capacity as opposed to an implicit numerical solution. Using this new mechanism, we accurately quantify and compare the reachable space of different subjects in order to effectively compare the functionality of the fingers. We evaluate the performance of our proposed method against the kinematic feed-forward approach in calculating the reachable space. An execution time of 0.3 seconds is observed for the proposed method in comparison to 25 minutes for the standard kinematic feed-forward method in forming the reachable space. In another benchmarking exercise, the method based on proposed explicit solutions in finding the capacity of the reachable space is three times faster than the well-known Scan-Line Fill algorithm.

In the current phalangeal terminology, the normal ROM is the standard or average range over a given population. The functional ROM is the ROM measured when people are engaged in daily activities, which is the common position of fingertips when normal day-to-day activities are performed. Task-specific declination angles are commonly used for measuring finger joint movements when performing specific tasks such as power grip, precision grip, key pinch and tip pinch. Subspaces associated with a ROM for specific tasks are called task-specific subspaces and essentially constitute a subset in the complete reachable space.

In the next section, we present a closed-form solution to find the reachable space accurately and quantitatively. The reachable space is defined by a set of six equations. We will present the main key points as follows:

- Mathematically determine closed-form descriptions explicitly for the boundary and the capacity of the reachable space.
- Experimental verification of the researchable space and daily activity subspaces.
- Comparison of mathematical (explicit) assertions with numerical computations in terms of both accuracy and computational cost.

We have organised the next sections as follows. Section 4.9 describes the feed-forward kinematic method in finding the reachable space. We mathematically present the underlying idea that forms the basis for the subsequent developments. Section 4.10 presents the closed-form solutions for the reachable space while the quantification is given in Section 4.11. Section 4.12 consists of two experiments: the first experiment compares and demonstrates the advantage of using reachable space rather than simple declination angles; the second experiment compares the computational cost in finding reachable space through our exclusive description against the numerical kinematic feed-forward method and compares the computational cost in finding the capacity of the reachable space between our method (presented in Section 4.12) and the numerical Scan-Line Fill algorithm followed by concluding remarks.

4.9 Reachable Space

As the concept of reachable space is closely related to the fingertip positions, a mathematical model can be used to find the fingertip positions. Denote wrist, metacarpophalangeal joint, proximal interphalangeal joint, distal interphalangeal joint and tip with the letters W, M, P, D and T, respectively, as indicated in Figure 4.9. Assume that lengths of the phalangeal bones and the angles of the phalangeal joints are known and let l_1, l_2, l_3, l_4 denote the lengths of the metacarpus, proximal, middle and distal bones and $\alpha, \hat{\alpha}, \beta, \gamma$ denote the angles of the phalangeal joints. Coordinates of joints of the hand are with respect to the origin positioned on the wrist. Thus, the x axis lies on the metacarpus and its direction is toward the metacarpophalangeal joint, the positive direction of the z axis is toward the left of the left hand, and the positive direction of the y axis is toward the hollow of the palm. In this chapter, we use homogeneous coordinates to ensure calculation simplicity. The Denavit-Hartenberg (D-H) convention, which is widely used [70, 203, 314], is similar to our representation and uses the same homogeneous coordinates. The fingertip position presented as a homogeneous vector notation is computed as follows:

$$p_T = T_{WM} R_y(\hat{\alpha}) R_z(\alpha) T_{MP} R_z(\beta) T_{PD} R_z(\gamma) T_{DT} \, p_W. \tag{4.11}$$

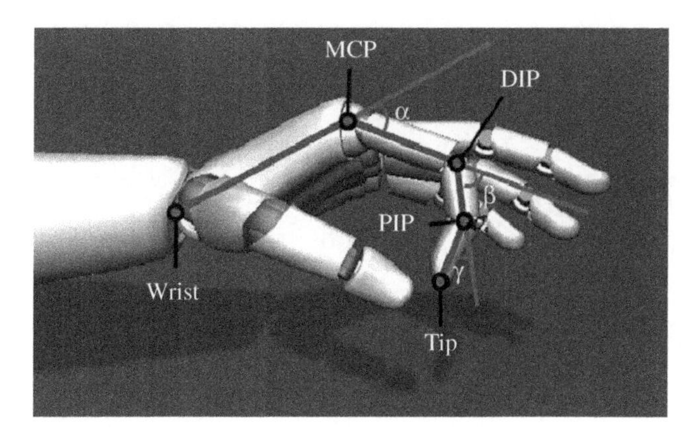

Figure 4.9 Parametric description for the index finger.

Here the notation is such that $T_{X,Y}$ is a translation matrix with $X, Y \in \{W, M, P, D\}$ and $R_x(\theta)$ is a rotation matrix with a rotation angle θ, rotated around the x axis in the counterclockwise direction. The translation operator used in equation (4.11) can be given in homogeneous coordinates as

$$T_{WM} = \begin{bmatrix} 1 & 0 & 0 & l_\theta \\ 0 & 1 & 0 & 0 \\ 0 & 0 & 1 & 0 \\ 0 & 0 & 0 & 1 \end{bmatrix}. \tag{4.12}$$

The MCP has two degrees of freedom: extension/flexion and adduction/abduction. The extension/flexion angle of MCP is denoted by α. The adduction/abduction angle of MCP is denoted by $\hat{\alpha}$. $R_z(\alpha)$ and $R_y(\hat{\alpha})$ are the rotation operators with respect to extension/flexion and adduction/abduction. $R_z(\beta)$ is the rotation operator with respect to extension/flexion of PIP. DIP has only an extension/flexion movement. $R_z(\gamma)$ is the rotation operator with respect to extension/flexion of the distal joint. In homogeneous coordinates, the rotation operator can be given as follows:

$$R_y(\hat{\alpha}) R_z(\alpha) = \begin{bmatrix} C_\alpha C_{\hat{\alpha}} & -C_{\hat{\alpha}} S_\alpha & S_{\hat{\alpha}} & 0 \\ S_\alpha & C_\alpha & 0 & 0 \\ -C_\alpha S_{\hat{\alpha}} & S_\alpha S_{\hat{\alpha}} & C_{\hat{\alpha}} & 0 \\ 0 & 0 & 0 & 1 \end{bmatrix},$$

$$R_z(\beta) = \begin{bmatrix} C_\beta & -S_\beta & 0 & 0 \\ S_\beta & C_\beta & 0 & 0 \\ 0 & 0 & 1 & 0 \\ 0 & 0 & 0 & 1 \end{bmatrix}, R_z(\gamma) = \begin{bmatrix} C_\gamma & -S_\gamma & 0 & 0 \\ S_\gamma & C_\gamma & 0 & 0 \\ 0 & 0 & 1 & 0 \\ 0 & 0 & 0 & 1 \end{bmatrix}.$$

In this approach, we do not use the widely used [87, 237, 259] constraints between PIP and DIP [137]. Our approach is fundamentally distinct as we primarily aim at therapy and in this case we cannot assume a constraint between PIP and DIP for the impaired hand. These constraints are only for a healthy hand. Inclusion of constraints limits the assessment of the

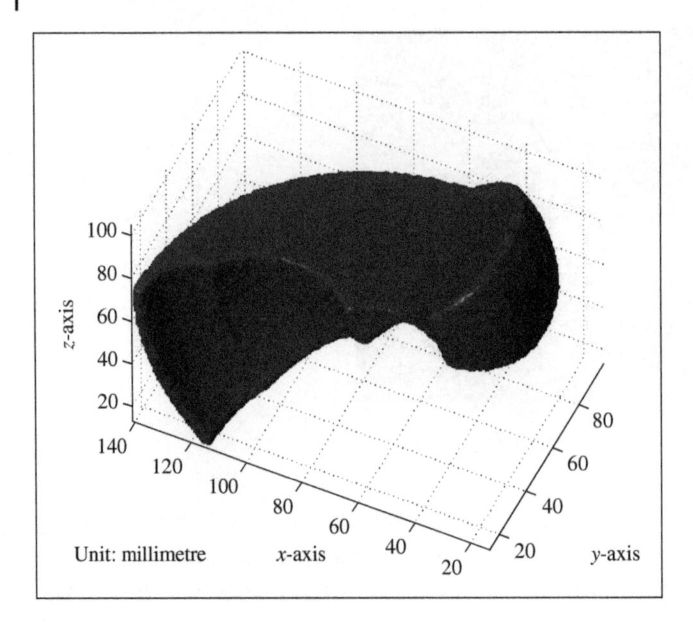

Figure 4.10 Simulation of a reachable space of fingertips for a normal hand in three dimensions based on figures in Table 4.6.

hand to healthy hands. However, impaired hands are major objects in therapy and assessment. Therefore, with the intention of quantifying the functional ability of the hand, we remove this constraint in our approach.

The obvious approach is to simulate the full reachable space exhaustively and compute the fingertip positions for various values of $\alpha, \hat{\alpha}, \beta, \gamma$ within the specific allowable limits. The limiting values of these parameters for a normal person are well documented and widely known. The normal range of motion values according to the American Society for Surgery of the Hand are given in Table 4.6. Reachable space derived from computer simulations for a normal hand is depicted in Figure 4.10. This section introduces the idea of reachable space with better visualisation compared to research works by Darling *et al.* [87] and Cruz *et al.* [82]. However, this approach of implicit numerical formations does not provide sufficient strength to our argument of using the reachable space as a performance metric, especially when it is aimed at therapeutic rehabilitation. Therefore, in Section 4.10, we present a more

Table 4.6 Normal active ROM of a finger according to the American Society for Surgery of the Hand.

Joint	Range (degrees)
MCP	0–100
PIP	0–105
DIP	0–85
Total arc:	0–290

rigorous mathematical formulation to compute six curves defining the boundary of the reachable space explicitly.

4.10 Boundary of the Reachable Space

In providing a generic approach covering the case of the impaired hand as well as the normal hand, the important role of the boundary of the reachable space was evident. Although the trajectory of a finger is always inside a specific shape as acknowledged in many research works [82, 203, 363], the exact boundary of the reachable space of the finger has not been clearly defined. In other words, the boundaries of the reachable space have been defined by other models, but not with the explicit mathematical formulation. These studies used a tracking system to localise the finger position and orientation providing distinct points inside the reachable space. Therefore, a full and exact reachable space of a finger is unattainable without exhaustively covering the dataset spanning the complete space. Moreover, in the comparison of two reachable spaces, either from different people or from the same person at different times, the expensive computational cost of using the kinematic feed-forward method in other studies is impracticable. Our work is intended to directly address this problem. Now we present the mathematical formulation underpinning the description of the boundary of the reachable space and the associated proof of our claims.

For simplicity, we consider finger movements in a 2D plane with the assumptions given in Section 4.9 (Figure 4.11). Known parameters include the angles of the phalangeal joints, denoted by α, β, γ. The angle at MCP is bounded by a_1 and a_2; the angle at PIP is bounded by b_1 and b_2; and the angle at DIP is bounded by c_1 and c_2, as given below:

$$a_1 \leq \alpha \leq a_2, \tag{4.13}$$

$$b_1 \leq \beta \leq b_2, \tag{4.14}$$

$$c_1 \leq \gamma \leq c_2. \tag{4.15}$$

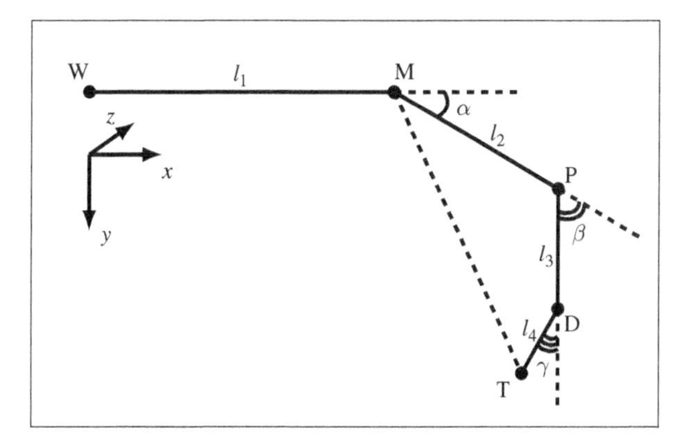

Figure 4.11 Finger model in the finger plane.

For the purpose of simplicity and clarity, we shift the origin from W to M. Then, the position of PIP, DIP and the fingertip can be represented in the complex plane as follows:

$$z_P = \lambda(\alpha) = l_2 e^{i\alpha}, \tag{4.16}$$

$$z_D = \eta(\alpha, \beta) = l_2 e^{i\alpha} + l_3 e^{i(\alpha+\beta)}, \tag{4.17}$$

$$\begin{aligned} z_T &= \zeta(\alpha, \beta, \gamma) \\ &= l_2 e^{i\alpha} + l_3 e^{i(\alpha+\beta)} + l_4 e^{i(\alpha+\beta+\gamma)}. \end{aligned} \tag{4.18}$$

The reachable space D of the finger can now be defined as

$$RS = \left\{ \zeta(\alpha, \beta, \gamma) \middle| \alpha : [a_1, a_2],\ \beta : [b_1, b_2],\ \gamma : [c_1, c_2] \right\}. \tag{4.19}$$

We also state certain properties (trivial proofs have been omitted) of the function $\epsilon(\cdot,\cdot,\cdot)$, useful for our subsequent proofs with α^i, β^i and γ^i in $[a_1, a_2]$, $[b_1, b_2]$ and $[c_1, c_2]$, respectively, for all $i \in \{1, 2\}$.

Property 1. $\left|\zeta(\alpha^1, \beta^1, \gamma^1)\right| = \left|\zeta(\alpha^2, \beta^1, \gamma^1)\right|$

Property 2. $\beta^1 \geq \beta^2 \implies \left|\zeta(\alpha^1, \beta^1, \gamma^1)\right| \leq \left|\zeta(\alpha^1, \beta^2, \gamma^1)\right|$

Property 3. $\gamma^1 \geq \gamma^2 \implies \left|\zeta(\alpha^1, \beta^1, \gamma^1)\right| \leq \left|\zeta(\alpha^1, \beta^1, \gamma^2)\right|$

Property 4. $\alpha^1 \geq \alpha^2 \implies \left|\zeta(\alpha^1, \beta^1, \gamma^1) - \lambda(\alpha^1)\right| \leq \left|\zeta(\alpha^1, \beta^1, \gamma^1) - \lambda(\alpha^2)\right|$

Property 5. $\gamma^1 \geq \gamma^2 \implies \left|\zeta(\alpha^1, \beta^1, \gamma^1) - \lambda(\alpha^1)\right| \leq \left|\zeta(\alpha^1, \beta^1, \gamma^2) - \lambda(\alpha^1)\right|$

Property 6. $\alpha^1 < \alpha^2 \implies \left|\zeta(\alpha^1, \beta^1, \gamma^1) - \eta(\alpha^1, \beta^1)\right| > \left|\zeta(\alpha^1, \beta^1, \gamma^1) - \eta(\alpha^2, \beta^1)\right|$
Proof for property 6: if $\alpha^1 < \alpha^2 \implies \arg\eta(\alpha^1, \beta^1) < \arg\eta(\alpha^2, \beta^1) \implies \arg\zeta(\alpha^1, \beta^1, \gamma^1) - \arg\eta(\alpha^1, \beta^1) > \arg\zeta(\alpha^1, \beta^1, \gamma^1) - \arg\eta(\alpha^2, \beta^1)$. Using trigonometry property in triangle ΔDTM, we have $\left|\zeta(\alpha^1, \beta^1, \gamma^1) - \eta(\alpha^1, \beta^1)\right| > \left|\zeta(\alpha^1, \beta^1, \gamma^1) - \eta(\alpha^2, \beta^1)\right|$. □

Property 7. $\beta^1 > \beta^2 \implies \left|\zeta(\alpha^1, \beta^1, \gamma^1) - \eta(\alpha^1, \beta^1)\right| < \left|\zeta(\alpha^1, \beta^1, \gamma^1) - \eta(\alpha^1, \beta^2)\right|$

Proof for property 7: if $\beta^1 > \beta^2 \implies \arg\eta(\alpha^1, \beta^1) > \arg\eta(\alpha^1, \beta^2) \implies \arg\zeta(\alpha^1, \beta^1, \gamma^1) - \arg\eta(\alpha^1, \beta^1) < \arg\zeta(\alpha^1, \beta^1, \gamma^1) - \arg\eta(\alpha^1, \beta^2)$. Using trigonometry property in triangle ΔDTM, we have $\left|\zeta(\alpha^1, \beta^1, \gamma^1) - \eta(\alpha^1, \beta^1)\right| < \left|\zeta(\alpha^1, \beta^1, \gamma^1) - \eta(\alpha^1, \beta^2)\right|$. □

Now we introduce the following theorem as the key point of this section.

Theorem 1. *Let the boundary of the reachable space RS be denoted by* Ξ. *Then* $\Xi := \xi_1 \cup \xi_2 \cup \xi_3 \cup \xi_4 \cup \xi_5 \cup \xi_6$ *where the* $\xi_i, i \in [1, \ldots, 6]$ *are given by the following continuous curves:*

$$\xi_1 := \zeta(\alpha, b_1, c_1) \Big| \alpha : [a_1, a_2]$$

$$\xi_2 := \zeta\left(a_2, \beta, c_1\right) \Big| \beta : \left[b_1, b_2\right]$$
$$\xi_3 := \zeta\left(a_2, b_2, \gamma\right) \Big| \gamma : \left[c_1, c_2\right]$$
$$\xi_4 := \zeta\left(\alpha, b_2, c_2\right) \Big| \alpha : \left[a_1, a_2\right]$$
$$\xi_5 := \zeta\left(a_1, \beta, c_2\right) \Big| \beta : \left[b_1, b_2\right]$$
$$\xi_6 := \zeta\left(a_1, b_1, \gamma\right) \Big| \gamma : \left[c_1, c_2\right]$$

Now we prove that $\xi_1, \xi_2, \xi_3, \xi_4, \xi_5, \xi_6$ are boundary parts individually. The point **T** is a boundary point of *RS* if, and only if, in every neighbourhood there exists at least one point in the set and at least one point not in the set. This is because **T** is clearly in *RS* and **T** is a boundary point if we can find a point in the neighbourhood that is not in the set.

For any disk Ω centred at **T** with a radius $\epsilon > 0$, any point E in Ω has the form $z_E = z_T + \epsilon e^{i\theta},\ \theta \in [0, 2\pi]$. Let $\alpha_0, \alpha' \in \left[a_1, a_2\right], \beta_0, \beta' \in \left[b_1, b_2\right]$ and $\gamma_0, \gamma' \in \left[c_1, c_2\right]$. If E is a reachable point, E can be denoted by $z_E = \zeta\left(\alpha', \beta', \gamma'\right)$ for the purpose of the following subproofs.

Subproof 1 ***(ξ_1 is a boundary curve of RS)*** *Let **T** be a point on ξ_1:*

$$z_T = \zeta\left(\alpha_0, b_1, c_1\right). \tag{4.20}$$

Choose θ such that E, a specific point in Ω, is

$$\theta = \arg z_T \tag{4.21}$$
$$\Rightarrow z_E = \left|z_T\right| e^{i \arg z_T} + \epsilon e^{i \arg z_T} \tag{4.22}$$
$$\Rightarrow z_E = \left|z_T + \epsilon\right| e^{i \arg z_T} \tag{4.23}$$
$$\Rightarrow \left|z_E\right| > \left|z_T\right|. \tag{4.24}$$

For $\alpha_0 \in \left[a_1, a_2\right]$, using property 1,

$$\left|\zeta\left(\alpha_0, \beta', \gamma'\right)\right| = \left|\zeta\left(\alpha', \beta', \gamma'\right)\right|, \tag{4.25}$$

For $\beta' > b_1$, using property 2,

$$\left|\zeta\left(\alpha_0, \beta', \gamma'\right)\right| < \left|\zeta\left(\alpha_0, b_1, \gamma'\right)\right|, \tag{4.26}$$

For $\gamma' > c_1$, using property 3,

$$\left|\zeta\left(\alpha_0, b_1, \gamma'\right)\right| < \left|\zeta\left(\alpha_0, b_1, c_1\right)\right|. \tag{4.27}$$

Then, from equations (4.25), (4.26), (4.27),

$$\left|\zeta\left(\alpha', \beta', \gamma'\right)\right| < \left|\zeta\left(\alpha_0, b_1, c_1\right)\right| \tag{4.28}$$

$$\Rightarrow \left|z_E\right| < \left|z_T\right|. \tag{4.29}$$

*This result (equation 4.29) contradicts our initial assumption (equation 4.24) in choosing the reachable point E. Therefore, the point E, which we chose, is not in RS and hence **T** is a boundary point and ξ_1 is a boundary curve of RS.*

Subproof 2 **(ξ_2 is a boundary curve of RS)** *Let* ***T*** *be a point on* ξ_2*:*

$$z_T = \zeta\left(a_2, \beta_0, c_1\right). \tag{4.30}$$

Choose θ *such that E, a specific point in* Ω*, is*

$$\theta = a_2 + \beta_0 + \widehat{TPD} \tag{4.31}$$

$$\Rightarrow P, T \text{ and } E \text{ are in a straight line} \tag{4.32}$$

$$\Rightarrow \left|z_E - z_P\right| = \left|z_E - z_T\right| + \left|z_T - z_P\right| \tag{4.33}$$

$$\Leftrightarrow \left|\zeta\left(\alpha', \beta', \gamma'\right) - \lambda\left(a_2\right)\right| = \epsilon + \left|\zeta\left(a_2, \beta_0, c_1\right) - \lambda\left(a_2\right)\right|. \tag{4.34}$$

Consider $a_2 > \alpha'$*, using property 4,*

$$\left|\zeta\left(\alpha', \beta', \gamma'\right) - \lambda\left(a_2\right)\right| < \left|\zeta\left(\alpha', \beta', \gamma'\right) - \lambda\left(\alpha'\right)\right|. \tag{4.35}$$

For $\beta_0 \in \left[b_1, b_2\right]$*,*

$$\left|\zeta\left(\alpha', \beta', \gamma'\right) - \lambda\left(\alpha'\right)\right| = \left|\zeta\left(\alpha', \beta_0, \gamma'\right) - \lambda\left(\alpha'\right)\right|. \tag{4.36}$$

For $\gamma' > c_1$*, using property 5,*

$$\left|\zeta\left(\alpha', b_0, \gamma'\right) - \lambda\left(\alpha'\right)\right| < \left|\zeta\left(\alpha', b_0, c_1\right) - \lambda\left(\alpha'\right)\right|. \tag{4.37}$$

Using property 5,

$$\left|\zeta\left(\alpha', b_0, c_1\right) - \lambda\left(\alpha'\right)\right| = \left|\zeta\left(a_2, b_0, c_1\right) - \lambda\left(a_2\right)\right|. \tag{4.38}$$

From equations (4.35), (4.36), (4.37), (4.38),

$$\left|\zeta\left(\alpha', \beta', \gamma'\right) - \lambda\left(a_2\right)\right| < \left|\zeta\left(a_2, b_0, c_1\right) - \lambda\left(a_2\right)\right|. \tag{4.39}$$

This result (equation 4.39) contradicts our initial assumption (equation 4.34) in choosing the reachable point E. Therefore, the point E, which we chose, is not in RS and hence ***T*** *is a boundary point and* ξ_2 *is a boundary curve of RS.*

Subproof 3 **(ξ_3 is a boundary curve of RS)** *Let* ***T*** *be a point on* ξ_3*:*

$$z_T = \zeta\left(a_2, b_2, \gamma_0\right). \tag{4.40}$$

Choose θ *such that E, a specific point in* Ω*, is*

$$\theta = a_2 + b_2 + \gamma_0 \tag{4.41}$$

$$\Rightarrow D, T, \text{ and } E \text{ are in a straight line} \tag{4.42}$$

$$\Leftrightarrow \left|z_E - z_D\right| = \left|z_E - z_T\right| + \left|z_T - z_D\right| \tag{4.43}$$

$$\left|\zeta\left(\alpha', \beta', \gamma'\right) - \eta\left(a_2, b_2\right)\right| = \epsilon + \left|\zeta\left(a_2, b_2, \gamma_0\right) - \eta\left(a_2, b_2\right)\right|. \tag{4.44}$$

Consider $a_2 > \alpha'$*, using property 6,*

$$\left|\zeta\left(\alpha', \beta', \gamma'\right) - \eta\left(a_2, b_2\right)\right| < \left|\zeta\left(\alpha', \beta', \gamma'\right) - \eta\left(\alpha', b_2\right)\right|. \tag{4.45}$$

Consider $b_2 > \beta'$*, using property 7,*

$$\left|\zeta\left(\alpha',\beta',\gamma'\right)-\eta\left(\alpha',b_2\right)\right| < \left|\zeta\left(\alpha',\beta',\gamma'\right)-\eta\left(\alpha',\beta'\right)\right|. \tag{4.46}$$

For $\gamma_0 \in \left[c_1, c_2\right]$*, and the link* l_4 *is constant,*

$$\left|\zeta\left(\alpha',\beta',\gamma'\right)-\eta\left(\alpha',\beta'\right)\right| = \left|\zeta\left(a_2,b_2,\gamma_0\right)-\eta\left(a_2,b_2\right)\right|. \tag{4.47}$$

From equations (4.45), (4.46), (4.47),

$$\left|\zeta\left(\alpha',\beta',\gamma'\right)-\eta\left(a_2,b_2\right)\right| < \left|\zeta\left(a_2,b_2,\gamma_0\right)-\eta\left(a_2,b_2\right)\right|. \tag{4.48}$$

This result (equation 4.48) contradicts our initial assumption (equation 4.44) in choosing the reachable point E. Therefore, the point E, which we chose, is not in RS and hence $\boldsymbol{T}$ *is a boundary point and* ξ_3 *is a boundary curve of RS.*

The proof for ξ_4, ξ_5, ξ_6 is similar to ξ_1, ξ_2, ξ_3 so we do not present it here.

4.11 Area of the Reachable Space

In this section, we present a mechanism to quantitatively represent the planar reachable space explicitly employing Green's theorem. In Green's theorem, if C is a simple closed curve and $\mathbf{F}(x,y)$ is defined everywhere inside C, then we can convert the line integral into a double integral (Figure 4.12). Therefore, we find the line integral $\int_C \mathbf{F} ds$ by means of calculating the double integral:

$$\iint_D \left(\frac{\partial \mathbf{F}_2}{\partial x} - \frac{\partial \mathbf{F}_1}{\partial y}\right) dA. \tag{4.49}$$

Indeed, this is in the form

$$\iint_D f(x,y)\, dA \tag{4.50}$$

when a vector field $\mathbf{F}(x,y)$ exists such that

$$f(x,y) = \frac{\partial \mathbf{F}_2}{\partial x} - \frac{\partial \mathbf{F}_1}{\partial y}. \tag{4.51}$$

Taking the area of the region RS to be equal to the double integral of $f(x,y) = 1$ over RS gives

$$Area\ of\ RS = \iint_D dA = \iint_D 1 dA. \tag{4.52}$$

There are many such vector fields F, but we choose $\mathbf{F}(x,y) = (-y/2, x/2)$. Then the area of the region RS bounded by $C = \partial RS$. Therefore,

$$Area\ of\ RS = \int_C \mathbf{F} ds = \frac{1}{2}\int_C x\ dy - y\ dx, \tag{4.53}$$

where $\mathbf{F}(x,y) = (-y/2, x/2)$.

Now we apply Green's theorem to calculate the area bounded by the curve ξ which was presented in the previous section. Since ξ is a counterclockwise-oriented boundary of RS,

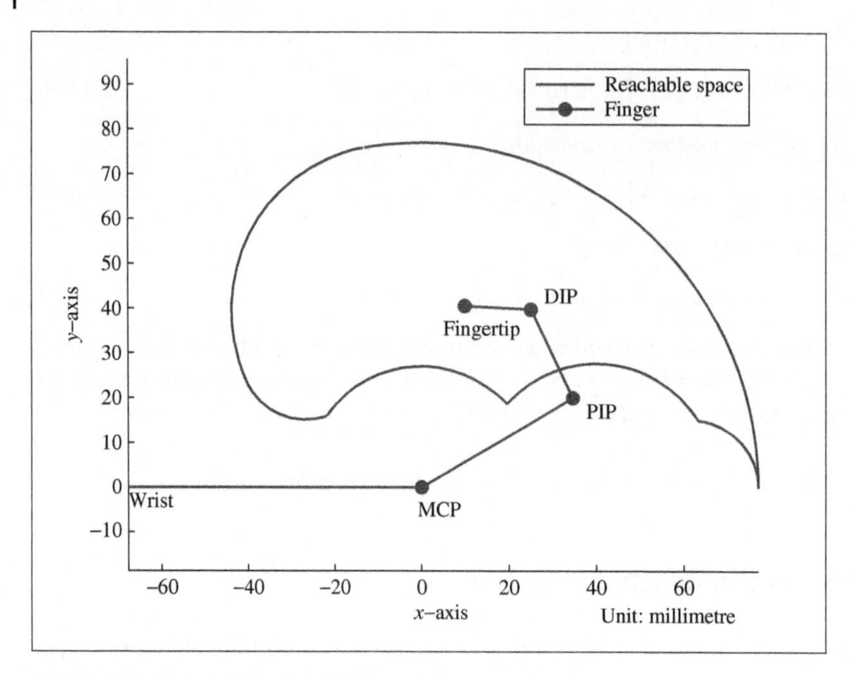

Figure 4.12 Reachable space of fingertips in two dimensions is built using boundary formulae with a normal active ROM of finger given in Table 4.6.

the area is just the line integral of the vector field $\mathbf{F}(x, y) = \frac{1}{2}(-y, x)$ around the curve ξ parameterized by α, β, γ:

$$\begin{aligned} \textit{Area of RS} &= \iint dA \\ &= \int_{\xi} F ds \\ &= Q_1 + Q_2 + Q_3 + Q_4 + Q_5 + Q_6, \end{aligned}$$

where

$$Q_i = \int_{\xi_i} F^{(i)} ds. \tag{4.54}$$

We rewrite the boundary curve ξ in a counterclockwise direction here:

$$\xi_1 = \zeta(\alpha, b_1, c_1) \quad \Big| \quad \alpha : [a_1, a_2], \tag{4.55}$$

$$\xi_2 = \zeta(a_2, \beta, c_1) \quad \Big| \quad \beta : [b_1, b_2], \tag{4.56}$$

$$\xi_3 = \zeta(a_2, b_2, \gamma) \quad \Big| \quad \gamma : [c_1, c_2], \tag{4.57}$$

$$\xi_4 = \zeta(\alpha, b_2, c_2) \quad \Big| \quad \alpha : [a_2, a_1], \tag{4.58}$$

$$\xi_5 = \zeta(a_1, \beta, c_2) \quad \Big| \quad \beta : [b_2, b_1], \tag{4.59}$$

$$\xi_6 = \zeta(a_1, b_1, \gamma) \quad \Big| \quad \gamma : [c_2, c_1]. \tag{4.60}$$

In order to use the line integral of a vector field, we transform our presentation from a polar form to a trigonometric form:

$$\begin{aligned}\xi_1 &= l_2\cos\alpha + l_3\cos(\alpha+b_1) + l_4\cos(\alpha+b_1+c_1) + \\ &\quad i\left(l_2\sin\alpha + l_3\sin(\alpha+b_1) + l_4\sin(\alpha+b_1+c_1)\right),\\ \xi_1' &= -\left(l_2\sin\alpha + l_3\sin(\alpha+b_1) + l_4\sin(\alpha+b_1+c_1)\right)\\ &\quad i\left(l_2\cos\alpha + l_3\cos(\alpha+b_1) + l_4\cos(\alpha+b_1+c_1)\right).\end{aligned}$$

The expression for Q_1 is computed as follows:

$$\begin{aligned}Q_1 &= \frac{1}{2}\int_{a_1}^{a_2}\Big(\left(l_2\cos\alpha + l_3\cos(\alpha+b_1) + l_4\cos(\alpha+b_1+c_1)\right)^2\\ &\quad + \left(l_2\sin\alpha + l_3\sin(\alpha+b_1) + l_4\sin(\alpha+b_1+c_1)\right)^2\Big)\,d\alpha\\ &= \frac{1}{2}\left(2l_3l_2\cos b_1 + 2l_4l_2\cos(b_1+c_1)\right.\\ &\quad \left.+ 2l_3l_4\cos c_1 + l_2^2 + l_3^2 + l_4^2\right)\int_{a_1}^{a_2} d\alpha\\ &= \frac{1}{2}\left(2l_3l_2\cos b_1 + 2l_4l_2\cos(b_1+c_1)\right.\\ &\quad \left.+ 2l_3l_4\cos c_1 + l_2^2 + l_3^2 + l_4^2\right)(a_2-a_1).\end{aligned}$$

The remaining curves can be derived in a similar way with ξ_1. Similarly, expressions for Q_2 to Q_6 can easily be computed. In summary, the area of reachable space is calculated using the following formula:

$$\begin{aligned}Area &= \frac{1}{2}\left(2l_3l_2\cos b_1 + 2l_4l_2\cos(b_1+c_1)\right.\\ &\quad \left.+ 2l_3l_4\cos c_1 + l_2^2 + l_3^2 + l_4^2\right)(a_2-a_1)\\ &\quad + \frac{1}{2}\left(l_2l_4\left(\sin(b_2+c_1) - \sin(b_1+c_1)\right)\right.\\ &\quad \left.+ l_2l_3\left(\sin b_2 - \sin b_1\right) + \left(2l_4l_3\cos c_1 + l_3^2 + l_4^2\right)(b_2-b_1)\right)\\ &\quad + \frac{1}{2}\left(l_2l_4\sin(b_2+c_2) - l_2l_4\sin(b_2+c_1)\right.\\ &\quad \left.+ l_3l_4\sin(c_2) - l_3l_4\sin(c_1) + l_4^2(c_2-c_1)\right)\\ &\quad + \frac{1}{2}\left(2l_3l_2\cos b_2 + 2l_4l_2\cos(b_2+c_2)\right.\\ &\quad \left.+ 2l_3l_4\cos c_2 + l_2^2 + l_3^2 + l_4^2\right)(a_1-a_2)\\ &\quad + \frac{1}{2}\left(l_2l_4\left(\sin(b_1+c_2) - \sin(b_2+c_2)\right)\right.\\ &\quad \left.+ l_2l_3\left(\sin b_1 - \sin b_2\right) + \left(2l_4l_3\cos c_2 + l_3^2 + l_4^2\right)(b_1-b_2)\right)\\ &\quad + \frac{1}{2}\left(l_2l_4\sin(b_1+c_1) - l_2l_4\sin(b_1+c_2)\right.\\ &\quad \left.+ l_3l_4\sin(c_1) - l_3l_4\sin(c_2) + l_4^2(c_1-c_2)\right)\end{aligned}$$

In order to use this formula to compute the area, it is necessary to know explicitly the equation of the reachable space. As it is a closed-form solution, the computational cost of a reachable space calculation is constant over different trials. In the following section, we validate our formula with respect to the numerical approach of calculating the reachable space using declination angles, in terms of accuracy as well as the computational cost.

4.12 Experiments

Our proposed approach was examined through two experiments. In the first experiment, we collected data from 10 subjects and demonstrated the advantages of using reachable space instead of declination angles as well as how convenient the proposed approach is in the construction of the reachable space of fingers. The second experiment utilised data from Hume *et al.*'s article [150] focused on the performance of the proposed approach in comparison with other existing approaches. The characteristics of the two sets of data are described below.

The data collection procedure was approved by Deakin University Human Ethics Advisory Group (HEAG) and all participants provided their written informed consent to participate in this study. Ten subjects (aged from 20 to 40) participated in the experiment. One subject had a history of a fractured left arm. Another subject previously had a broken left wrist and "jarred" fingers. In the experiment, finger movement ranges for both hands were measured using a goniometer by the therapist. Three angles at MCP, PIP and DIP were measured where the hand with the finger was extended and then in the flexed position (composite fixed). The measurement results of three declination angles of index fingers are shown in Figure 4.13. While variance angles of finger extension of the hand are low ($\sigma_{MCP} = 4$, $\sigma_{PIP} = 5.9$, $\sigma_{DIP} = 2.5$), we observed large differences between angles corresponding to the flex position ($\sigma_{MCP} = 11.4$, $\sigma_{PIP} = 7.1$, $\sigma_{DIP} = 20.3$). The declination angles of index fingers are represented in 3D space in Figure 4.14, where each dimension is associated with one joint angle. Each line stands for a measured result of the index finger of one hand of one subject. There are a total of 20 lines associated with 20 index fingers (10 subjects, left and right hands). The coordinate of two end points of the line is associated with two sets of values of three declination angles in the extension position and the flexion position. Subsequently, we calculated the reachable space from the data captured. The reachable spaces of 20 index fingers of both hands from 10 subjects are shown in Figure 4.15. The correlation between declination angles and reachable spaces is discussed in more detail in the following section, which considers the documented angles for daily activity tasks given in the literature.

In Hume's article, 35 right-handed men, aged 26 to 28 years, and none with a history of antecedent hand injuries, were studied. The trial recorded the maximum active ROM, daily

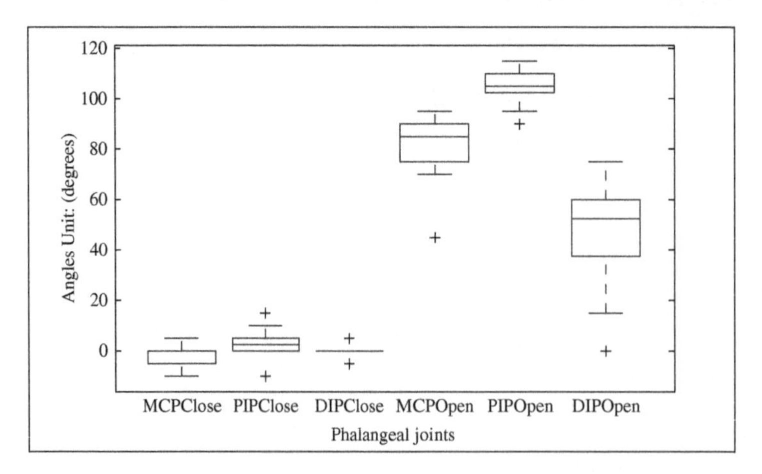

Figure 4.13 Measurement results of three declination angles of the index finger of both hands.

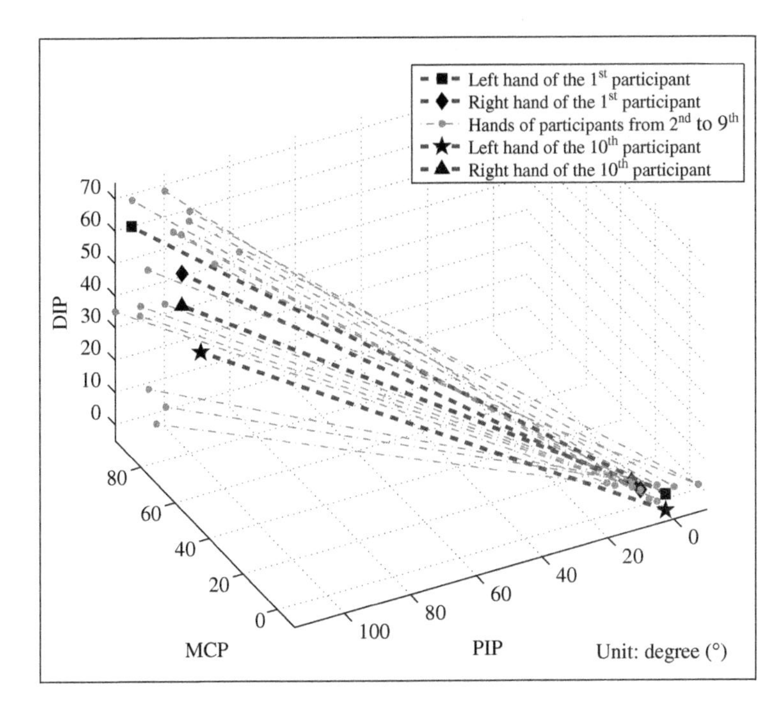

Figure 4.14 Measurement results of three declination angles are represented in the 3D space. Each axis stands for one joint.

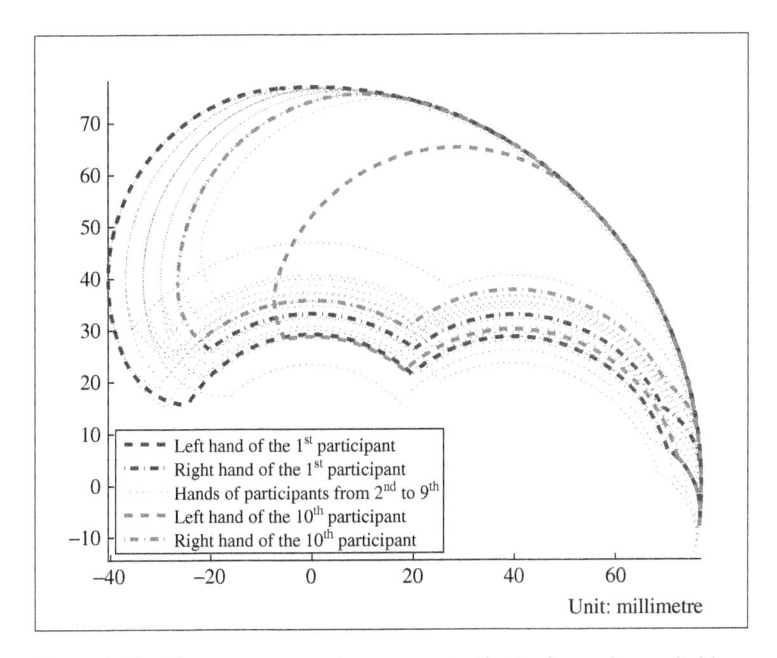

Figure 4.15 Measurement results represented in the form of a reachable space description.

Table 4.7 Task-specific positions of the joints of the hand: fingers.

	Task-specific positions (fingers)				
	Key pinch	Tip pinch	Grasp	Grip	Functional
MCP	62 (±8)	58 (±7)	33 (±6)	72 (±12)	61 (±12)
PIP	76 (±8)	76 (±13)	39 (±7)	28 (±5)	60 (±12)
DIP	46(±8)	33 (±12)	26 (±5)	50 (±5)	39 (±14)

Unit: degree (°)

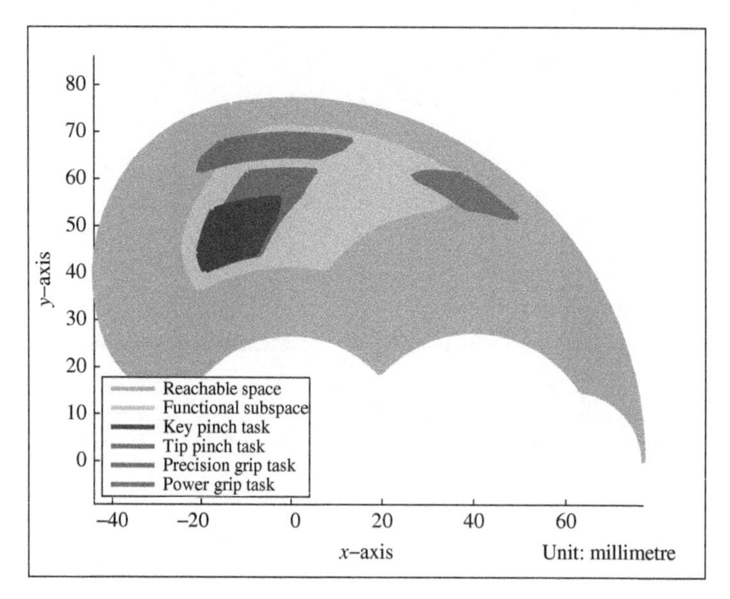

Figure 4.16 The task-specific sub-spaces associated with the task-specific positions of the finger joints, as shown in Table 4.7.

activities (functional) ROM (Table 4.6) and the position of finger joints when subjects were engaged in these specific tasks (Table 4.7). Figure 4.16 shows the reachable space associated with values in Table 4.7. In the daily activity test, there were 11 activities recorded: holding a telephone, holding a can, using a zipper, holding a toothbrush, turning a key, using a comb, writing with a pen, holding a fork, holding scissors, unscrewing a jar and holding a hammer. The task-specific tests included key pinch, tip pinch, precision grip (grasp) and power grip.

4.13 Results and Discussion

From the collected data of 10 subjects, we selected three special cases to compare with Hume's documented data. The data were from the left hands of the sixth, seventh and tenth participants. Participants 6 and 7 had normal hands while participant 10 had a broken wrist

Table 4.8 Range of movement of participants 6, 7 and 10.

	MCP	PIP	DIP
Participant 6	[0, 90]	[5, 110]	[0, 35]
Participant 7	[0, 85]	[0, 110]	[0, 15]
Participant 10	[−5, 45]	[0, 115]	[−5, 50]
Functional range	[33, 73]	[36, 86]	[20, 61]
			Unit: degree (°)

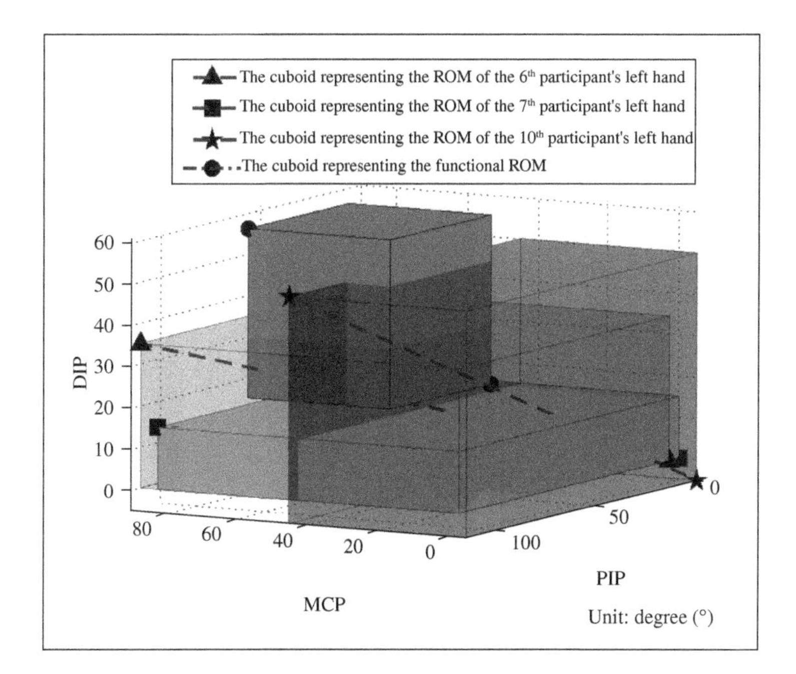

Figure 4.17 Range of movements of participants 6, 7 and 10 and the functional ROMs are represented in three dimensions.

where all the fingers were jarred. The ROMs for the index fingers of these subjects are listed in Table 4.8. Notice that none of these three subjects possessed the accepted functional ROM for a normal hand with respect to the DIP when employing the declination angle approach (Figure 4.17). Figure 4.9 depicts the ROM in the declination angle approach via rectangular cuboids. This is typical of a subject with a fractured wrist and not for the other two subjects with normal hands. This demonstrates the unique profile that individuals' hands will demonstrate, particularly following a hand injury. The ability to assess the reachable space of the individual and obtain a profile of a hand is manifested in this example and goes beyond that of the joint declination concept. The reachable space approach (Figure 4.18) using the same numerical values provides a more descriptive reasoning as two in three reachable spaces cover the functional subspace (shown in solid

Table 4.9 Comparison of execution time between the explicit method and kinematic feed-forward (KFF) in finding a reachable space and between the explicit method and Scan-line Fill (SLF) in computing areas of reachable space.

	Find reachable space		Compute area	
	Explicit (second)	**KFF (second)**	**Explicit (millisecond)**	**SLF (millisecond)**
Full area	0.27	1500	3.1	6.3
Functional	0.23	142	0.2	0.7
Grasp	0.46	4.6	0.06	0.17
Grip	0.28	5	0.06	0.19
Key	0.29	8	0.06	0.19
Pinch	0.27	16.2	0.06	0.17

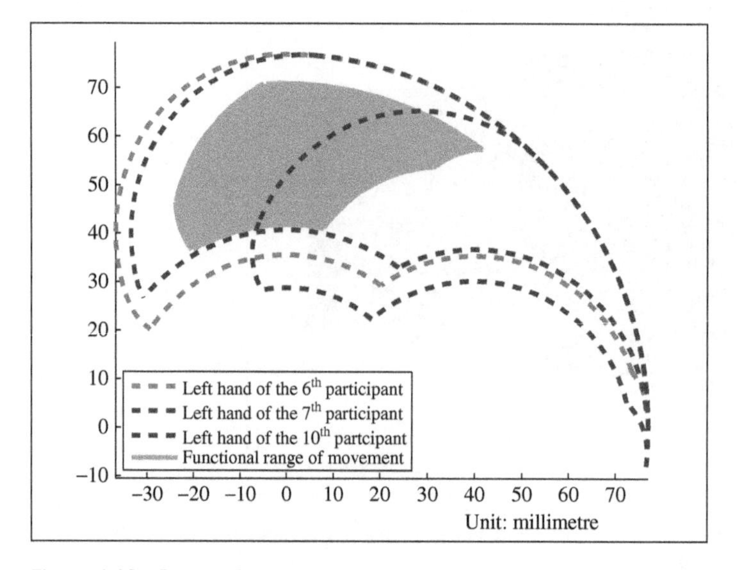

Figure 4.18 Range of movements of the 6th, 7th and 10th participants and the functional ROMs are represented in reachable space in two dimensions.

grey colour). The left reachable space from the subject with a fractured wrist can only cover part of the functional sub-space. This experiment shows that a reachable space approach is better than the traditional approach of using declination angles to represent patients' hand status.

We used Hume's data to illustrate the underlying concept of the reachable space for the general case. We used the bone length values to measure a subject's index finger with a ruler as follows: the metacarpal bone length is 68 mm, the proximal phalanx bone length is 40 mm, the medial phalanx bone length is 22 mm and the distal phalanx bone is 15 mm. From the real-time reachable space and task-specific sub-space computations, we visualised a specific task (task-specific sub-space) in comparison to the full functionality of the fingers (reachable space). Figure 4.16 shows the reachable space associated with values in Table 4.7. Hume's experimental data represented under this new approach provides a better insight into the finger flexibility and the generic status with respect to previous analysis of the injured fingers or the relative status with respect to the normal finger functionality. For example, if the reachable space of a finger encompasses the task-specific sub-spaces, then we can expect that the subject is able to perform daily activities after passing the strength tests. Otherwise, daily activities cannot be performed and the strength tests are redundant. In another example, if one subject has the reachable space that encompasses the key pinch task sub-space, we can expect that the patient can perform the key pinch task.

With numerous advantages when representing a range of finger movements, a reachable space approach is a good candidate as an alternative to the traditional approach, which simply uses declination angles to describe the ROM of the hand. However, kinematic feed-forward methods to obtain reachable spaces can be costly in terms of time and computational resources. A system with an Intel Core i7-3740QM 2.7GHz CPU, 8GB RAM, Nvidia NVS 5400M graphic card and Windows 7 64-bit operating system takes approximately 25 minutes to find a reachable space of a normal hand using the kinematic feed-forward method with a resolution of 0.5 degrees. This implementation can be improved with the Monte Carlo random sampling approach or by exploiting the computational power of GPU, but the execution time is still hard to compare with our proposed method (in Section 4.10) with only 0.3 seconds for the resolution of 0.01 degrees. Further, our proposed method to find the area of the reachable space (in Section 4.11) performs better in terms of the execution time compared to that of the Scan-line Fill algorithm, which is a general algorithm to find an arbitrary area [386]. The reason for this significant difference in the execution time is that, given the bone lengths and the ROM, our approach can find the capacity in a deterministic time $O(1)$ with explicit computation. If we use the kinematic feed-forward method mentioned in other research work to calculate the reachable space capacity, the computation cost is at least cubic time $O\left(n^3\right)$ to generate the reachable space in addition to at least quadratic time $O\left(n^2\right)$ for the Scan-line Fill algorithm. In order to ensure fairness in the comparison, both methods used the same boundary lines computed by the formula in Section 4.10. Without our contribution in Section 4.10, the Scan-line Fill algorithm consumes much more resources in finding the area of reachable space. A more detailed time benchmark exercise between our proposed method, the kinematic feed-forward technique and the Scan-line Fill algorithm in finding reachable spaces, subspaces and calculating the area is depicted in Table 4.9.

Table 4.10 shows the area of the reachable space that we computed from the subspaces using both the Scan-line Fill algorithm with Euclidean metrics and our explicit solution. The Scan-line Fill algorithm is a numerical method that is used to compute the area of an arbitrary shape. The results in Table 4.10 confirmed our calculations for the description of the reachable space. As the Scan-line Fill algorithm computes one pixel at a time, it would not consider a pixel if the boundary line crossed that pixel. Therefore, the area calculated by the Scan-line Fill algorithm is always smaller than or equal to the real area of the reachable space. This fact is confirmed in Table 4.10, where all of the values calculated by the Scan-line Fill method are smaller than the values calculated by the explicit method. Therefore, finding the capacity of the reachable space is then straightforward, accurate and faster.

Table 4.10 Reachable space areas computed by the Scan-line Fill method (SLF) and the explicit method.

	SLF (mm^2)	Explicit (mm^2)
Full reachable space	5081	5089.7
Functional space	1274.7	1275.5
Key pinch space	181.57	181.6
Tip pinch space	263.19	263.21
Grasp space	86.68	87.07
Grip space	140.86	141.69

In this work, we provide an alternate approach to quantify finger functionality using the concept of reachable space. Although this research was developed in a 2D setting, the underlying approach can easily be extended to the "real" 3D physiological space. We have considered this case to present the core ideas, as the angles of abduction/adduction at the MCP joint are often small and not sufficiently emphasised in daily activities. Another limitation in this research work was the absence of the Monte Carlo random sampling implementation, which can provide a better perspective on the performance of our approach. However, the difference in performance between the proposed approach and the Scan-line Fill algorithm was deemed sufficient to confirm that the approach is more relevant and appropriate than existing approaches.

4.14 Conclusion and Future Work

Building a more descriptive finger functionality profile using the reachable space idea requires the bone lengths and the ROM, making it specific to the individual. By focusing only on the fingertip capabilities regardless of declination angles, the reachable space can avoid redundancies in angle configurations due to the non-uniqueness of the declination angles for a given fingertip position. An experiment is conducted to confirm the advantages of using reachable space rather than simply using declination angles. Our contributions to the systems design arena are explicit formulae to find the reachable space via boundary curves and the area of the reachable space, while the computational cost compared to other methods such as the kinematic feed-forward method and Scan-line Fill algorithm is significantly less. In the future, we intend to extend this work to find the reachable spaces from a set of fingertip coordinates using a depth sensor such as the Leap Motion Controller or the Creative Senz3D Camera as well as a closed-form solution for 3D space. The underlying work is focused directly on the characterisation of finger performance for phalangeal rehabilitation.

5

Non-contact Measurement of Respiratory Function via Doppler Radar

5.1 Introduction

The non-contact detection characteristic of Doppler radar provides an unobtrusive means of respiration detection and monitoring. This avoids additional preparations such as physical sensor attachment or special clothing that can be useful for certain healthcare applications. Non-contact detection of basic human functions such as respiration using Doppler radar is particularly useful in comparison to respiration belts, which are simply inconvenient or even impractical; for instance, in long-term sleep monitoring (sleep studies) and respiration monitoring of burn patients or patients with dermatological conditions. As the belt is strapped to the chest, the natural breathing process is somewhat interfered with and therefore the measurements are likely to be affected. A non-contact form of measurement penetrating clothing would facilitate the gathering of breathing data that has not been available in the past. This will provide greater insights into conditions such as sleep apnoea with enhanced patient comfort. Furthermore, as Doppler radar is relatively robust against environmental factors such as ambient temperature, light interference and other electromagnetic signals such as WiFi, it offers less practical limitations for long-term monitoring and detection compared to, for example, computer vision-based systems or chest straps.

Typically, recording and monitoring of vital signs such as blood pressure, temperature, pulse rate and respiration rate are considered standard hospital procedures. Often, these are recorded only when patients experience respiratory problems or they are in certain critical conditions where abnormal respiratory rates are one of the key predictors of such events [77, 80]. Normal breathing rates for resting adults vary from 12 to 20 breaths/min [393] and some studies suggest a rate of over 20 breaths/min is abnormal and one over 24 breaths/min is critical [80]. As we gain more insight into the respiratory function in its natural form, one interest among breathing and sleep researchers is, can long-term respiratory signatures potentially be used in the diagnosis of respiratory disorders? In particular, can the identification of different respiratory patterns lead to detecting specific respiratory conditions? George *et al.* [393] have already discussed the importance of respiratory rate as well as associating certain types of breathing patterns with certain respiratory disorders. Therefore, a non-contact mechanism that accurately captures respiratory function under various breathing conditions is destined to support research leading to new clinical practices in many areas relevant to respiratory physiology.

Human Motion Capture and Identification for Assistive Systems Design in Rehabilitation, First Edition.
Pubudu N. Pathirana, Saiyi Li, Yee Siong Lee and Trieu Pham.

The of use of microwave Doppler radar for capturing different dynamics of breathing patterns in addition to the respiration rate has been widely investigated. Although finding the respiration rate is essential, identifying abnormal breathing patterns in real time could be used to gain further insights into respiratory disorders and refine diagnostic procedures. Several known breathing disorders were professionally role-played and captured in a real-time laboratory environment using a non-contact Doppler radar to evaluate the feasibility of this non-contact form of measurement in capturing breathing patterns under different conditions associated with certain breathing disorders. In addition to that, inhalation and exhalation flow patterns under different breathing scenarios were investigated to further support the feasibility of Doppler radar to accurately estimate the tidal volume. The results obtained for both studies were compared with the gold standard measurement schemes, such as respiration belt and spirometry readings, yielding significant correlations with the Doppler radar-based information. In light of the above, Doppler radar is highlighted as an alternative approach not only for determining respiration rates but also for identifying breathing patterns and tidal volumes as a preferred non-wearable alternative to the conventional contact sensing methods.

Real-time respiratory measurement with Doppler radar has an important advantage in the monitoring of certain conditions such as sleep apnoea, sudden infant death syndrome (SIDS) and many other general clinical uses requiring fast non-wearable and non-contact measurement of the respiratory function. Various types of respiratory dynamics, such as normal breathing, rapid breathing, slow inhalation-fast exhalation and fast inhalation-slow exhalation, can be directly measured by Doppler radar and respective respiratory frequencies can be derived using the fast Fourier transform and the results are highly correlated with conventional contact measurement devices via a respiration strap. Advancement in signal processing techniques allows more information to be obtained from the Doppler radar measurements, which potentially can be used to facilitate additional insights into breathing activity and are likely to trigger a number of new applications in respiratory medicine.

Respiration monitoring is essential in the diagnosis and treatment of conditions such as chronic obstructive pulmonary disease, heart disease and a number of sleep-related conditions [51]. Furthermore, dysfunctional respiratory patterns such as rapid or shallow breathing [117] or high-frequency breathing rates have also been associated with certain psychosomatic conditions [124], all of which, at present, are typically measured via respiration rates alone. However, a more detailed analysis of breathing patterns [17, 79, 240, 247, 335, 393] will provide physicians with new insights into diagnostic medicine, particularly if this can be performed non-invasively. Non-contact Doppler radar has already been considered in a variety of patient monitoring and measurement scenarios in healthcare, including heartbeat and respiration monitoring in place of conventional methods such as the chest strap, photoplethysmograph [250] and ECG [118]. Research reported using Doppler radar in measuring human physiological activity [40, 199, 200, 269, 341, 342, 387] has predominantly demonstrated the feasibility of Doppler radar in obtaining breathing frequency or heart rate using FFT, wavelet analysis or time-frequency analysis [40, 47, 347].

A complete respiration cycle is typically defined by inhalation (inspiration) and exhalation (expiration) states accompanied by a pause [14]. Breathing rates are predominantly calculated independent of the inhalation to exhalation ratio (*I:E*) for each breathing cycle.

For normal and spontaneous breathing, there is an *abundance of time* for the exhalation process from the inspired tidal volume, but in certain pathological states, for instance, asthma and COPD (chronic obstructive pulmonary disease), reduced expiratory flow would need a longer time to empty the inspired lung volume [81]. Typically, for adults, a normal *I:E* ratio is in the range of 1:2, but this varies between individuals depending on the health and the physiological state of the individual [162]. Consequently, more information about each component is extremely important as it can be useful in early detection of several respiratory disorders.

Another important parameter associated with breathing is respiratory tidal volume [198, 230], which can also be derived from microwave radar due to the relationship between the chest wall displacement and the tidal volume. Different types of breathing can potentially be deduced from such chest wall or abdomen displacement information during inhalation and exhalation. For instance, the displacement of the chest wall or abdomen in shallow breathing is expected to be small and the complete breathing cycle would occur in a shorter time period compared to normal breathing. Doppler radar operates by transmitting a radio wave signal and receiving the modulated version of the signal due to the motion triggered by the target [230, 278]. The reflected wave is in the modulated form, where it undergoes a frequency shift proportional to the radial velocity that can be described using the Doppler effect. When a target has a quasi-periodic motion, the time-varying position of the target can be represented as a phase-modulated signal and the phase shift is directly proportional to the object's movement. Thus, the movement of the chest wall/abdomen for respiration due to the inhalation, exhalation and the pause states can be detected and modelled using the reflected Doppler-shifted signal.

5.2 Fundamental Operation of Microwave Doppler Radar

5.2.1 Velocity and frequency

The Doppler effect occurs when there is a shift in the frequency of the wave, either reflected or radiated, received by an object in motion [384]. Consider a transmitted sine wave signal with an angular frequency ω_0,

$$T_x = \sin(\omega_0 t + \phi_0), \tag{5.1}$$

where T_x is the transmitted signal, t is the time and ϕ_0 is the arbitrary phase shift. Assume that the target is stationary at a distance of r_0 from the radar and the transmission time from radar to target is r_0/c, where c is the wave propagation velocity. The target range at time t is given by $r(t) = r_0 + \dot{r}(t - t_0)$, where r is the range of the target from the radar, $\dot{r}$ (velocity) is the rate of change of r, and t_0 is the time at $r = r_0$. The received signal at the stationary target is the same as the transmitted signal at the time r_0/c, which can be given as

$$R_{\text{target}} = \sin\left(\omega_0 t - \frac{\omega_0 r_0}{c} + \phi_0\right). \tag{5.2}$$

The received signal from the target at time t would have been sent Δt seconds prior to time t. This can be represented as $\Delta t = 2r_0/c$. Referring to equation (5.1), the signal can be depicted in the same formulation given as

$$R_x = \sin(\omega_0(t - \Delta t) + \phi_0). \tag{5.3}$$

Substituting $\Delta t = 2r_0/c$ into equation (5.3), the received signal is further represented as

$$R_x = \sin\left(\omega_0 t - \frac{2\omega_0 r_0}{c} + \phi_0\right). \tag{5.4}$$

For a target moving (radially) with respect to the radar, the distance will vary and by using $r(t) = r_0 + \dot{r}(t - t_0)$ and $\omega_d = 2\omega_0\dot{r}/c$, the received signal can be further derived as

$$\begin{aligned} R_x &= \sin\left(\omega_0\left(t - \frac{2r(t)}{c}\right) + \phi_0\right) \\ &= \sin\left(w_0\left(t - \frac{2r_0}{c} - \frac{2\dot{r}(t - t_0)}{c}\right) + \phi_0\right) \\ &= \sin\left(\omega_0\left(1 - \frac{2\dot{r}}{c}\right)t - \frac{2\omega_0}{c}(r_0 - \dot{r}t_0) + \phi_0\right) \\ &= \sin\left((\omega_0 - \omega_d)t - \frac{2\omega_0 r_0}{c} + \omega_d t_0 + \phi_0\right), \end{aligned} \tag{5.5}$$

where the frequency of the reflected signal is shifted by ω_d and the phase angle by $\omega_d t_0$. Therefore, the Doppler shift ω_d can also be denoted as $\omega_d = 2\pi f_d$, where $f_d = 2\dot{r}f_0/c$ is the Doppler shift in Hertz and f_0 is the transmitted frequency. Using $\lambda = c/f_0$, f_d can be written as $f_d = -2\dot{r}/\lambda$, where the negative sign accounts for the fact that if $\dot{r}$ is negative (when the target is approaching), the Doppler frequency will be positive or vice versa [384]. From equation (5.5), the phase angle Φ of the received signal is given as $\omega_d t_0$. Therefore, the transmitted wave from the radar to the target will be reflected to the receiver with some phase shifting and can be represented as phase modulation, given as

$$\Phi = \frac{2\omega_0\dot{r}}{c}t_0 = \frac{4\pi(r)}{\lambda}. \tag{5.6}$$

The measurement model for human respiration using Doppler radar can be derived as follows. Generally, the Doppler shift in frequency is given by

$$f_d(t) = 2fv(t)/c = 2v(t)/\lambda, \tag{5.7}$$

where $v(t)$ is the velocity of the target, λ is the wavelength of the transmitted signal and c is the velocity of the propagating wave. Assuming the target to be stationary or undergoing a periodic movement of $x(t)$ with no net velocity, the Doppler frequency shift can be represented in the form of non-linear phase modulation as the phase signal $\Phi_r(t)$ given by $\Phi_r(t) = 4\pi x(t)/\lambda$, where $x(t)$ is the displacement of the chest wall or abdomen. Using a continuous wave (CW) radar, the transmitted signal is represented by

$$T(t) = \cos(\omega_0 t + \phi_0(t)), \tag{5.8}$$

where $T(t)$ is the transmitted signal and ϕ_0 is the arbitrary phase shift or the phase noise of the signal source. If the transmitted wave $T(t)$ is reflected by the target/subject at a nominal distance d_o, a time-varying displacement of $x(t)$ is caused by the movement of the torso (abdomen). Thus, the distance [100] between the transmitter and the target is given as $d(t) = d_0 + x(t)$. The measurement of the time delay between the transmitter and the target is denoted as the distance travelled over the signal's propagation velocity, given as $d(t)/c$. Thus, due to the movement of the abdomen during the process of respiration, the distance between the antenna and the abdomen at the time of reflection is denoted as $d(t - d(t)/c)$ and the round trip time can be further derived as $t_d = 2(d_0 + x(t - d(t)/c))/c$.

Using the similar formulation shown in the equation (5.3) along with $\omega_0 = 2\pi f$ and $c = f\lambda$, the received signal $R(t)$ can be represented as

$$R(t) = \cos\left[\omega_0(t - t_d) + \phi(t - t_d)\right]$$
$$= \cos\left[\omega_0\left(t - \frac{2d_0 + 2x\left(t - \frac{d(t)}{c}\right)}{c}\right) + \phi\left(t - \frac{2d_0 + 2x\left(t - \frac{d(t)}{c}\right)}{c}\right)\right] \quad (5.9)$$

and further approximated as

$$R(t) \approx \cos\left(2\pi ft - \frac{4\pi d_0}{\lambda} - \frac{4\pi x(t)}{\lambda} + \phi\left(t - \frac{2d_0}{c}\right)\right). \quad (5.10)$$

Demodulation of the phase is used to determine the motion signature, which can be detected at the receiver. In the direct conversion system, the received signal will be mixed with the local oscillator to obtain the baseband output, given as

$$B(t) = \cos\left(\theta + \frac{4\pi x(t)}{\lambda} + \Delta\phi(t)\right). \quad (5.11)$$

In a quadrature receiver system, the received signal will be split into two forms, an in-phase $[I_B(t)]$ and a quadrature phase $[Q_B(t)]$ signal, where the phase difference will be $\pi/2$. Therefore, a general two orthogonal baseband outputs of the quadrature receiver system can be denoted as

$$I_B(t) = \cos\left(\theta + \frac{4\pi x(t)}{\lambda} + \Delta\phi(t)\right), \quad (5.12)$$

$$Q_B(t) = \sin\left(\theta + \frac{4\pi x(t)}{\lambda} + \Delta\phi(t)\right). \quad (5.13)$$

Here, $\theta = 4\pi d_0/\lambda$ is the constant phase shift dependent on the nominal distance to the target and $\Delta\phi(t)$ is the residual phase noise. The benefit of using a quadrature receiver is to overcome the null problem [118] where at least one output (either I/Q) is not null when the other is null.

In the past few decades, different types of radar system structures have been proposed for non-contact respiratory function detection and continuous monitoring. Due to simplicity and a high level of integration, a direct conversion Doppler radar (homodyne radar system) is considered [128] for this application. In respiration detection using Doppler radar, the Doppler shift is due to the movement of the chest/abdomen during respiration. In a non-contact Doppler radar, a transmitted monotone signal is modulated by the movement of the chest/abdomen wall (due to the breathing activity) to a frequency proportional to the radial velocity [277].

Referring to [197, 198], using a quadrature receiver, under the influences of random body movements and hardware imperfections, the reflected signal due to respiration activity and noise (interference/ artefacts) can be represented as

$$I(t) = \cos\left(\theta + \frac{4\pi x(t)}{\lambda} + r_m(t) + \Delta\phi(t)\right) + N_I, \quad (5.14)$$

$$Q(t) = \sin\left(\theta + \frac{4\pi x(t)}{\lambda} + r_m(t) + \Delta\phi(t)\right) + N_Q, \quad (5.15)$$

where θ is the constant phase shift dependent on the nominal distance to the target, $\Delta\phi(t)$ is the residual phase noise, N_i and N_Q refer to the DC offsets and $r_m(t)$ is the interference resulting from random body movements, i.e. jerking of limbs or movement of the body. In order to recover the respiration signal obtained from the Doppler radar measurements, filtering or an advanced signal processing method such as a wavelet can be applied to minimise or filter off the $r_m(t)$ component.

5.2.2 Correction of I/Q amplitude and phase imbalance

Two orthogonal outputs (I and Q) are obtained from a quadrature receiver system but in practice (due to the imperfection of components in the hardware design) suffer from amplitude and phase imbalance, which affects the accuracy of the recovered data at the output [275]. Consequently, phase and amplitude correction is necessary to increase accuracy. There are a number of approaches used to correct the amplitude and phase imbalance [147, 336]. In Steila [336], a final form of two orthonormal vectors using a method similar to the Gram Schmidt orthogonalisation (GSO) [275] has been proposed, as shown later in equation (5.21). The derivation of this is as follows. The ideally received signal $R_x(t)$ is defined by

$$R_x(t) = X_I\cos(w_0t) + X_Q\sin(w_0t), \tag{5.16}$$

where X_I and X_Q are the in-phase and quadrature phase of the information signal respectively. In our approach, with the presence of amplitude imbalance and phase offset, the received signal at the mixer can be represented as

$$R'_x(t) = R_x(t) * \cos(w_0t) + R_x(t) * A_e * \sin(w_0t + \phi), \tag{5.17}$$

where A_e and ϕ are the amplitude and phase imbalance. Demodulation of the received signal is as follows:

$$I' = R_x(t) * \cos(w_0t), \tag{5.18}$$

$$Q' = R_x(t) * A_e * \sin(w_0t + \phi). \tag{5.19}$$

Expanding the derivation:

$$\begin{aligned} I' &= X_I\cos(w_0t)\cos(w_0t) + X_Q\sin(w_0t)\cos(w_0t), \\ Q' &= X_I\cos(w_0t) * A_e(\sin(w_0t)\cos(\phi) + \cos(w_0t)\sin(\phi)) \\ &\quad + X_Q\sin(w_0t) * A_e(\sin(w_0t)\cos(\phi) + \cos(w_0t)\sin(\phi)). \end{aligned} \tag{5.20}$$

After the low-pass filtering and ignoring the term $\frac{1}{2}$, representation of orthogonal X_I and X_Q in matrix form is

$$\begin{bmatrix} X_I \\ X_Q \end{bmatrix} = \begin{bmatrix} 1 & 0 \\ -\tan(\phi) & \dfrac{1}{A_e\cos(\phi)} \end{bmatrix} \begin{bmatrix} I' \\ Q' \end{bmatrix}. \tag{5.21}$$

Using equation (5.21), correction on amplitude and phase imbalance can be performed.

From equation (5.21), the imbalance factors of A_e and ϕ need to be estimated for I/Q correction. This procedure is similar to the GSO procedure as the quadrature-phase signal

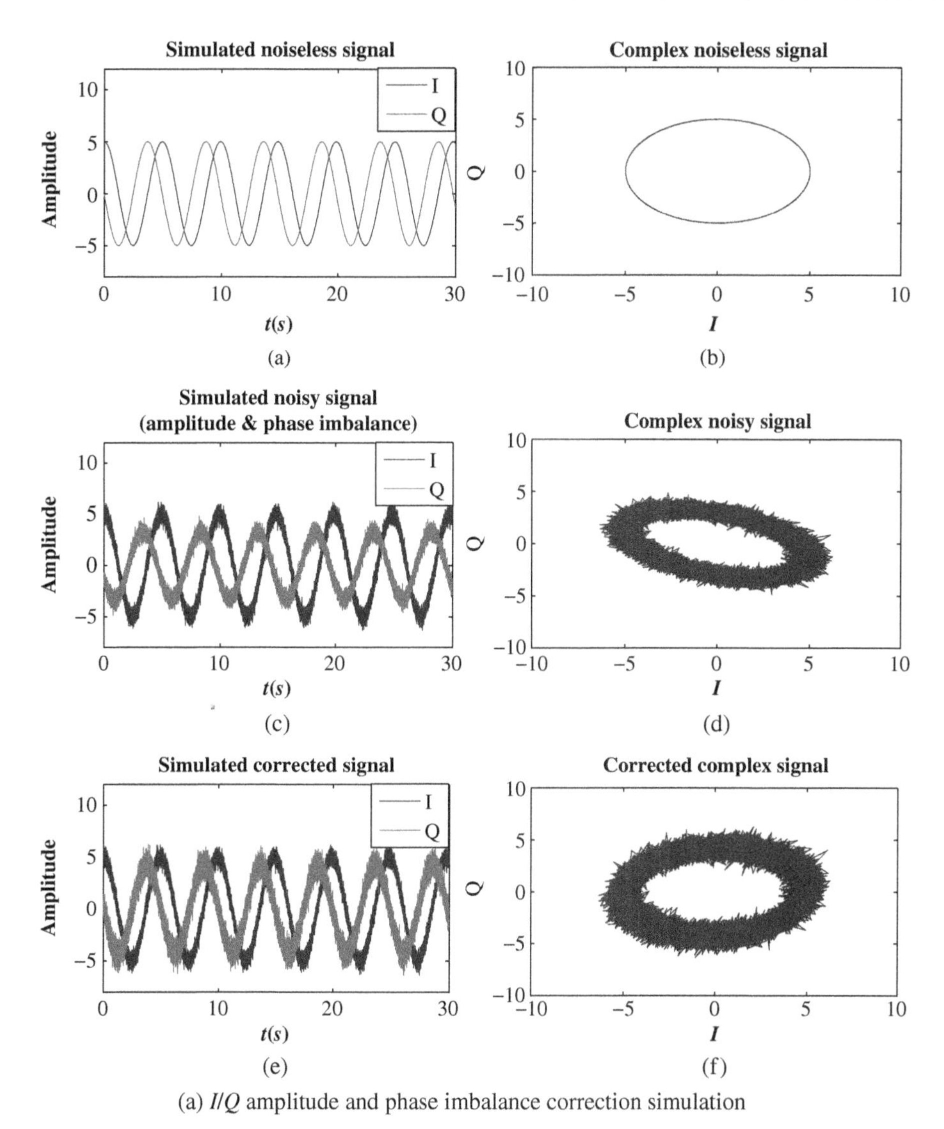

(a) I/Q amplitude and phase imbalance correction simulation

Figure 5.1 I/Q Imbalance simulation and results evaluation.

is orthogonal to the in-phase signal. The simulation was performed by assuming that the breathing frequency is in the vicinity of 0.2 Hz in the I and Q representation. In the simulation results shown in Figure 5.1(a,c), the phase offset of 25° with amplitude imbalance in the quadrature signal was simulated in the noisy signal. We have estimated that the amplitude imbalance ratio and phase offset between the I and Q signal have corrected the signal using equation (5.21), as shown in Figure 5.1(a). The amplitude imbalance was obtained by taking the average ratio of Q/I while the phase offset was estimated by computing the phase difference between the I and Q signals.

Estimated parameters would be slightly different from the real value due to the noise in the signal, but it will be adequate to correct the Q signal based on the I signal. From the

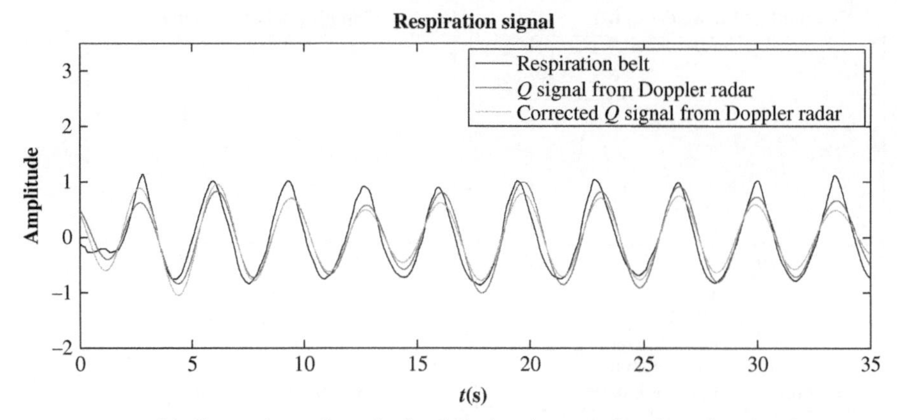

(b) Comparison of respiration belt signal versus doppler radar signal

Figure 5.1 (*Continued*)

Table 5.1 Quantitative evaluation of the Doppler radar signal with the reference respiration belt.

Dataset	Mean square error	Correlation coefficient
1	0.017	0.968
2	0.094	0.938
3	0.009	0.965
4	0.005	0.942
5	0.015	0.975

results shown in Figure 5.1(a,e), the corrected Q signal is similar to the simulated noiseless signal (Figure 5.1(a)) in the amplitude and the phase offset. The same approach was used with the real data and subsequently compared with the respiration belt signal. The corrected Q signal is slightly better than the uncorrected Q signal as the mean squared errors are "0.041 651" and "0.050 928", respectively (see Figure 5.1b). To further compare, five datasets (a minute of recording for each dataset) were collected from the subject (random breathing) where the mean square error (MSE) and correlation coefficient were computed. Results are shown in Table 5.1 and a good correlation was obtained between the Doppler signals and the respiratory belt signals.

5.3 Signal Processing Approach

5.3.1 Respiration rate

Capturing the respiration rate is a basic function of any respiratory monitoring system and is achieved primarily by Fourier transform and wavelet transform techniques.

Fast Fourier transform (FFT)

Fourier transform is used for characterising linear systems as well as to identify the frequency components that make up a continuous waveform [35]. FFT is an efficient and fast approach in computing the discrete Fourier transform (DFT). In this book, the FFT (using the Cooley-Tukey DFT algorithm) [35] was used to determine the related frequency component of breathing and the resultant spectra are shown in Section 5.6.3 via spectral peak extraction.

The extracted peak frequency can be used to approximate the breathing rates, especially under a normal breathing condition. As for abnormal breathing patterns, it is found that FFT is no longer suitable for capturing the respiration rate as it does not give the information on the time instance at which the frequency occurs. This is due to the non-periodic characteristic in an abnormal breathing cycle and led us to explore using a continuous wavelet transform.

Continuous wavelet transform (CWT)

Computation of the FFT over the complete time period of the signal only returns the corresponding frequency components of the signal but not the information on the time instance at which a particular frequency occurs. However, the short time Fourier transform (STFT) using a sliding window provides information on both time and frequency with limited resolution on both dimensions simultaneously [71]. The wavelet transform is destined to address this issue. In the continuous wavelet transform (CWT), the mother wavelet is dilated in such a way as to cater for temporal changes of different frequencies [333]. A mother wavelet function is defined as a function of $\psi(t)\epsilon L^2(\mathbf{Re})$ with a zero mean and localised in both frequency and time. Properties of the wavelet can be summarised as follows:

$$\int_{-\infty}^{\infty} \psi(t)dt = 0, \tag{5.22}$$

$$\| \psi(t) \|^2 = \int_{-\infty}^{\infty} \psi(t)\psi^*(t)dt = 1. \tag{5.23}$$

Through dilation and translation of a mother wavelet $\psi(t)$, a family of wavelets is produced. We can denote this as

$$\psi_{s,u}(t) = \frac{1}{\sqrt{s}}\psi\left(\frac{t-u}{s}\right), \tag{5.24}$$

where $(u, s)\ \epsilon\ \Re$, u is the translating parameter specifying the region of interest and s is the dilation parameter that is greater than zero. Continuous wavelet transform is the coefficient of the basis $\psi_{u,s}(t)$ [71], which is the inner product of the family of the wavelet $\psi_{u,s}(t)$ with the signal $f(t)$ denoted as

$$F_W(s, u) = < f(t), \psi_{s,u}(t) > \tag{5.25}$$

$$= \int_{-\infty}^{\infty} f(t)\frac{1}{\sqrt{s}}\overline{\psi}\left(\frac{t-u}{s}\right) dt, \tag{5.26}$$

where $\overline{\psi}$ is the complex conjugate of ψ and $F_W(s, u)$ is the time scale map. Using equation (5.25), the one-dimensional signal of $f(t)$ can be mapped into a two-dimensional coefficient of $F_W(s,u)$ and a time frequency analysis can be performed where we could locate a particular frequency (s) at that particular time instance u.

If $f(t)$ is an $L^2(\Re)$ function, the inverse transform of the wavelet can be denoted as

$$f(t) = \frac{1}{C_\psi} \int_0^\infty \int_{-\infty}^\infty Wf(s,u) \frac{1}{\sqrt{s}} \psi \frac{t-u}{s} du \frac{ds}{s^2}, \tag{5.27}$$

where C_ψ is

$$C_\psi = \int_0^\infty \frac{|\Psi(\omega)|^2}{\omega} d\omega < \infty \tag{5.28}$$

and $\Psi(\omega)$ is the Fourier transform of the mother wavelet $\psi(t)$.

5.3.2 Extracting respiratory signatures

A complete breathing cycle is comprised of inhalation (I), exhalation (E) and pause components, where the ratio of I:E can certainly be asymmetric [162] for normal breathing patterns. Therefore, computation of breathing rates purely based on simple single frequency signatures computed via FFTs is not sufficient to provide detailed breathing pattern features – particularly for the identification and analysis of respiratory conditions. For this purpose, two different approaches were demonstrated. In the first approach, the pre-processed raw data were modelled using a piecewise linear least-squares approach [205]. In the second approach, the raw data were processed using a SG (Savitzky-Golay polynomial least square) [319] smoothing filter and further analysed using Fourier filtering [270]. The first approach offers a simple method applicable for real-time processing while the second approach offers more accurate identification of the respiration cycle components and their properties.

The Piecewise Linear Fitting Method

This method fits non-linear, typically noisy, waveforms by choosing an optimal segmentation of the waveform and then fitting each segment with a linear function [205]. Here the segmentation process is critical and, in this case, appropriate lengths of non-overlapping segments were used, which were obtained via trial and error.

The Savitzky-Golay Method and Fourier Filtering

The Savitzky-Golay filter is a least-squares polynomial filter [319]. By applying the filter to the noisy data obtained from the chemical spectrum analysers, Savitzky and Golay demonstrated how it reduces noise while preserving the shape and height of waveform peaks. Savitzky and Golay demonstrated the use of this filter in removing the noisy data obtained from the chemical spectrum while preserving the shape and height of waveform peaks. Referring to [319], a least-squares polynomial $p(n)$ of order N is fitted to the signals with a moving window of size $2M + 1$ centred at $n = 0$. This can be explicitly stated as

$$p(n) = \sum_{k=0}^{N} a_k n^k, \tag{5.29}$$

where a_k is the kth coefficient of the polynomial function. The mean squared approximation error (ε_N) for the underlying group of data point samples centred at $n = 0$ can be represented as

$$\varepsilon_N = \sum_{n=-M}^{M} (p(n) - x[n])^2 \tag{5.30}$$

$$= \sum_{n=-M}^{M} \left(\sum_{k=0}^{N} a_k n^k - x[n] \right)^2. \tag{5.31}$$

Thus, the output value is smoothed and derived from the central point of $n = 0$ of the moving window and the whole procedure is repeated over the stream of data by convolution [319]:

$$y[n] = \sum_{m=-M}^{M} h[m]x[n-m] \tag{5.32}$$

$$= \sum_{m=n-M}^{n+M} h[n-m]x[m]. \tag{5.33}$$

The SG filter was used to smooth the raw data from the Doppler Radar while preserving the shape and reducing the noise. Here, the SG filter was used to smooth the input raw data after the DC components were removed. The output from the SG filter improved the shape of the signal significantly where noise and redundancy were filtered extensively, as shown later in Figure 5.5 (Data Set 1, sub-figure: a,c).

The signals were smoothed by an SG filter and then reconstructed using Fourier filtering. This was to extract absolute maxima and minima points of the breathing curve that denotes each of the inhalation and exhalation components. Fourier filtering [270] has already been used as one of the processing algorithms to further eliminate noise and to reconstruct the signals. It is a filtering function that manipulates specific frequency components of a signal by taking the Fourier transform of the corresponding signals that later either attenuate or amplify frequencies of interest. The Fourier filter was used to eliminate noise employing a bandpass filter depending on the desired breathing frequency range while not distorting the signal significantly. The shape of the Fourier filtered signal was quite similar to the signal resulting from the piecewise linear fitting but was smoother and local minima and maxima were prominent.

Breathing signal decomposition

For the breathing cycles obtained from Doppler radar we assumed that the transition from *local minima to local maxima* on the curve represents the inhalation component and vice versa for the exhalation component, respectively. A peak detection algorithm was then used to determine the maximum and minimum points of each transition, defining the inhalation and exhalation components, respectively. These components were extracted separately and represented by a fourth-order polynomial. We then computed the average representation for normal and fast breathing components (inhalation and exhalation) to be used as a model for component identification, as discussed later in Section 5.5.

Identification of dynamic time warping

Dynamic time warping (DTW) is used to optimally align two time series where one time series is transformed to best fit the other [311]. This technique has been extensively used in speech recognition to identify the similarity of spoken phases from two waveforms as the duration of each spoken sound can vary but with similar overall waveform shapes. DTW has also been used in other areas such as data mining and gait recognition [170]. Typically, similarity between time series or for time series classification often required a distance measurement between them. Computation of Euclidean distance between the two time series will not yield accurate results if one of the two identical time series is slightly shifted along the time axis. To overcome this limitation, DTW was introduced, as previously described [311].

Referring to [69, 311], given two time series of A and B of length m and n, respectively,

$$A = a_1, a_2, a_i, \ldots, a_n, \quad B = b_1, b_2, b_j, \ldots, b_m. \tag{5.34}$$

and to align these two time series using DTW, a matrix of n by m is constructed where the ith and jth elements of the matrix contain the distance $d(a_i, b_j)$ between two points in two time series (a_i and b_j). Typically, using square distance, $d(a_i, b_j) = (a_i - b_j)^2$, the alignment between points a_i and b_j can be represented in the constructed matrix of $n \times m$ of each element (i, j). Mapping between A and B is represented by a warping path W, which is defined by

$$W = w_1, w_2, \ldots, w_K, \quad max(|A|, |B|) \leq K \leq |A| + |B|, \tag{5.35}$$

where K is the warp path length, kth element of the warp path, $w_k = (i, j)_k$ and i is the index from time series A while j is the index for time series B. The warp path begins at each time series at $w_1 = (1, 1)$ and terminates at the end of both time series at $w_K = (|A|, |B|)$. Every index of both time series will be covered and the i and j increased monotonically in the warp path. This is given as

$$w_k = (i, j), w_{k+1} = (i', j'), i \leq i' \leq i + 1, j \leq j' \leq j + 1. \tag{5.36}$$

The minimum distance of the warp path is the optimal warp path, given as

$$Dist(W) = \sum_{k=1}^{k=K} Dist(w_{ki}, w_{kj}), \tag{5.37}$$

where $Dist(W)$ is the Euclidean distance of the warp path W and w_{ki}, w_{kj} are the distances between two data point indexes between two time series in the kth element of the warp path. Here, DTW was used for registering and comparing breathing components to determine the temporal features extracted from the breathing component model.

Each inhalation and exhalation component was extracted to obtain the polynomial coefficients from normal and fast breathing data, respectively, and results indicate that a fourth-order RMSE (root mean square error) and Corr (correlation coefficient) polynomial was sufficient to fit these components (e.g. randomly chosen inhalation and exhalation component), as shown in the Table 5.3, polynomial modelling and DTW performance evaluation (a). Subsequently, using the same approach, the computed fourth-order polynomial model was used to characterise two different types of inhalation and exhalation breathing components (normal and fast). This model was then used to identify the experimental breathing scenario, as discussed later in Section 5.5.

5.3.3 Low-pass filtering (LPF)

As mentioned in Yuan *et al.* [393], the temporal pattern is essential in addition to the respiration rate and naturally provides information pertaining to breathing anomalies that allows medical researchers to progress in a number of research directions. The breathing rate can be estimated from the LPF output using fast Fourier transform (FFT) while the shape of the breathing curve does not reflect real breathing patterns if it is altered by large motion artefacts, as shown in the LPF signal in Figure 5.22(a,b). In contrast, if the Doppler return signal due to breathing activity is not affected by any motion artefact, LPF could be performed effectively as shown in Figure 5.22(c). Referring to [100], the proposed LPF employed to process the Doppler received signal is a 600 order Kaiser window FIR (finite impulse response) with $\beta = 6.5$ and a cut-off frequency of 1.0 Hz to account for the normal and abnormal respiratory conditions.

Table 5.7 in Section 5.7.2 depicts the performance comparison of different LPF schemes in filtering off the motion artefacts/noises (dataset from experiment (1–3)).

5.3.4 Discrete wavelet transform

As discussed by John *et al.* [367], wavelet analysis has received significant attention, especially among the signal and image processing communities. Wavelet transformation has been widely used in numerous applications as previously reported [185, 261], while the underlying principles are described in Daubechies [88]. In the wavelet transformation, the original time domain is transformed to the time-scale domain where a given signal is decomposed into several types of other signals with varying levels of resolution [392].

One-dimensional discrete wavelet transform described in terms of a filter bank is shown in Figure 5.2(a) while (b) depicts the multi-level decomposition architecture using wavelet transformation. As this involved multi-level decomposition, the approximation coefficient cA_1 serves as the input at the second-level filter bank where the analysis stage is depicted in Figure 5.2(a). Based on the number of levels (L) of decomposition, the process is iteratively repeated with a reference signal of $cA_L(n)$ and detail coefficient of $cD_L(n), cD_{L-1}(n), ..., cD_1(n)$.

Typically, DWT employs a dyadic grid that consists of integer power of two scalings in a and b, zero redundancy and having orthonormal wavelet basic functions [16]. In practice, the transform integral remains continuous in DWT but is only derived on a discretised grid of a scales and b locations which have the form of

$$\psi_{m,n} = \frac{1}{\sqrt{a_0^m}}\psi\left(\frac{t - nb_0a_0^m}{a_0^m}\right),$$

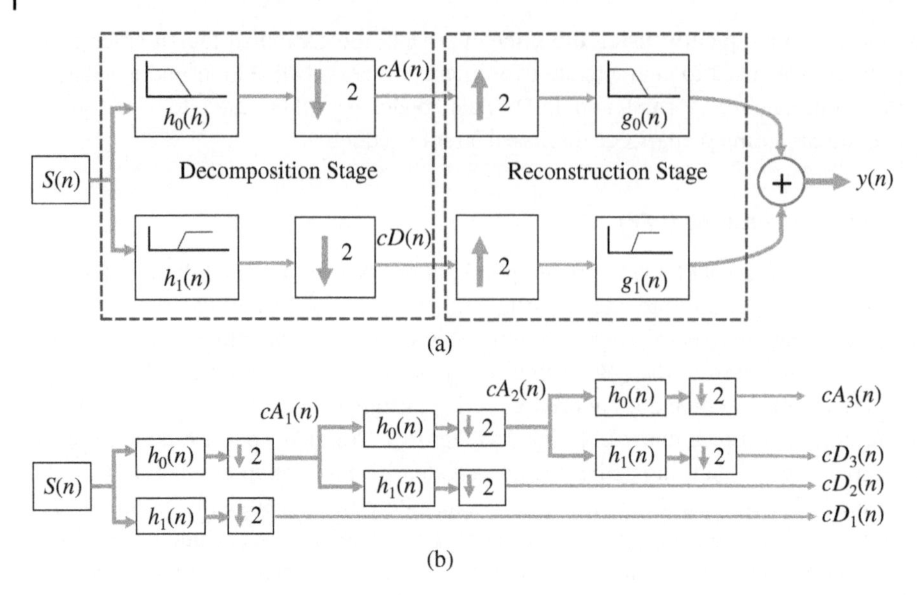

Figure 5.2 (a) One-dimensional filter bank for wavelet transformation; (b) multi-level (three-level) transformation.

where m and n control the wavelet dilation and translation, respectively, a_0 is a specified fixed dilation step parameter (value > 1) and b_0 is the location parameter (value > 0). Using $a_0 = 2$ and $b_0 = 1$, the dyadic wavelet transform of signal $s(t)$ is given as follows:

$$\text{DWT } s(m,n) = 2^{-m/2} \int_{-\infty}^{\infty} s(t)\varphi * \left(\frac{t - 2^m n}{2^m}\right) dt, \tag{5.38}$$

where m and n are scale and time shift parameters, respectively, * is the complex conjugate and $\varphi(t)$ is the given basis function (mother wavelet).

As shown in Figure 5.2, generally, the output of the low-pass filter is known as a wavelet approximation (scaling) coefficient cA_m and the output of the high-pass filter is called wavelet detail (wavelet) coefficient cD_m. The approximation and the detail coefficient at the mth level can be denoted as $cA_m[n] = \sum_{-\infty}^{\infty} l_d[k]cA_{m-1}[2n-k]$ and $cD_m[n] = \sum_{-\infty}^{\infty} h_d[k]cA_{m-1}[2n-k]$.

The taps of the high-pass filter h_d and the low pass filter l_d are derived from the scaling and wavelet function of a chosen mother wavelet family, i.e, Daubechies, Haar, Coiflets. In this experiment, we have explored the use of the Daubechies (Db10) mother wavelet in 12 levels of decomposition and the approximation of the breathing signal shows a high level of correlation at the ninth level in comparison to the respiration belt (*respiband*).

5.4 Common Data Acquisitions Setup

A simple Doppler radar system (Figure 5.3a) has a continuous wave (CW) that operates at 2.7 GHz with 2.14 dBm, two panel antennae where one is (Tx) and the other (Rx), an I/Q demodulator (Analog Device AD8347) and a data acquisition module(NI-DAQ) that can be

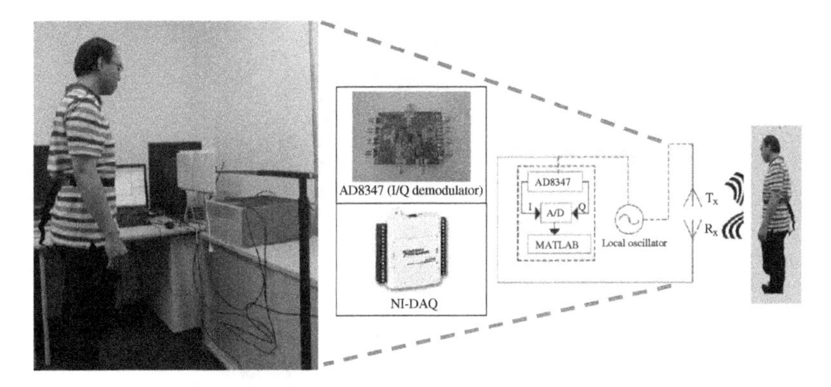

(a) Simple Doppler radar system configuration

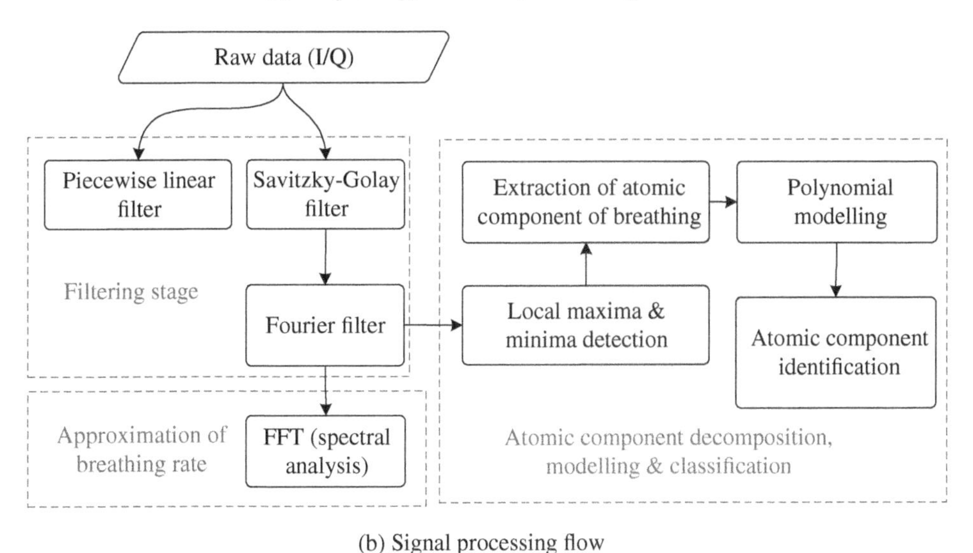

(b) Signal processing flow

Figure 5.3 Doppler radar system and signal processing flow. Source: Yee Siong Lee.

built easily with commercially available, off-the-shelf electronic components. The received signals were directly converted into *I/Q* decomposition using AD8347, where the demodulated signal was then sent to a DAQ for further processing using MATLAB. Alternatively, a research-based Doppler radar can be obtained from Gu [126].

A typical experimental setup would require the subjects to be positioned at a certain range away from the antenna (transmitter, Tx, and receiver, Rx). The distance apart depends on the frequency and operating power of the Doppler radar. The panel antennae were aligned to focus on the abdomen to capture a better Doppler effect due to respiration. The subject, with normal clothing (see Figure 5.3a), was asked to stand in front of the antenna and breathe in specific ways for a determined period of time. The measured signal from Doppler radar can be computed using the proposed signal processing techniques shown in Figure 5.3(b).

Figure 5.4 Environment setup for seated position experiments using Gu [126]. Source: Yee Siong Lee.

Alternatively, a 2.4 GHz (Figure 5.4) Doppler radar module [126] was also being used in the experiments in this book. The system transmits a 2.4 GHz continuous wave at 0 dBm (1 mW) and was attached to a two panel antenna (transmitter and receiver) having a data acquisition module (DAQ: NI-USB6009). The received signals were then sent to DAQ (sampled at 1000 Hz) for further processing using MATLAB. For this experiment, the subject was positioned 0.8 m away from the panel antenna where the antenna was aligned to focus on the abdomen (belly) rather than the chest in order to capture a more reliable Doppler signal. This setup varies for different subjects depending on belly breathing or chest breathing. Data are collected from the subject with normal clothing and the person was instructed to follow specific breathing patterns, as discussed in Section 5.5. For each experiment, an external respiration belt (MLT1132 Piezo Respiratory Belt Transducer, sampled at 1000 Hz) attached to PowerLab (ADInstruments) was used as a reference signal to evaluate the performance of the Doppler radar.

For validation purposes, a respiband (MLT1132 -Piezo Respiratory Belt Transducer) attached to PowerLab (ADInstruments) was used as a reference signal to compare with the Doppler measurements. Results in Figure 5.1(b) show the normalised raw respiration signal obtained from the respiration belt and normalised filtered Doppler radar signals. For the decomposition of the breathing signal into inhalation and exhalation components, it is necessary to calculate the transition time of each breathing component independent of the breathing amplitude. In addition to this, preliminary measurements were obtained from a voluntary subject (asthmatic) to understand if there were any detectable breathing pattern differences in his breathing compared to normal patterns (see Figure 5.5 and Table 5.2).

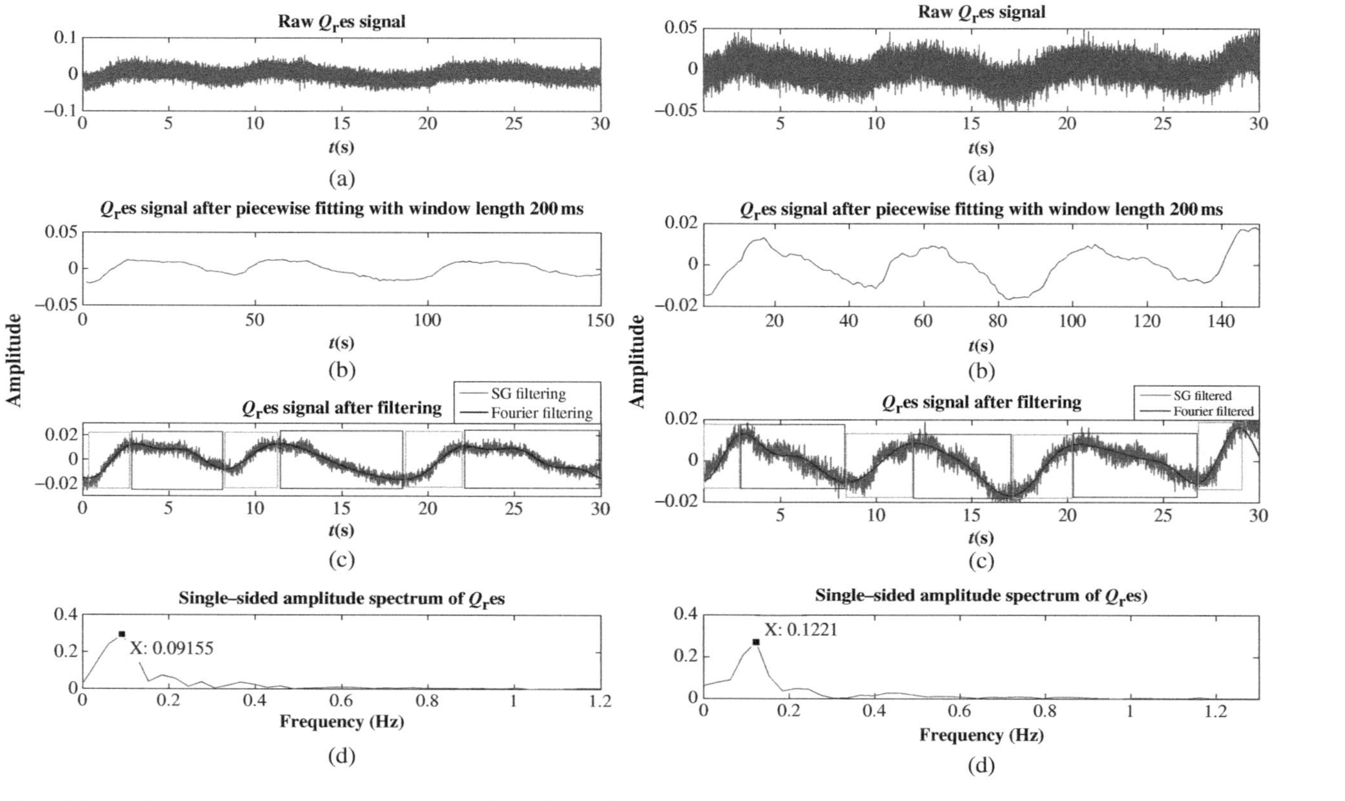

Figure 5.5 Breathing patterns from a voluntary asthmatic subject (datas 1 and datas 2). Source: Yee Siong Lee.

Table 5.2 Doppler radar signals from various types of breathing scenarios.

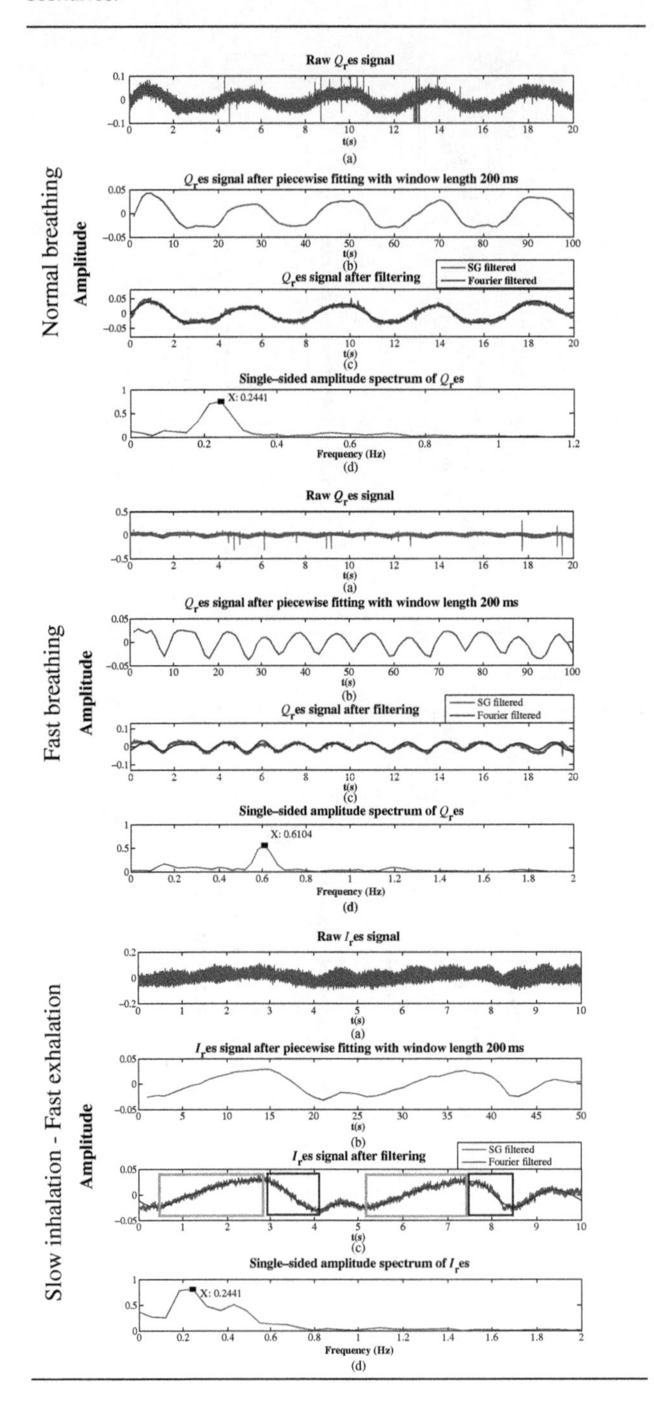

Table 5.2 (Continued)

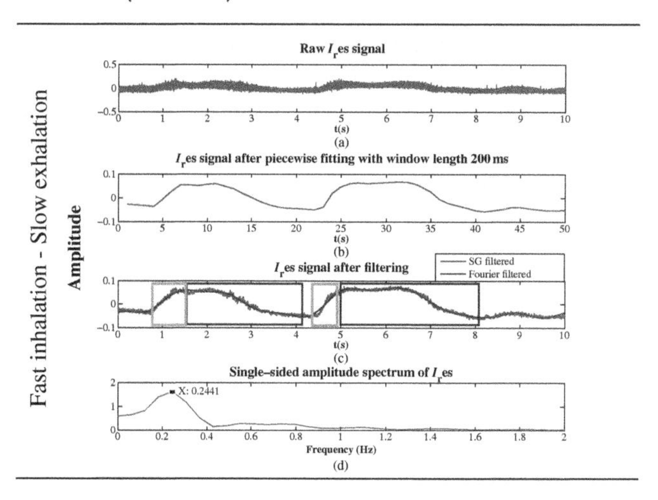

5.5 Capturing the Dynamics of Respiration

The results were based on choosing the *I*/*Q* baseband signal closest to the optimum point [49] and best matched with the independent breathing measurements. Only a portion of observations were displayed in this book where the output consisted of Doppler radar-based measurements for different types of inhalation and exhalation patterns collected over a specific period of time.

5.5.1 Normal breathing

Normal adult breathing rates range from 12 to 20 cycles (inhalation, exhalation, pause) in a minute [393]. Table 5.2 (first row) represents the normal breathing pattern. It can be seen that for a period of 20 seconds there were 5 breaths which corresponds to 0.25 Hz ($\approx$15 breaths per minute) and the FFT of the signal shows a constant peak at 0.2441 Hz or 14.646 breaths per minute. The patterns and extracted rate correlated with the independent breathing counts.

5.5.2 Fast breathing

Rapid breathing is typically defined as above 20 breaths per minute for resting adults and this is called Tachypnoea [245]. In this experiment, our aim was to establish if breathing at different rates can be detected robustly and the feasibility of subsequent classification. Table 5.2 (second row) represents the fast breathing pattern with different dynamics. Here, the subject was inhaling and exhaling at a faster rate, resulting in a shorter breathing cycle. Results show the occurrence of 12 breathing cycles in a period of 20 seconds (36 per minute). The FFT also shows a peak at 0.6104 Hz, corresponding to 36.6 breaths per minute, which is similar to the independent breathing cycle counts.

5.5.3 Slow inhalation–fast exhalation

Another type of breathing scenario where the inhalation is slower than the exhalation rate was mimicked. Data were collected for a period of 10 seconds and from Table 5.2 (third row), a longer inhalation time (mark in light colour box) and a shorter exhalation time (mark in dark colour box) are evident. This is as expected as the subject inhales slowly and exhales at a faster pace. Results shows that there were two clear breathing cycles in a period of 10 seconds. Observed results show an average of 2.5:1 for the *I:E* ratio, where the FFT computation approximated the breathing rate to be 14.65 (0.2441 Hz) breaths per minute and the expected breathing rate was 12 breaths per minute, found from independent measurements. For these particular experiments, an average of 2.5 seconds was required for inhalation compared to 1 second needed for exhalation.

5.5.4 Fast inhalation–slow exhalation

Table 5.2 (fourth row) shows the signal representation for fast inhalation and slow exhalation. Measurements clearly show that two breathing cycles with an average *I:E* ratio of 1:2.5 occurred. The breathing rate was expected to be 12 breaths per minute and from the FFT, the breathing rate was estimated as 14.65 (0.2441 Hz) breaths per minute. Results from both observations clearly show that the exhalation is longer than the inhalation. Both of the cases discussed in Sections 5.5.3 and 5.5.4 further prove that the respiration rate alone is not adequate in describing the respiratory activities of the subjects. More descriptive information could be obtained through the breathing cycle decomposition approach from the non-contact Doppler radar measurements.

5.5.5 Capturing abnormal breathing patterns

It is clear that simply recording breathing frequencies, measured as angular frequency using spectral methods, is inadequate for analysing asymmetric breathing patterns [162], although useful for extracting the fundamental cycle for breathing periods. The evidence so far is that decomposing the breathing cycle into its inhalation and exhalation components offers a more accurate and insightful approach to detecting and interpreting breathing and can be performed reliably using Doppler radar. In this particular experiment, the breathing patterns of a voluntary subject (age 23, height 180 cm and weight 95 kg) who has *asthma* were collected within the duration of 30 seconds *but not during an asthma attack*. Results are shown in Figure 5.5. Notice that the inhalation component (marked in the light colour box) is of a shorter duration compared to the exhalation component (marked in the dark colour box) where the approximated *I:E* ratio for that subject is 1:2.5. Both results showed a longer duration recorded for exhalation compared to inhalation, where the implications are such that the subject could be having difficulties in exhaling [358], which enforces the value in the analysis by decomposition.

In future work, experiments from Sections 5.5.1 to 5.5.5 will be extended with an increased number of subjects (normal and abnormal) in a clinical trial to further support the qualitative and quantitative evaluations. This can facilitate finding a more accurate and insightful way to describe the respiratory functions using a non-contact form of measurements. Furthermore, additional analysis could be performed, including the amplitude variation and the shape of each decomposed breathing component pertaining to different types of subjects. For instance, amplitude variations in the voluntary subject with asthma were observed to be less than those of the subject with normal breathing. Consideration of respiratory effort, breathing patterns and other related factors (e.g. respiratory function such as tidal volume) would be essential in future studies evaluating the potential use of Doppler radar in respiratory research that includes sensing, detections, analysis and qualitative assertions.

5.5.6 Breathing component decomposition, analysis and classification

Although a complete breathing cycle comprises inhalation and exhalation, short and even long pauses can also exist between these states depending on the regularity of breathing and other factors such as the need for oxygen, the surrounding environment, etc. A long pause, for instance, of more than 10 seconds [21] is defined as an abnormal event and is known as apnoea, relevant for detecting sleep apnoea and even SIDS. Breathing patterns can also potentially be used together with the analysis of tidal volume [230] to diagnose other aspects of breathing problems such as shallow breathing and to detect apnoea. These have been reported in Lee *et al.* [200] using microwave Doppler radar.

The main purpose of decomposing the breathing cycles is to gain useful information about the breathing activity. For instance, an abnormal breathing rate of 8 breaths/min could be analysed with more information, such as inhalation and exhalation rates, etc. This can be particularly useful in the early diagnosis of specific breathing conditions or in a pulmonary rehabilitation [63, 176, 213], especially if it could be performed in a non-contact form.

I:E ratio analysis

The ratio between the inhalation or exhalation components was computed from the average time duration in considerations of the entire set. Using the collected data, there were 15 fast and seven normal components extracted from the datasets and the ratios of each of the components (in comparison to the average time of respective inhalation/exhalation components) are shown in Figure 5.6. It was seen that there were two distinct groups corresponding to two different breathing dynamics in two separate events, where this could not be estimated from the respiration rate estimation (spectral analysis):

1. Extraction of inhalation and exhalation components based on normal and fast breathing criteria.
2. Computation of fourth-order polynomial models for each breathing condition (normal and fast) from the extracted components respectively.
3. Using dynamic time warping to find the optimal alignment between the pre-defined model from (2) and the randomly picked breathing component.
4. Using the correlation method to identify the similarity of the aligned results from (3) for identification. Compute the MSE between the curves.

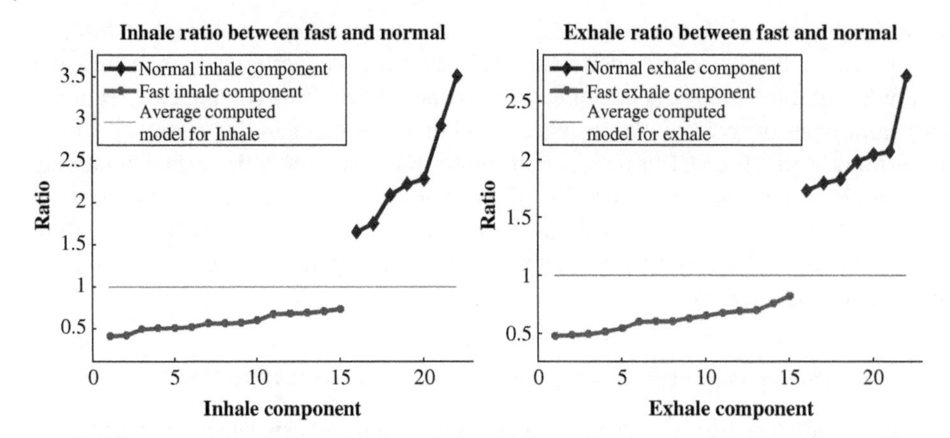

Figure 5.6 Ratio of breathing components.

Two different polynomials for inhalation and exhalation in normal and fast breathing were modelled from the datasets (Procedures 1 and 2). For validation, dynamic time warping was performed between randomly chosen components (any dataset) with the model based on polynomial representation (Procedures 3 and 4).

The purpose of performing this experiment was to use the derived model as a reference and to classify each breathing component based on two different classes. In brief, by deriving a model based on the rate of breathing, we can, in fact, identify and correlate the extracted breathing components with the derived model to distinguish different respiratory classes. For validation purposes, the experiments were performed as follows:

- Fast inhalation component with normal and fast inhalation model.
- Normal inhalation component with normal and fast inhalation model.
- Fast exhalation component with normal and fast exhalation model.
- Normal exhalation component with normal and fast exhalation model.

Each of the breathing components were randomly picked from the datasets. It was then evaluated and represented in terms of mean square error (MSE) and correlation coefficient (Corr), as shown in the Table 5.3. For graphical representation, as an example, in Figure 5.7(a) we associate the normal inhalation component with the normal and fast inhalation model and in Figure 5.7(b) the fast exhalation component with the normal and fast exhalation model.

Here, the feasibility of breathing detection under varying conditions using Doppler radar was demonstrated. It was shown that non-invasive breathing detection using Doppler radar could potentially be used to detect different types of breathing patterns, such as rapid breathing and slow breathing. It is also demonstrated that by decomposing the respiratory cycle into inhalation, pause and exhalation, it is possible to extract additional information on the breathing activities. For this purpose, a fourth-order polynomial was proposed to represent each atomic component of breathing and demonstrated the use of DTW in classifying a breathing component independently into the corresponding class. In the derived model, each component is associated with a specific breathing scenario, which in particular are fast and normal breathing.

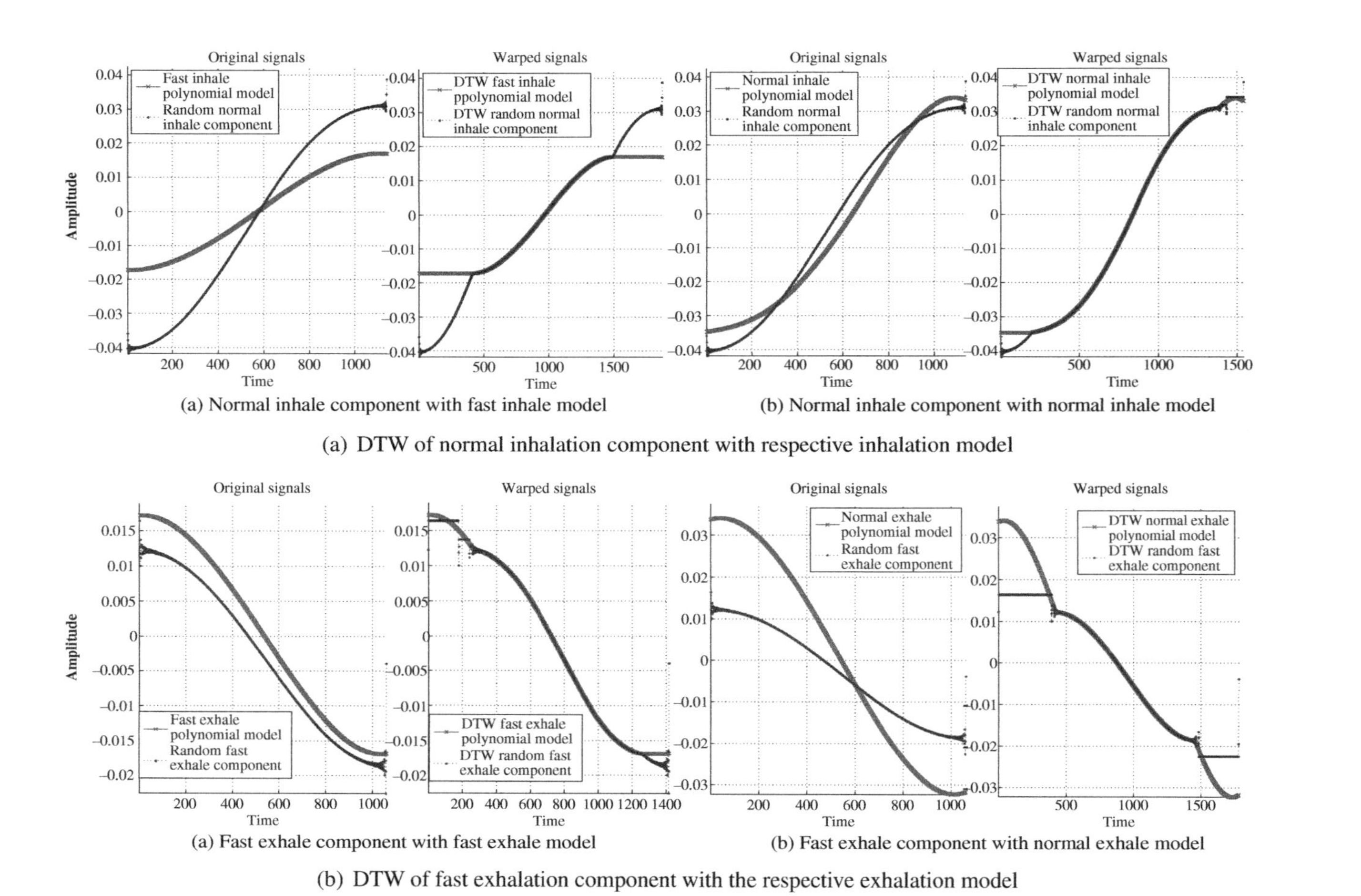

Figure 5.7 DTW evaluation.

Table 5.3 Polynomial modelling and DTW performance evaluation.

(a) Polynomial order evaluation

	Inhalation		Exhalation	
Order	**RMSE**	**Corr**	**RMSE**	**Corr**
1	2.14E-03	0.9912	2.04E-03	0.9918
2	2.02E-03	0.9921	2.03E-03	0.9919
3	3.15E-04	0.9998	5.39E-05	0.9999
4	1.97E-17	1	1.02E-17	1
5	3.26E-17	1	1.36E-17	1

(b) Performance evaluation of random breathing component with selected model

Breathing component	Polynomial model	MSE	Corr	Class
Fast inhalation	Normal	1.11e-04	0.933	**Fast**
	Fast	4.28e-06	0.989	
Normal inhalation	Normal	2.23e-06	0.999	**Normal**
	Fast	8.37e-05	0.954	
Fast exhalation	Normal	4.58e-05	0.972	**Fast**
	Fast	4.47e-07	0.999	
Normal exhalation	Normal	2.50e-06	0.999	**Normal**
	Fast	7.76e-05	0.958	

5.6 Capturing Special Breathing Patterns

The feasibility of using Doppler radar in capturing different types of breathing patterns was investigated. Various breathing types were professionally role-played as per the widely accepted descriptions [68, 393]. In addition, the relationship between the inspiration and expiration tidal volumes obtained from Doppler radar correlates with the measurements obtained from the experimental trials using spirometer readings for various breathing conditions. Analysis of the breathing pattern, independent of the respiration rate measurements, can often provide clues as to the existence of specific respiratory conditions [393]. For example, abnormal breathing, as in Cheyne-Stokes breathing, can be seen in normal people, altitude-related respiratory defects or patients with severe neurological or cardiac diseases [371]. Therefore, relying on the respiration rate alone will not provide sufficient information on the breathing pattern. Here, different types of breathing patterns were professionally role-played in real time based on the description of the patterns reported in the literature [68, 393] including:

1. Normal breathing
2. Kussmaul's breathing
3. Cheyne-Stokes respiration

4. Ataxic breathing and Biot's breathing
5. Cheyne-Stokes variant
6. Central sleep apnoea
7. Dysrhythmic breathing (non-rhythmic breathing).

The intention behind creating these scenarios is to establish that the measurements made using microwave Doppler radar correlate quite closely with the measurements made by clinically established techniques. Indeed, to use Doppler radar as a clinical tool, clinical trials with a sufficient number of subjects suffering from these conditions are required for an extensive analysis. This preliminary work lays a strong foundation and enforces the need for such extensive clinical trials.

5.6.1 Correlation of radar signal with spirometer in tidal volume estimations

To the best of our knowledge, Doppler radar ability to capture *different types of breathing patterns* is still an unresolved issue. Further to this, although respiratory tidal volume has been deducted using Doppler radar [229] for normal breathing conditions using the known relationship between chest wall movement and tidal volume, this has not been investigated for other forms of respiratory conditions.

Continuous and simultaneous measurement of the breathing patterns and tidal volume for longer terms is useful and could potentially be used to assess certain breathing disorders, especially if it can be performed via non-contact techniques. As mentioned in Kuratomi *et al.* [187], variation in breath by breath volume and temporal pattern distribution changes is dependent on the clinical condition and the pathophysiology, which in turn can be related to either restrictive or obstructive lung diseases. This signifies the importance of having tidal volume measurements in addition to the breathing patterns. By definition, tidal volume (TV) is the amount of air inspired or expired during regular breathing. It is a measurement of the amount of air flowing in and out of the lungs [245].

The linear relationship between the lung and air flow during regular breathing has been previously documented (see, for example, [181, 229]) and since the output of the Doppler radar is proportional to the abdomen/chest wall movement, the amount of air flow either in or out can be derived in a similar way. Firstly, calibration was performed in order to obtain the most reliable chest displacement measurements as follows:

1. Compute FFT for radar and spirometer signals.
2. Determine the maximum amplitude for both signals.
3. Compute the ratio from (2).
4. Reconstruct a new signal for radar from the calculated ratio by multiplying the ratio by the radar signal.

This will provide a functional relationship between the displacement of the chest wall (signals from Doppler radar) and the tidal volume to estimate the inhalation and exhalation flow rates.

5.6.2 Experiment setup

For each breathing pattern experiment, three datasets were collected (only one set of observations for each breathing pattern is shown in this book) from subject 1 where all the

Table 5.4 Evaluation of Doppler radar measurements with the respiration belt for six subjects.

	Dataset 1					Dataset 2		
Subject	**Gender**	**Mode**	**MSE**	**Error (%)**	**Correlation**	**MSE**	**Error (%)**	**Correlation**
1	Male	Abdomen	0.0236	1.18	0.9355	0.0764	3.82	0.9006
2	Male	Abdomen	0.0394	1.97	0.9610	0.0358	1.79	0.9065
3	Male	Abdomen	0.0102	0.51	0.9468	0.0172	0.86	0.9430
4	Female	Chest	0.0248	1.24	0.9718	0.0163	0.82	0.9668
5	Male	Chest	0.0544	2.72	0.8378	0.0140	0.70	0.9079
6	Female	Abdomen	0.0270	1.35	0.9601	0.0116	0.58	0.9505

datasets show a good correlation with the reference respiration belt signal. This is to ensure a higher level of accuracy for the Doppler radar in capturing the movements of the abdomen during breathing activity in this non-contact approach. Additionally, 10 datasets of random breathing activities were collected from five additional participants (age $= 25 \pm 5$; male and female) and the performance evaluation of the results is shown in Table 5.4.

Another set of experiments were performed to evaluate the air flow in and out during inhalation and exhalation using a Doppler radar system and a spirometer (MLT1000L respiratory flow-head attached to the Powerlab, sampled at 1000 Hz) as a reference. All the experiments were performed with the subject in a seated position (unless stated in a supine position) with a straight back and minimum movement of the body. In a seated position, the support from the back rest of the chair will minimise the movement of the body compared to a standing position and therefore further improvements could be expected with a subject in a supine position.

5.6.3 Results

The dynamics and shape of the breathing patterns are shown in the following sections. For each condition, the human subject was instructed to follow a certain breathing pattern to investigate the feasibility of using Doppler radar in capturing such conditions. Without concentrating on the duration of each breathing pattern, the subject was briefed on the characteristics of each type of breathing pattern before each experiment commenced. Each Doppler record was compared with the standard respiration belt measurement as a reference. To this end, both the results from Doppler radar and the respiration belt were normalised (to have a range of $[-1\ 1]$) to find the correlation of the breathing patterns obtained. The data are normalized to the range of $[a, b]$ using:

$$X' = a + \frac{X - X_{min}}{X_{max} - X_{min}}(b - a), \tag{5.39}$$

where $a = -1$ and $b = 1$.

Further, the respiration rate analysis was approximated from the FFT and a time frequency analysis was obtained from the continuous wavelet transform by computing the spectral density and the time-frequency distribution, respectively. The use of CWT instead

Table 5.5 Evaluation of Doppler radar measurements compared to respiration belts.

Types of breathing	MSE	Corr_Coeff
Ataxic	0.0038	0.9465
Biot	0.0063	0.9762
Central sleep apnoea	0.0121	0.9427
Cheyne-Stokes	0.0094	0.9376
Cheyne-Stokes variant	0.0089	0.9650
Dysrhythmic breathing	0.0236	0.9355
Normal breathing	0.0461	0.9198
Kussmaul's breathing	0.0764	0.9006

of purely relying on the spectral analysis for more information can be obtained from the time-frequency analysis to further understand how the breathing activity had taken place. The wavelet coefficients from the CWT can be used to extract features (i.e. energy, entropy, frequency distribution, power, etc.) along with pattern recognition techniques to characterise breathing disorders as well as to filter out the motion artefacts. We discuss this further in Section 5.6.4 and also in future work to potentially use this technique as a diagnostic tool for which strong and convincing evidence of reliability of Doppler radar in a range of breathing conditions is of paramount importance.

Table 5.5 shows the overall validation of the normalised Doppler radar measurements in comparison to the normalised respiration belt signals using the mean square error (MSE) and correlation coefficient (Corr_Coeff) as the average of three datasets.

Normal breathing

The normal breathing rate (see Figure 5.8) for an adult ranges from 12 to 20 breaths/min [393]. In this particular trial, the subject was asked to breathe normally and at ease, again, sitting on a chair, upright and facing the antenna. From Figure 5.9, the breathing rate estimated using spectral density is 14.6 breaths/min (0.2441 Hz), which agrees with the CWT where a dominant frequency is present at the 0.2441 Hz band.

Central sleep apnoea

For this experiment, data were collected with the subject in the supine position (sleeping on a bed) to emulate the typical environment of sleep. Central sleep apnoea is one type of sleep-related disorder. Typically, sleep apnoea occurs at least five times per hour of sleep and each apnoeic period lasts for at least 10 seconds. During central apnoea, there is a cessation of air flow with no diaphragmatic and intercostal muscle activity. This means that there is no air exchange either through the nose or the mouth. From Figure 5.10(a), there are approximately 8 breaths/min but from the spectral analysis in Figure 5.11(a), it is approximated that 15.12 breaths/min correspond to 0.2518 Hz. This is not accurate as a spectral density approximation

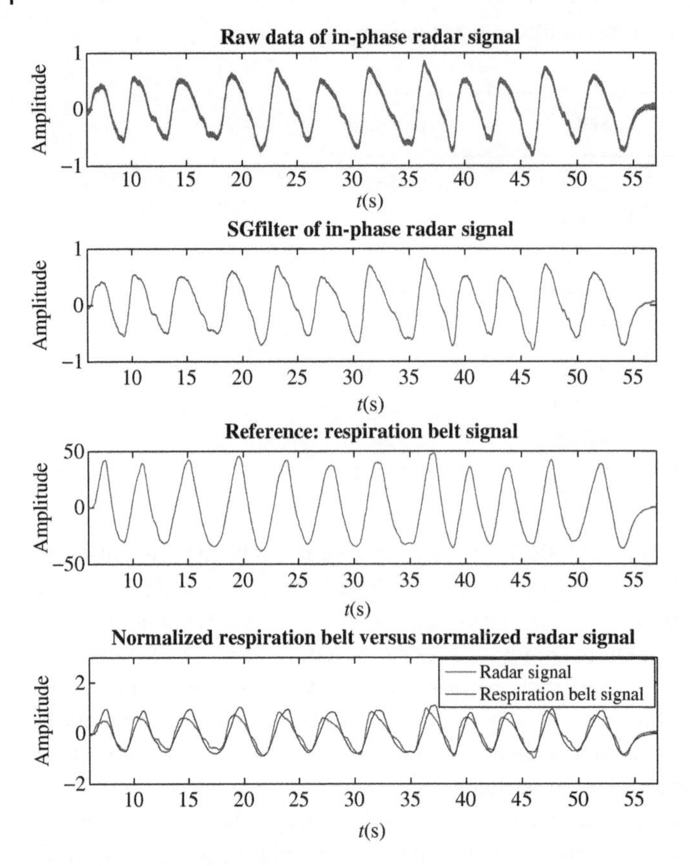

Figure 5.8 Normal breathing signal (from top to bottom): raw radar signal; filtered signal; reference respiration belt signal; and normalised respiration belt signal versus normalised filtered radar signal).

is not an appropriate approach to find the respiration rate for abnormal breathing patterns. From Figure 5.11(a), using the distribution of frequency versus time, we can clearly see that there is no sign of breathing activities (apnoea) for the first 8 seconds, 45–60 seconds and 90–110 seconds. This ability of Doppler radar to detect the apnoea state is beneficial for sleep monitoring, which includes sleep apnoea conditions and the more critical sudden infant death syndrome (SIDS).

Ataxic breathing and Biot's breathing

Ataxic breathing is often characterised by irregular cyclic breathing periods (irregular frequency and interspersed tidal volume) followed by unpredictable periods of apnoea or pauses in breathing. Biot's breathing is another special type of breathing abnormality with nearly regular cyclic breathing followed by an apnoea segment. These forms of breathing are sometimes associated with brain stem stokes, narcotic medications and patients with medullary lesions [68, 393]. For ataxic types of breathing, from Figure 5.10(b), it is difficult to approximate the breathing rate from the patterns, but the most frequent form is that of irregular cyclic breathing followed by a pause. Spectral density cannot be considered to

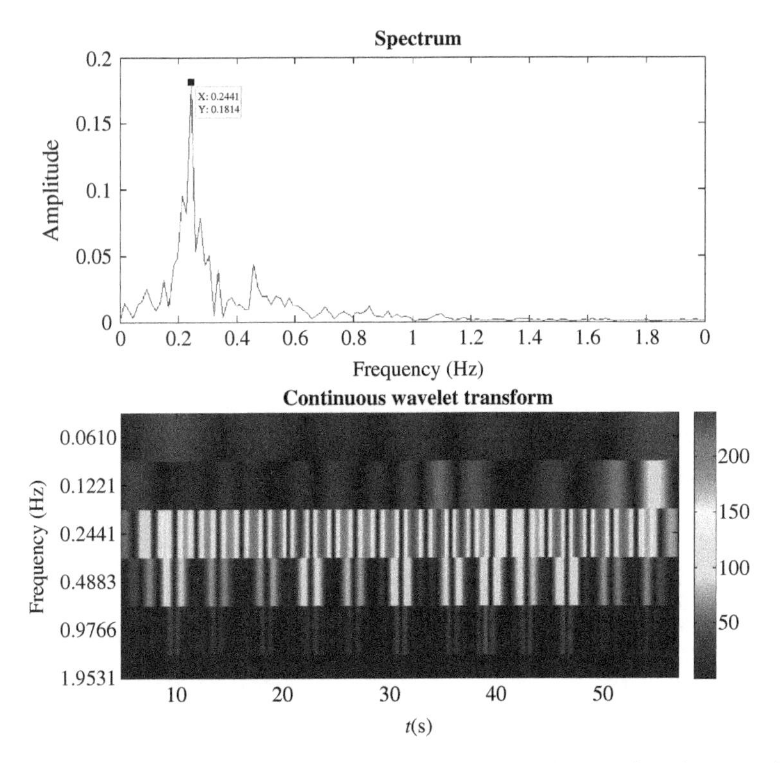

Figure 5.9 Spectral density (FFT) and continuous wavelet transform for normal breathing from Doppler radar.

provide a good approximation (see Figure 5.11b) as there is no dominant peak found, while from CWT, we can clearly see the distribution of signals from a time versus frequency analysis. For instance, from the first breathing segments, the subject breathes with a lower frequency of 0.1221 Hz followed by a period of rapid breathing at a frequency of 0.4883 Hz and eventually slowing down to 0.1221 Hz. This type of information is not available from spectral analysis and can further be improved with a higher frequency resolution.

As for Biot's breathing, from Figure 5.10(c), there were 8 breaths/min consisting of a long apnoea period in between regular breathing patterns. From the spectral analysis shown in Figure 5.13(a), it was approximated by 23.802 breaths/min which is not consistent with the patterns shown in Figure 5.10(c) due to the combination of both regular and irregular breathing patterns. In the CWT analysis, however, both are separated and therefore the results tend to be more accurate as the dominant frequency was clearly seen at 0.4883 Hz, which corresponds to 29.3 breaths/min occurring at relative time-stamp corresponding to the breathing patterns shown in Figure 5.10(c).

Cheyne-Stokes respiration

In Cheyne-Stokes breathing, there is a cyclic change in breathing with a crescendo-decrescendo type of sequence followed by pauses or central apnoea. In certain neurological disorders, this type of breathing is often encountered, i.e. bilateral cerebral hemispheric lesions [68], patients with stroke, brain tumour or traumatic brain injury. Carbon monoxide

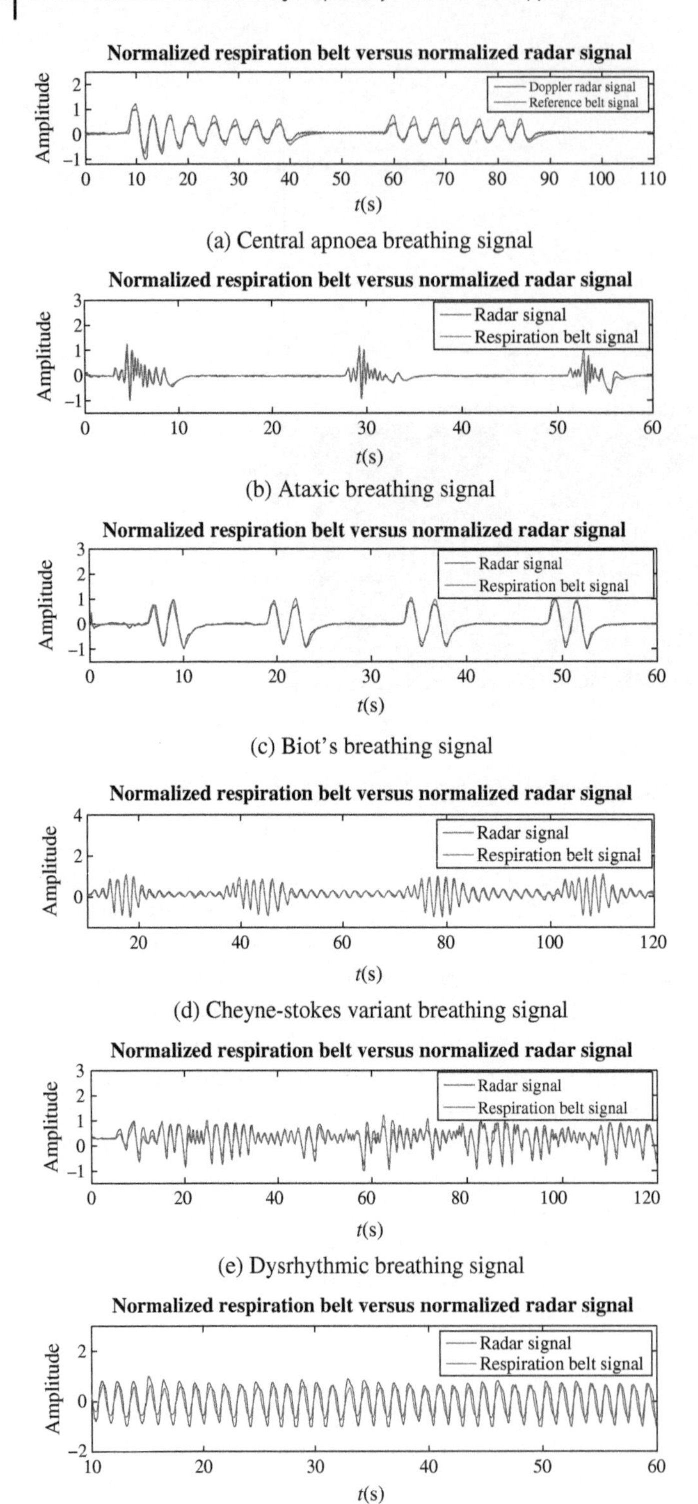

Figure 5.10 Normalised respiration belt signal versus normalised filtered radar signal.

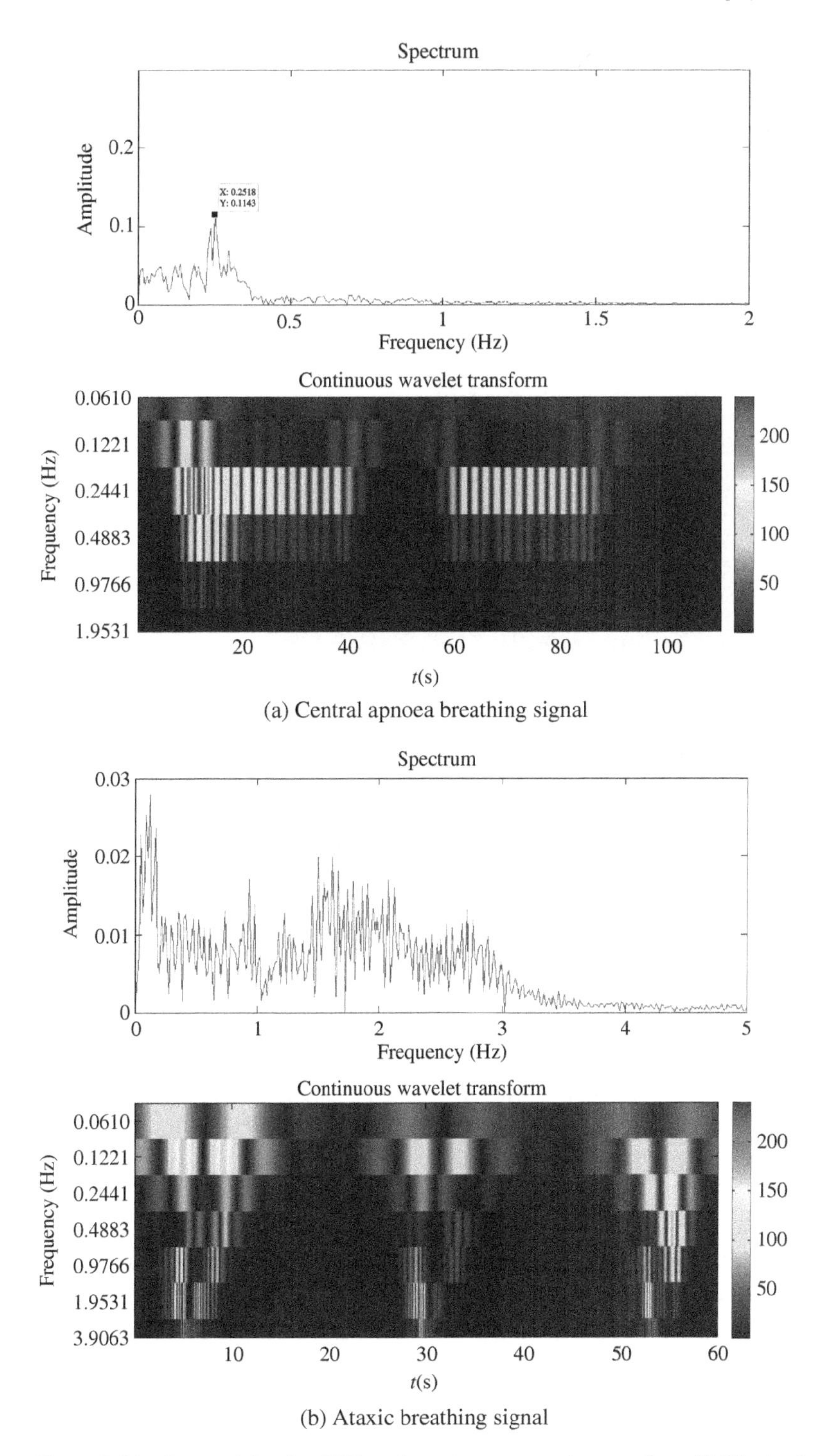

Figure 5.11 Spectral density (FFT) and continuous wavelet transform (CWT) plot for central apnoea and ataxic breathing.

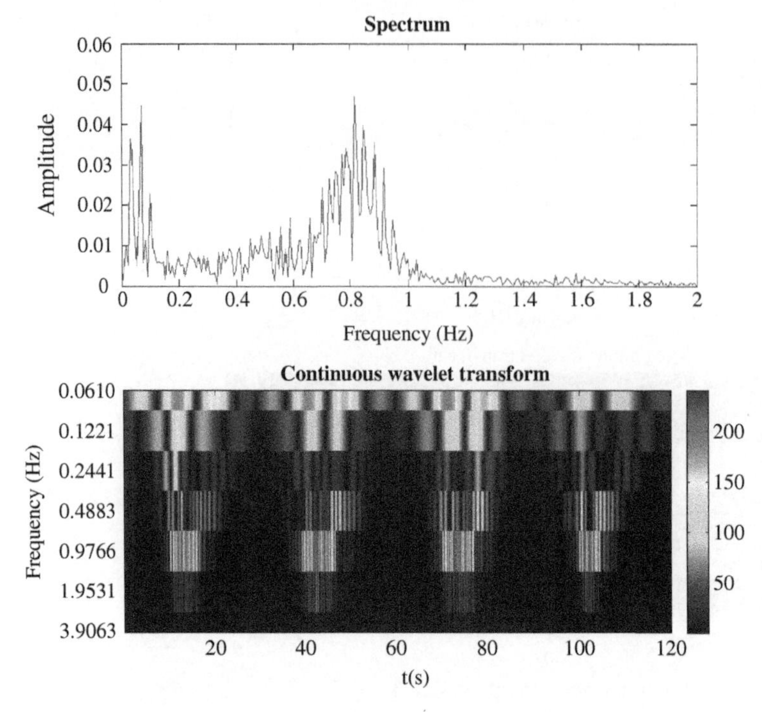

Figure 5.12 Spectral density (FFT) and continuous wavelet transform for Cheyne-Stokes breathing from Doppler radar.

poisoning, metabolic encephalopathy, altitude sickness and non-rapid eye movement of patients with congestive heart failure [393] can also exhibit this type of respiratory pattern.

From Figure 5.14, it can be clearly seen that the pattern changes with a crescendo-decrescendo sequence followed by an apnoea state. This further proves that Doppler radar is capable of capturing the changes of breathing in a non-contact method with good correlation to the one captured using the respiration belt. For a respiration rate analysis, from Figure 5.12, it is difficult to approximate the rate from the peak itself using the FFT method and it is prominent in the CWT. From the CWT, we can see the sequence of crescendo-decrescendo from the energy level with the time versus frequency distribution. This could possibly be used as one of the features in classifying types of breathing disorders, which will be our focus in the future.

Cheyne-Stokes variant

Cheyne-Stokes variant breathing is similar to Cheyne-Stokes respiration but the central apnoea stage is substituted by hypo-apnoea. This type of breathing is found in brain stem lesions as well as in bilateral cerebral hemispheric disease [68]. From Figure 5.10(d), the pattern is similar to the Cheyne-Stokes breathing pattern but is followed by a hypo-apnoea scenario where there is a small variation in breathing instead of a flat signal. Spectral analysis (see Figure 5.13b) shows few peaks and without the timing information and

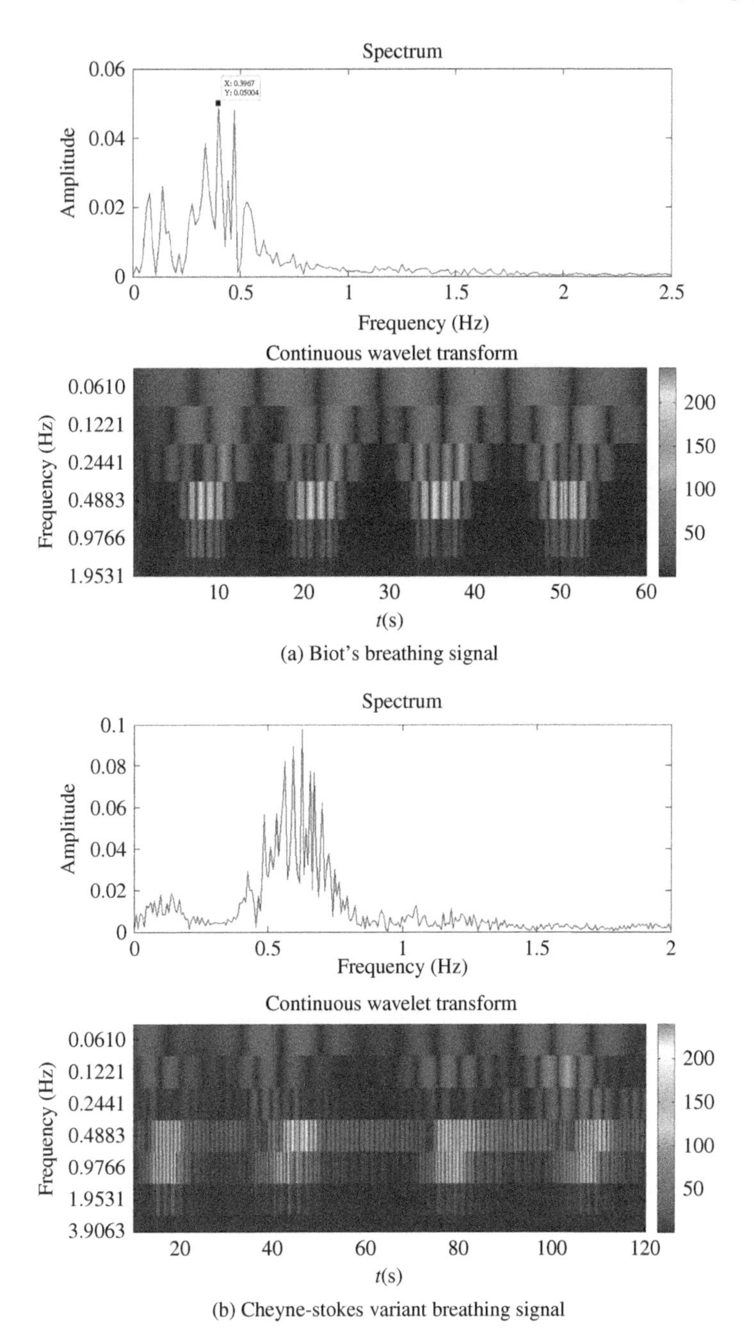

Figure 5.13 Spectral density (FFT) and continuous wavelet transform (CWT) plot for Biot's and Cheyne-Stokes variant breathing.

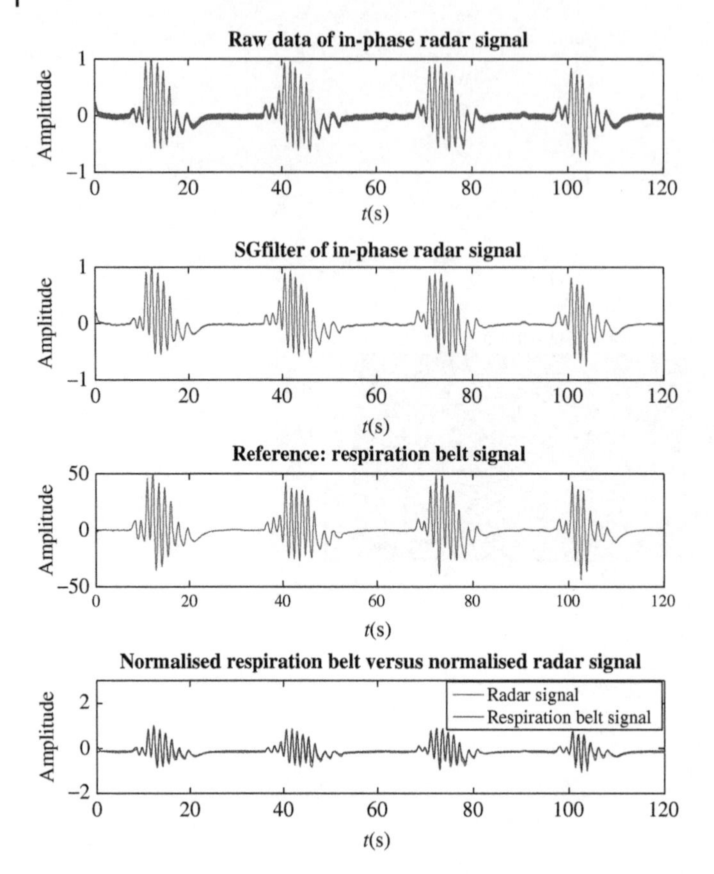

Figure 5.14 Cheyne-Stokes breathing signal (from top to bottom): raw radar signal; filtered signal; reference respiration belt signal; and normalised respiration belt signal versus normalised filtered radar signal.

it is difficult to interpret frequency information. CWT, on the other hand, provides adequate spectral-temporal information and results show two dominant frequency bands at 0.4883 Hz and 0.9766 Hz and a hypo-apnoea state lies in the 0.4883 Hz band at the time approximately between 20–40 s, 50–75 s and 85–100 s.

Dysrhythmic breathing

This type of breathing is characterised as non-rhythmic breathing with an irregular rhythm, rate and amplitude. It could be caused by an abnormality in the respiratory pattern generator invoked in the brain stem. The results of the pattern can be seen in Figure 5.10(e) where the pattern is non-rhythmic variation in rate and amplitude. In this case the Doppler radar was still able to capture the changes, yielding consistent correlations with the respiration belt readings. Again, it is difficult to approximate the breathing rate from the spectral representation, but with CWT it is possible to represent the breathing frequency corresponding to the occurrences in time, as shown in Figure 5.15(a). From the results, the dominant frequency is approximated in the range of 0.4883 Hz.

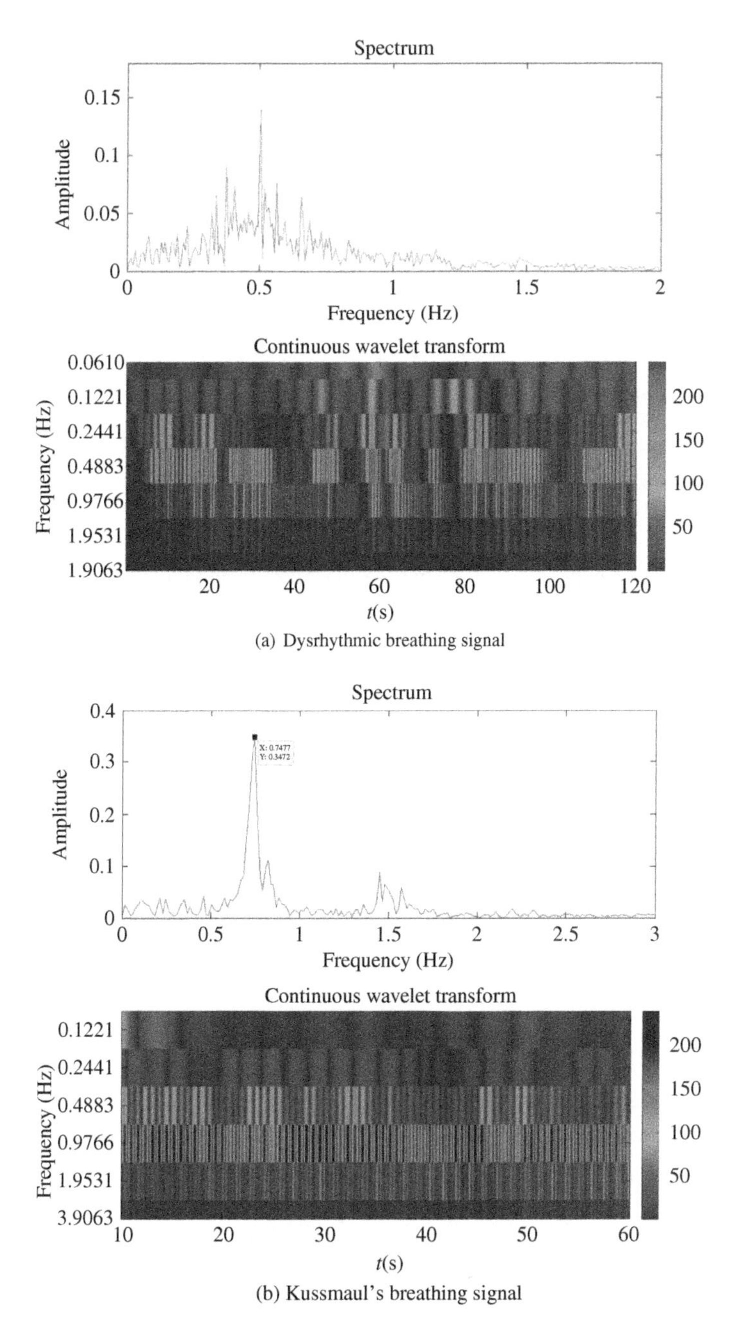

Figure 5.15 Spectral density (FFT) and continuous wavelet transform (CWT) plot for dysrhythmic and Kussmaul's breathing.

Kussmaul's breathing

This type of breathing is characterised by a regular increased rate and corresponds to tidal volume patterns consistent with "gasping for air" associated with a severe metabolic acidosis [393]. Since this type of breathing is considered as almost periodic, the spectral density shown in Figure 5.15(b) can be used to represent the breathing rate approximated at 0.7477 Hz corresponding to 44.9 breaths/min, which is quite accurate from the number of breath count approximated in Figure 5.10(f). CWT shows that the dominant frequency band is located at the 0.4883–0.9766 Hz band, but the results could be further improved with a higher frequency resolution.

5.6.4 Motion signature from Doppler radar

To further demonstrate the potential use of CWT, two experiments were performed where we capture the measurement of a breathing signal under the influence of motions related to turning of the body and small body movements when in the supine position to reflect any possible events during sleep. The results of the experiments are shown in Figure 5.16. From the time series representation, it is possible to identify the breathing, apnoea, turning of the body and body movement events with the naked eye but, with CWT, this process can be

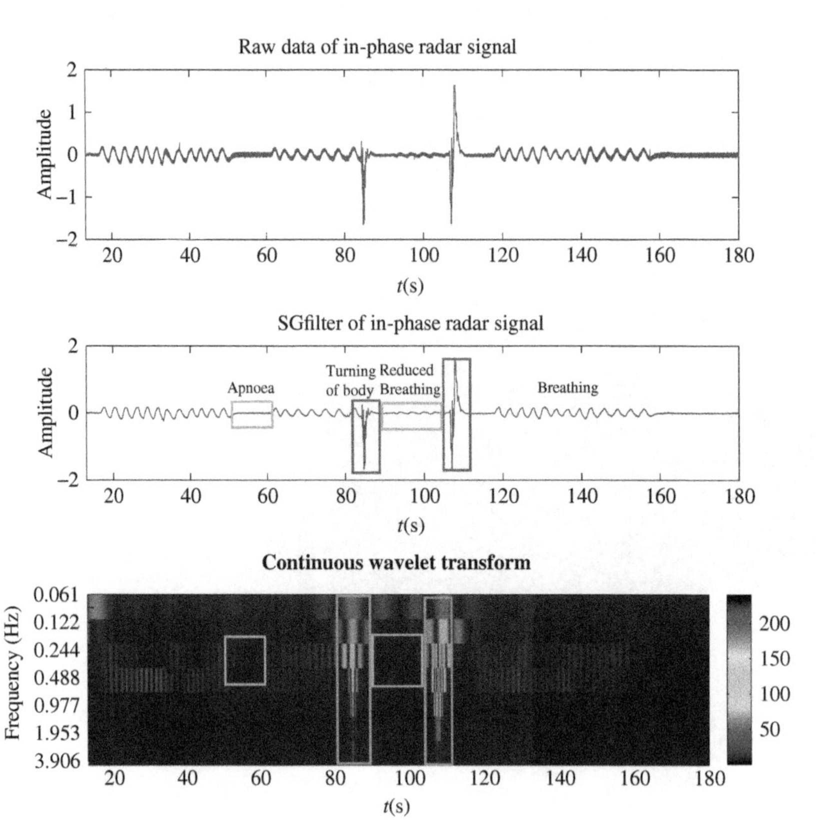

Figure 5.16 Mixture of signal characteristic from Doppler radar measurement in the supine position.

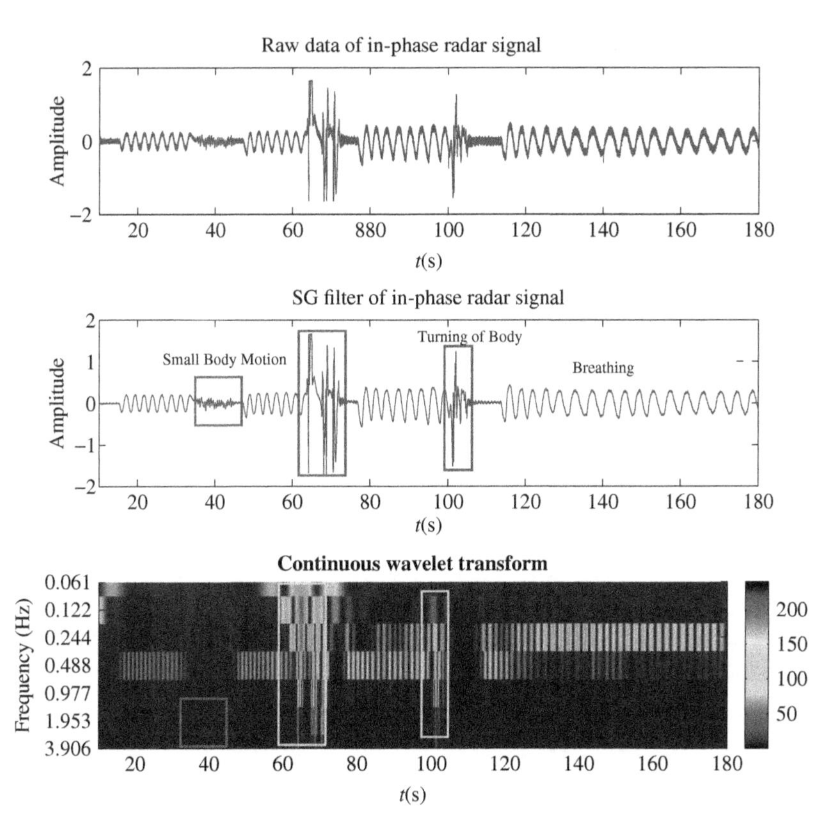

Figure 5.16 *(Continued)*

automated in a long-term monitoring analysis, especially in sleep studies involving a large amount of data. Additionally, using Doppler radar, body movement such as turning of the body could be identified from the signature pattern of the signals as shown in the results, whereas this information is not easily available from the respiration belt. This study will be extended to more subjects in a longer duration data collection exercise in the future for further evaluation and analysis, especially for sleep studies. Moreover, the type of features that can be extracted from CWT will be investigated to assist in signal analysis and diagnostics.

5.6.5 Measurement of volume in (inhalation) and volume out (exhalation)

An experiment was conducted to investigate the correlation between the Doppler radar and spirometer in capturing the respiratory function. For this purpose, a few different experimental conditions were designed to investigate the feasibility and consistency of Doppler radar in obtaining the flow in and out corresponding to the spirometric measurements. In this experiment, the subject was asked to:

1. Breathe deeply, pause and continue to breathe deeply.
2. Breathe normally followed by deep breaths.
3. Breathe deeply followed by normal breathing.

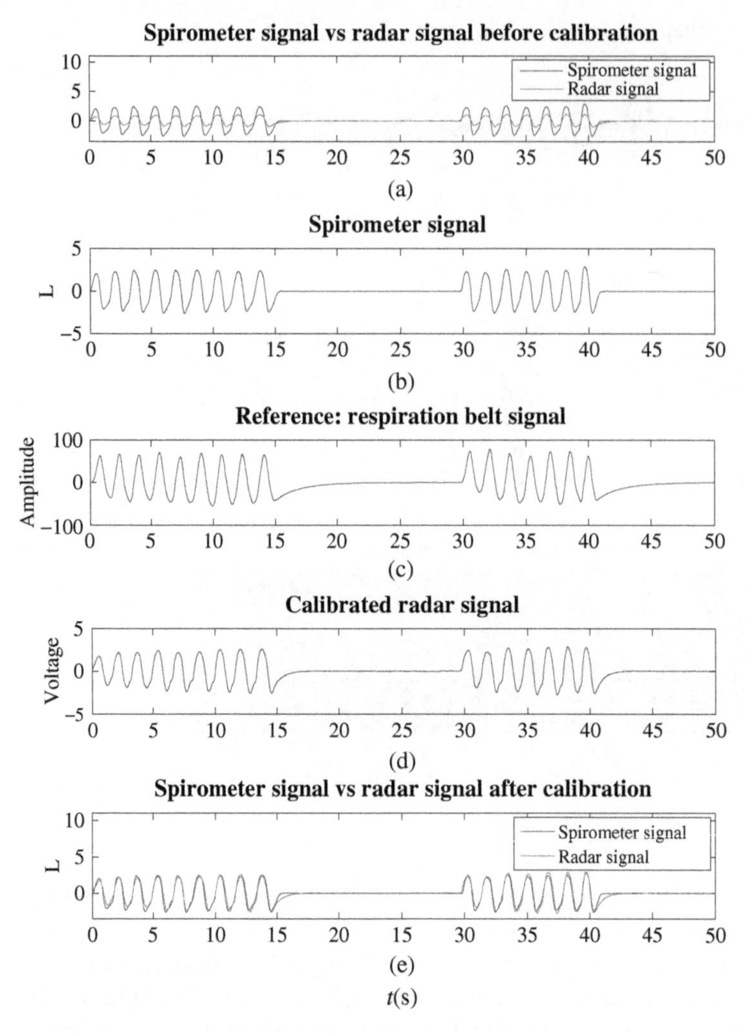

Figure 5.17 Experiment 1: deep breaths followed by apnoea: (a) filtered radar signal versus spirometer reading; (b) spirometer reading; (c) reference respiration belt signal; (d) calibrated radar signal and (e) spirometer reading versus calibrated radar signal.

The results are shown in Figures 5.17, 5.18 and 5.19, respectively, for each instance. From the results, the shape of the Doppler radar signal is similar to the spirometer readings but with an amplitude difference due to non-calibration. After calibration, the Doppler radar signals are almost identical to the spirometer readings. Assuming that the shifting of Doppler radar is caused by the movement of the chest/belly during respiration, a linear relationship between the change of flow has been derived based on the Doppler signal. We also see that the Doppler signal was sensitive enough to detect the differences in the breathing patterns. However, a calibration period of approximately 30 seconds is required and this can be automated.

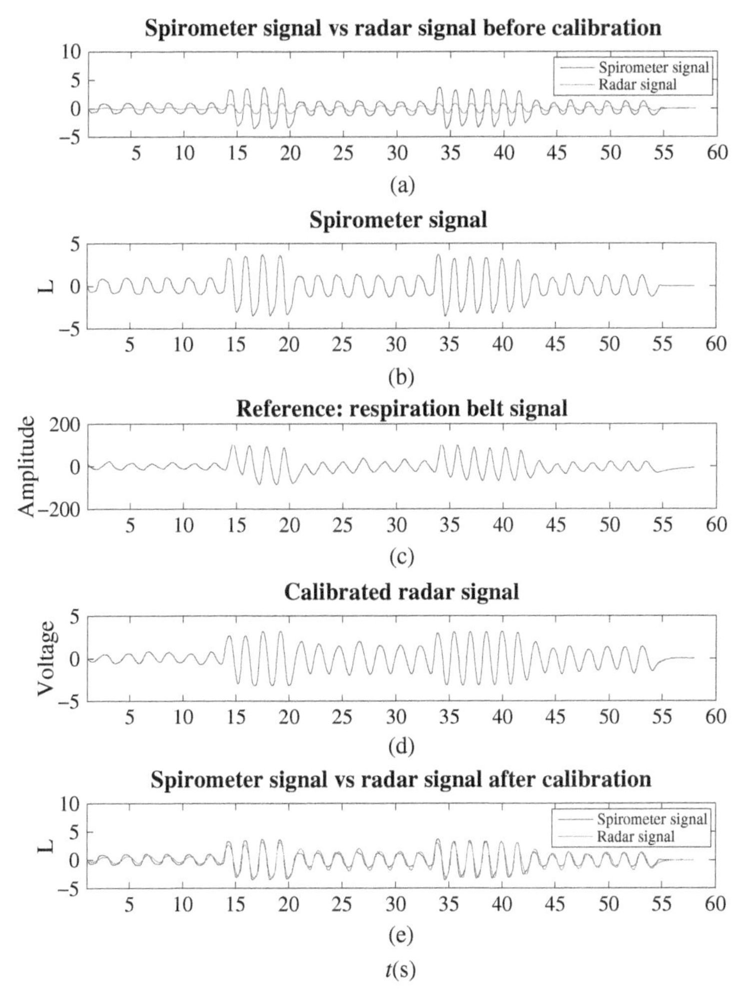

Figure 5.18 Experiment 2: normal breaths followed by deep breaths: (a) filtered radar signal versus spirometer reading; (b) spirometer reading; (c) reference respiration belt signal; (d) calibrated radar signal and (e) spirometer reading versus calibrated radar signal.

In all experiments, computation of the mean square error (MSE) and correlation coefficients was performed between Doppler radar signals and spirometer readings. Results show consistently high correlations between the Doppler radar and spirometer, as shown in the Table 5.6. The MSE provides a measure of the exact fit between the Doppler radar signal to the reference model (spirometer data) where Pearson's correlation coefficient shows how well the curves are related to each other, which is shown in Figure 5.20. If the correlation coefficient is one, the Doppler radar tidal volume would be equal to the spirometer reading and always lie on the red line.

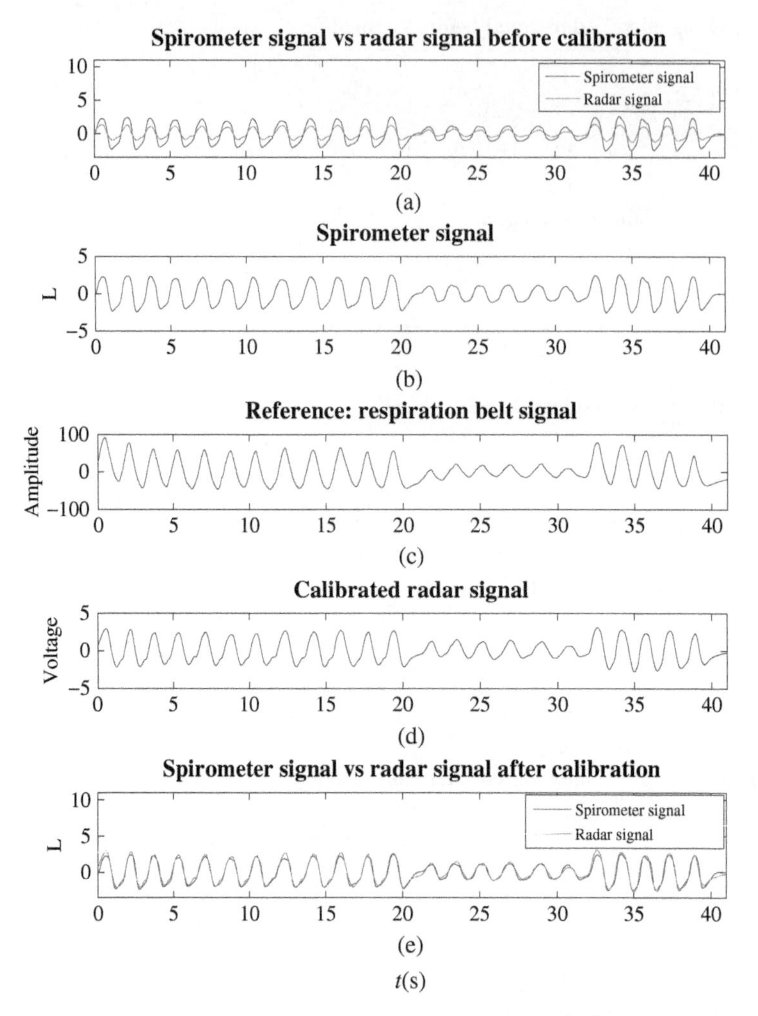

Figure 5.19 Experiment 3: deep breaths followed by normal breaths: (a) filtered radar signal versus spirometer reading; (b) spirometer reading; (c) reference respiration belt signal; (d) calibrated radar signal and (e) spirometer reading versus calibrated radar signal.

Table 5.6 Coefficient comparison on Doppler radar signal with spirometer.

	Comparison	MSE	Correlation coefficient
Experiment 1	Non-calibrated radar signal with spirometer	0.5867	0.9664
	Calibrated radar signal with spirometer	0.1064	0.9664
Experiment 2	Non-calibrated radar signal with spirometer	1.2306	0.9551
	Calibrated radar signal with spirometer	0.1939	0.9551
Experiment 3	Non-calibrated radar signal with spirometer	0.5023	0.9799
	Calibrated radar signal with spirometer	0.0739	0.9799

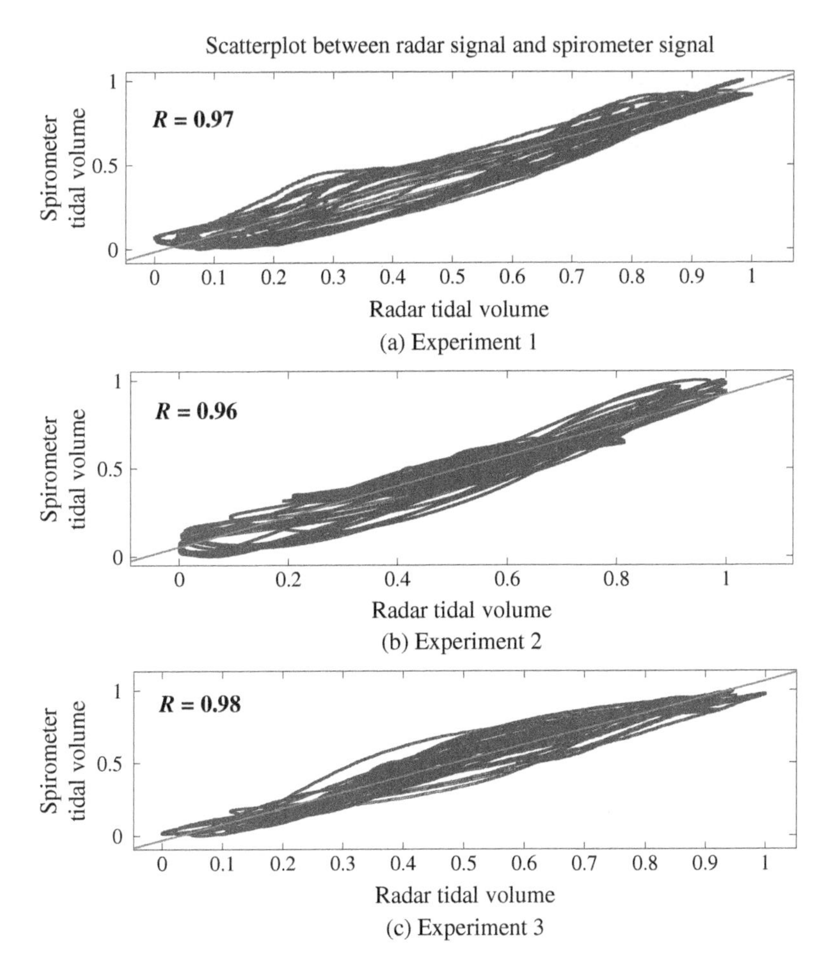

Figure 5.20 Scatterplot between a radar signal and spirometer reading: (from top to bottom: Experiment 1; Experiment 2; Experiment 3).

5.7 Removal of Motion Artefacts from Doppler Radar-based Respiratory Measurements

The measurement of respiratory function using Doppler radar is often affected by the motion artefacts, particularly in the case of long-term monitoring. Reducing the effect of these disturbances is vital before any attempt at data interpretation irrespective of the application. In this book, multi-level DWT (discrete wavelet transform) was used in filtering out motion artefacts typically caused by random limb or body movements when extracting the desired breathing signal from the DWT approximation components. The isolated artefacts can be identified from the DWT detailed components, which could be particularly useful in certain sleep studies. Experiments were conducted to ascertain the level of performance and the feasibility of the underlying approach for real-time non-contact respiratory monitoring applications. DWT is used in recovering the normal breathing patterns and apnoea states from the received Doppler radar signals that experienced motion artefacts for the cases of subjects in sitting as well as supine postures. The reconstructed Doppler-based measurements from DWT were highly correlated with the independent measurements captured by clinically used respiratory straps.

Detection and monitoring of human physiological functions such as respiration using Doppler radar offers the distinct advantage of non-contact measurement and tolerance to variations in environmental factors such as ambient light and temperature. Particularly for applications such as long-term respiration monitoring during sleep, the ability of Doppler radar signals to penetrate clothing allows their use as part of a complete remote sensing device. This improves patient comfort, which is vital for long-term monitoring, alleviating the need to use wearable devices such as respiration belts during sleep. Indeed it allows the capture of data in a more natural or home-based environment and simply provides clinicians and researchers with data that have not been available to date.

Most of the literature supports the use of Doppler radar in measuring respiratory function to an acceptable level of correlation and accuracy in comparison to the clinically used wearable respiration strap. Nevertheless, most results reported were focused on obtaining the Doppler-based measurements in a controlled environment where the received Doppler radar signals were assumed to be entirely due to the movement of the chest/abdomen from the respiration activity (breathing or apnoea). However, in practice, the received signal is inherently composed of other motion artefacts due to random body movements [128] such as jerking of limbs.

In long-term respiratory function monitoring applications such as sleep studies, the captured Doppler-shifted signal comprised breathing and cessation periods that could easily be detected [200], particularly when the subject is relatively stable and other bodily movements are not present. As the Doppler-based detections rely entirely on the movement of the chest wall for the observation of the respiratory function, any other body movement (motion artefacts) can adversely affect the measurements. This can be particularly significant when observing subjects with conditions such as periodic limb movement disorder (PLMD) [66, 265] where jerking of the legs or arms during sleep may occur as many as three times a minute [334]. Furthermore, any other normal movement of the body during sleep can completely obscure the breathing signal, at least for the duration of the interference. Having some knowledge of the signal characteristics can provide a deeper insight into the composition of the received Doppler signal that was subjected to these interferences. Therefore, separation or filtering of the breathing signal from certain undesirable motion artefacts is an essential precursor to any form of information extraction exercise involving respiration data.

In Vergara *et al.* [364], it was shown that blind source separation (BSS) could be used to separate certain human motion artefacts and the respiratory signal from the received signal. In a BSS context, it is necessary to have more measurements (m) than the actual number of sources (n) (i.e. $m \geq n$ [168]). Researchers [208, 391] have also used the radar array configuration to address the random body movement problem; the reported work mainly focused on the frequency of the vital sign instead of detailed analysis of the patterns, which is the main aim of this book. The use of the radar array configuration in suppressing/removing the motion artefacts increases the cost and the complexity of the overall system, compelling developers to look at implementations with one receiver element. This essentially is the case with our proposed approach while countering artefacts.

In capturing respiratory patterns, we show that multi-level decomposition using discrete wavelet transform (DWT) performs better than simply low-pass filtering (LPF) [48, 100] to exclude certain motion artefacts/noise. We demonstrate that the disturbances due to movements of the human body can be filtered out using multi-level DWT by using a single receiver element, unlike Vergara *et al.* [364] where multiple receivers were required.

5.7.1 Experimental verification

Experiments 1 to 4 (below) were performed on the subjects in a *seated* posture, as shown in Figure 5.21, while Experiment 5 was performed on a subject in a *supine position* at a distance of about 1 metre from the antenna. The panel antenna was focused on to the chest/abdomen of the subject to capture a more reliable Doppler signal with a higher signal-to-noise ratio (SNR). The subjects were wearing normal clothing and the external respiband (ADInstruments, MLT1132 Piezo Respiratory Belt Transducer with a sampling rate of 1000 Hz) was attached to the chest as the reference system for comparison purposes.

Figure 5.21 Environment setup for experiment trials.

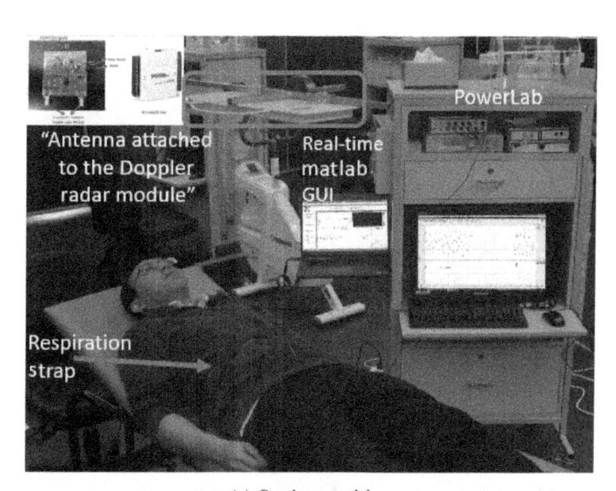

(a) Supine position

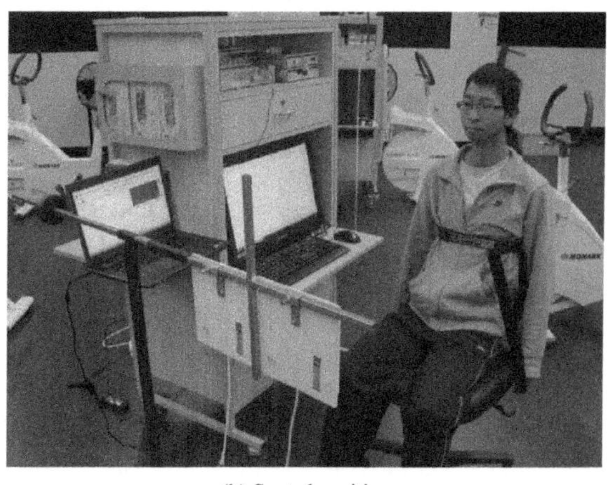

(b) Seated position

The subjects were asked to breathe normally along with some designated motion artefacts such as jerking of the upper right limb and small body movement as follows:

1. Experiment 1: Mixture of breathing, apnoea and movement of the upper limb (jerking of right arm).
2. Experiment 2: Mixture of breathing and movement of the upper limb (swing of the person's right arm).
3. Experiment 3: Mixture of breathing and small movements of the body (created from moving the shoulder forward and backward).
4. Experiment 4: Normal breathing activity with minimal motion artefacts.
5. Experiment 5: Mixture of breathing and movement of the upper limb (jerking of right arm) while in a supine position.

Furthermore, based on the specific application, location of the antenna around the body could be used to improve the SNR and hence the analysis of breathing function. Here, as the main focus of this book is on filtering out certain types of motion artefacts from the breathing signal using multi-level DWT, positioning the antenna at the right location was not investigated. The underlying concepts are independent of any possible optimisations in receiver configurations for signal conditioning and can readily be implemented in conjunction.

5.7.2 Results and discussion

The data captured cover different dynamics due to different breathing pattern experiments with motion artefacts as listed in Section 5.7.1. Figures 5.22 and 5.23 depict the results of the experiments conducted showing the comparison of the performance of DWT and LPF [100] with the reference respiration strap. The analysis and the reconstruction of the signal would be based on the approximation coefficient obtained from level 9 (Db10) as we observed that the reconstructed signal at this level yielded a higher correlation coefficient (Corr) and a lower mean square error (MSE), as depicted in Figure 5.24(a) for all the experimental datasets. As for the detailed components, the more prominent filtered artefacts/noises were observed to be present at level 6 and above. Therefore, throughout this book, detailed components at wavelet level 6 were used to represent the artefacts' signatures as the main focus of this book was on the recovery of the respiration signal from the noisy radar signals.

Daubechies (Db10) was chosen as the mother wavelet based on performance compared to other wavelet families. Using datasets acquired in the experiment, the performance of the proposed approach was compared to the reference respiration strap measurements in terms of MSE and Corr measures. The results are shown in Figure 5.24(b). Wavelets used in these comparison studies include *Daubechies (Db), Haar, Symlets (Sym), BiorSplines (bior), ReverseBior (rbior), Dmeyer (dmey), and coiflets (coif)*. Overall, a better Corr and a lower MSE are achieved with Db10.

In Experiment 1, a mixture of breathing, apnoea and motion artefacts were present, as shown in Figure 5.22(a). Due to the higher level of sensitivity in Doppler radar, when the breathing signal is affected by motion artefacts (jerking/movement of right arm), a normal breathing pattern (see Figure 5.22d) is no longer visible in the received signal. Using the DWT approximation component of the signal at level 9, the breathing signal was recovered

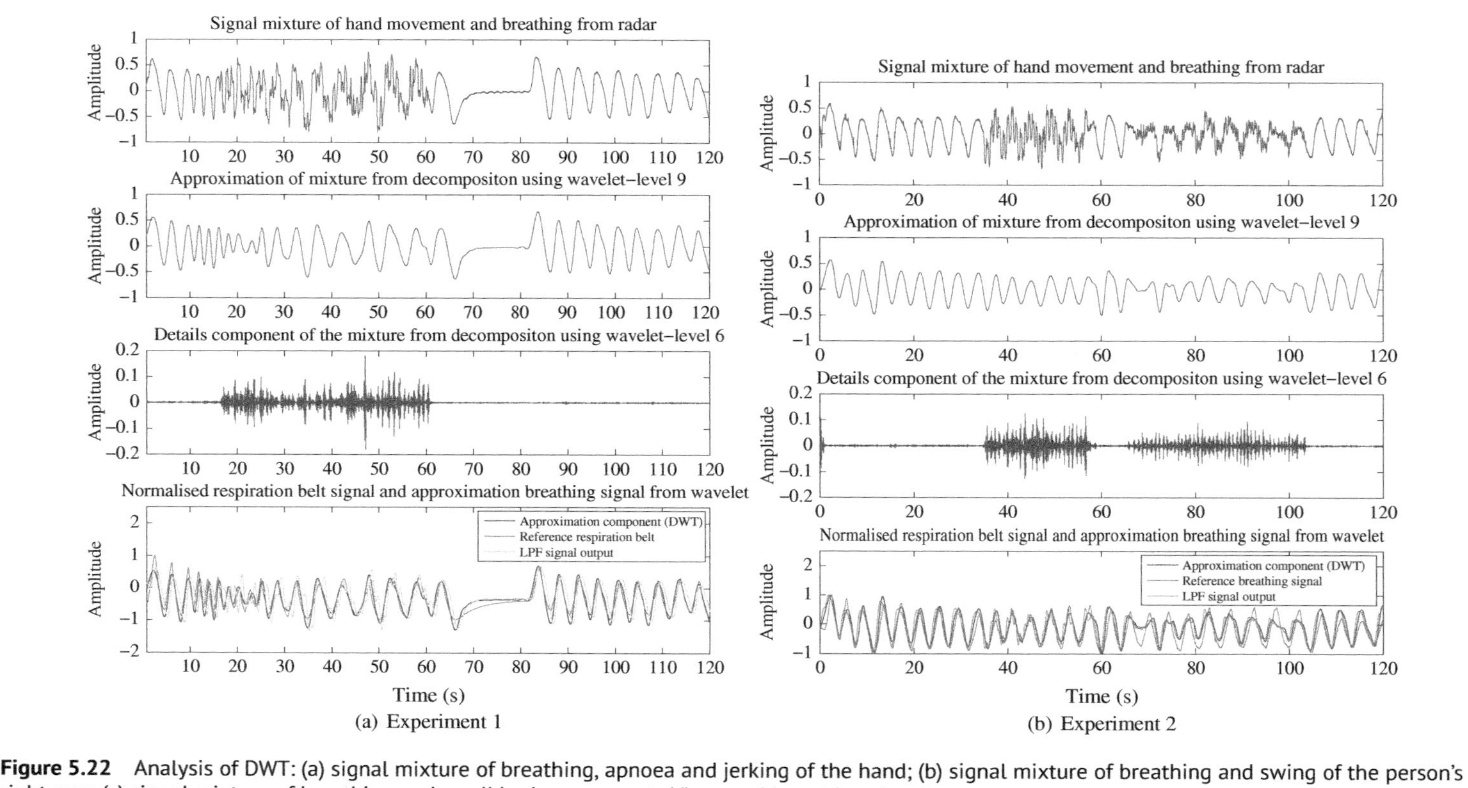

Figure 5.22 Analysis of DWT: (a) signal mixture of breathing, apnoea and jerking of the hand; (b) signal mixture of breathing and swing of the person's right arm; (c) signal mixture of breathing and small body movement; (d) normal breathing signal.

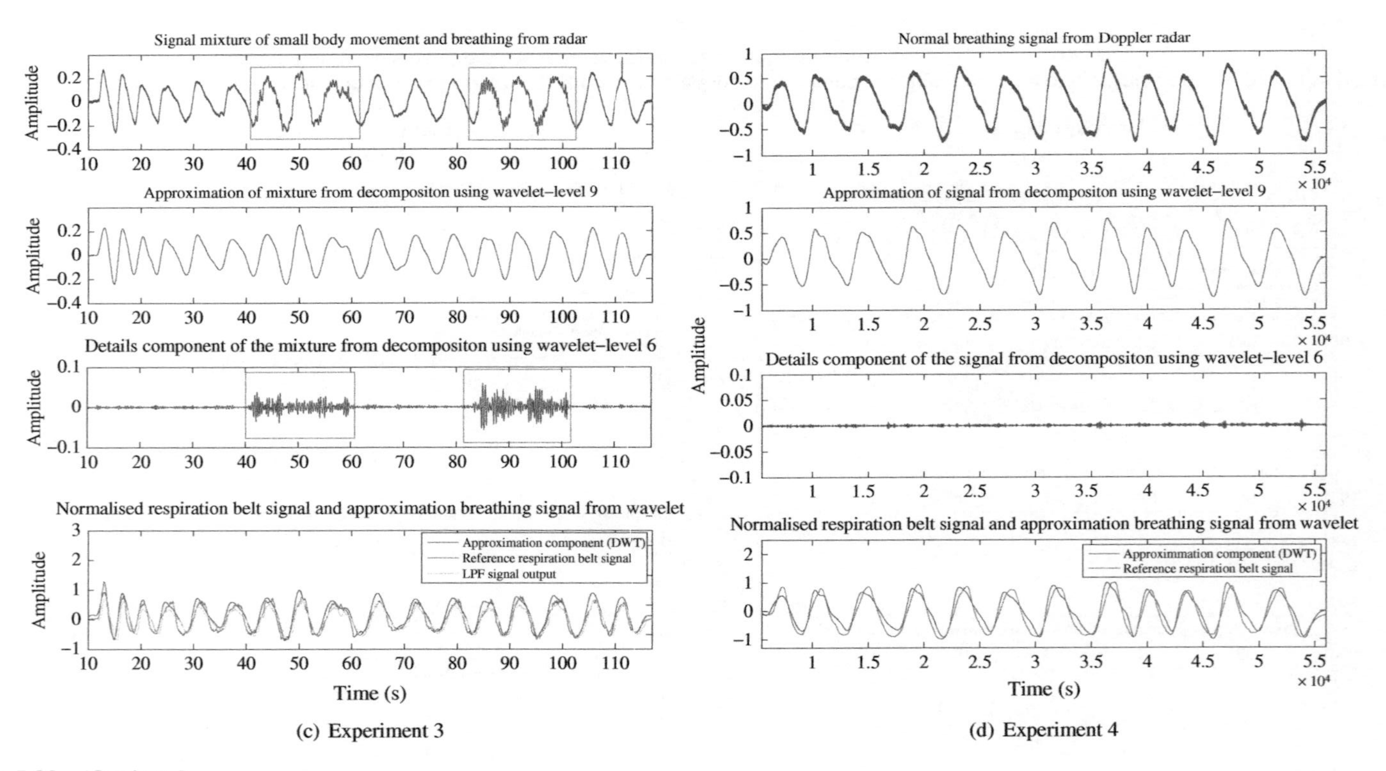

Figure 5.22 (*Continued*)

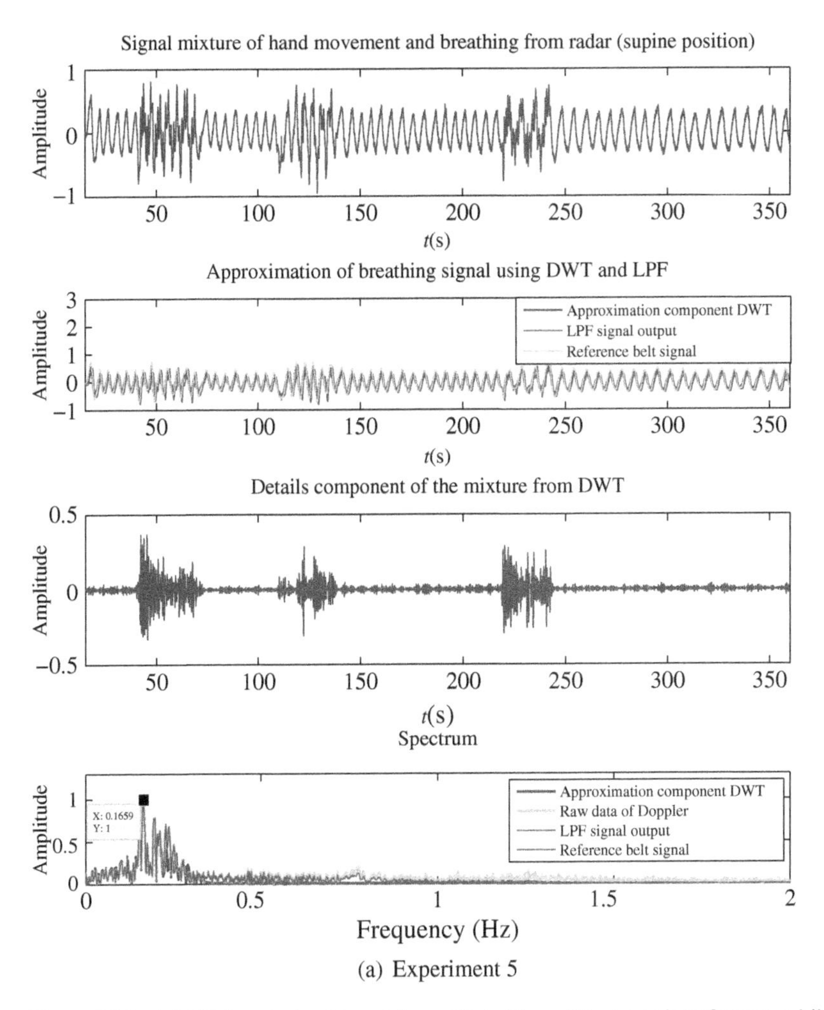

Figure 5.23 (a) DWT analysis of breathing with jerking of the hand interference while in a supine position; (b) spectrum analysis using FFT for Experiment 2.

by filtering out the motion artefacts due to the jerking of arms, which can clearly be observed in the detailed component (DWT) at level 6. The approximation signal is 91%, which is identical to the reference respiration strap and has an MSE (mean squared error) of 0.0335. Evaluation was based on normalised signals (DWT and the reference respiration strap) and is therefore independent of the captured amplitude of the signals. Further, by just using LPF as shown in the LPF signal output, the shape of the curve is still affected by the motion artefacts compared to the approximated (DWT) signal, as shown in the last sub-figure in Figure 5.22(a), where the performance of the comparison is listed in Table 5.7. From these observations, we can clearly see that the local maxima and minima from the approximation signals were quite close to the reference respiband. This also confirms the validity of the cycle by cycle variation (including the approximation of the *I:E* ratio).

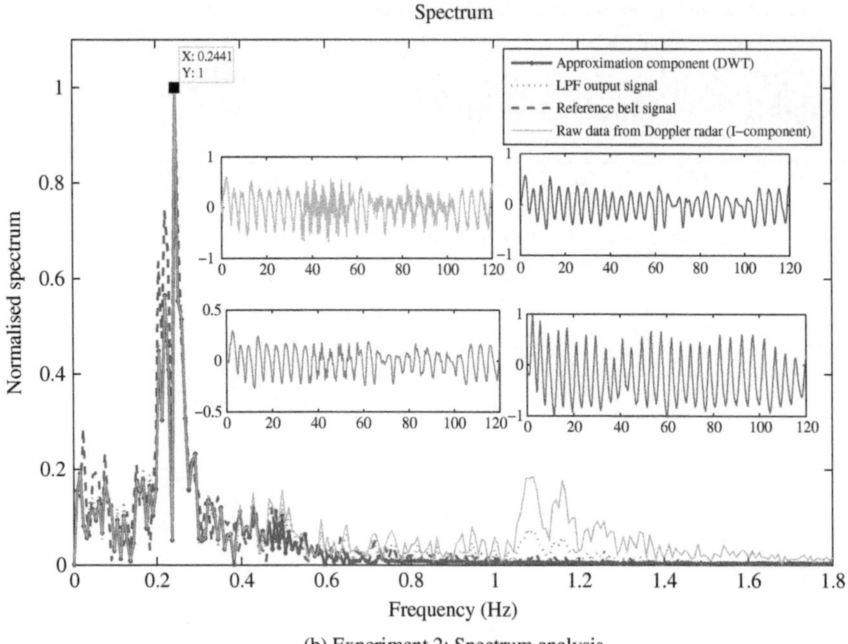

(b) Experiment 2: Spectrum analysis

Figure 5.23 *(Continued)*

In Experiment 2, as shown in Figure 5.22(b), a mixture of breathing and motion artefacts was present. From the received raw data, we can see that there are two different motion artefacts in terms of strength and frequency affecting the breathing curve. Using DWT at level 9, an approximation of the breathing signal was recovered while the motion artefacts and noise were visible in the detail component at level 6, similar to Experiment 1. Referring to Table 5.7, the approximation of breathing signal from DWT is 88% identical and has a smaller MSE compared to the LPF breathing signal in comparison to the reference respiration strap. The extracted detailed components can be used to indicate the presence of motion, which is particularly useful in sleep studies. Moreover, additional information could be derived from these signatures, e.g. the time-related information and frequency of occurance. All this information can be obtained from DWT analysis using a non-contact sensing approach.

Furthermore, as shown in Figure 5.22(c), for small movements of the body (such as moving the shoulders forward and backward), LPF and multi-level DWT were efficient in recovering the breathing signal, but DWT gives a better performance compared to LPF, as shown in Table 5.7. If the breathing signal is free from motion artefacts, the approximation components of breathing activities can clearly be seen at level 9, while the detail component would not have any observable signal present, as shown in Figure 5.22(d).

As discussed in [100, 197], the respiration rate could be estimated simply by performing FFT on the signals. We have used data from Experiment 2 to show the difference in the spectrum for each signal output from Doppler radar, DWT, LPF and reference respiband. In addition to estimating the breathing rate accurately from the occurrences of the peaks

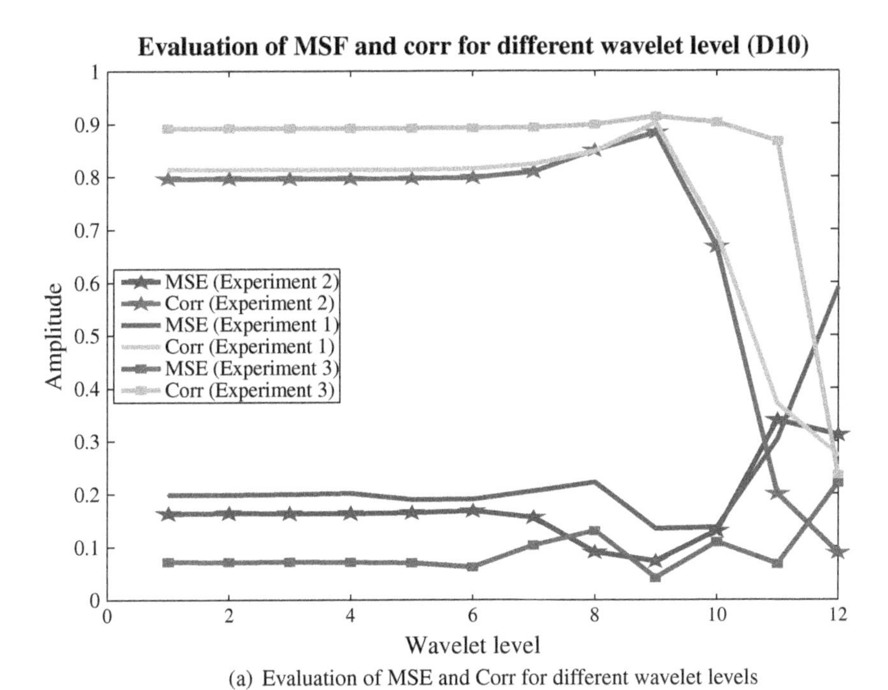

(a) Evaluation of MSE and Corr for different wavelet levels

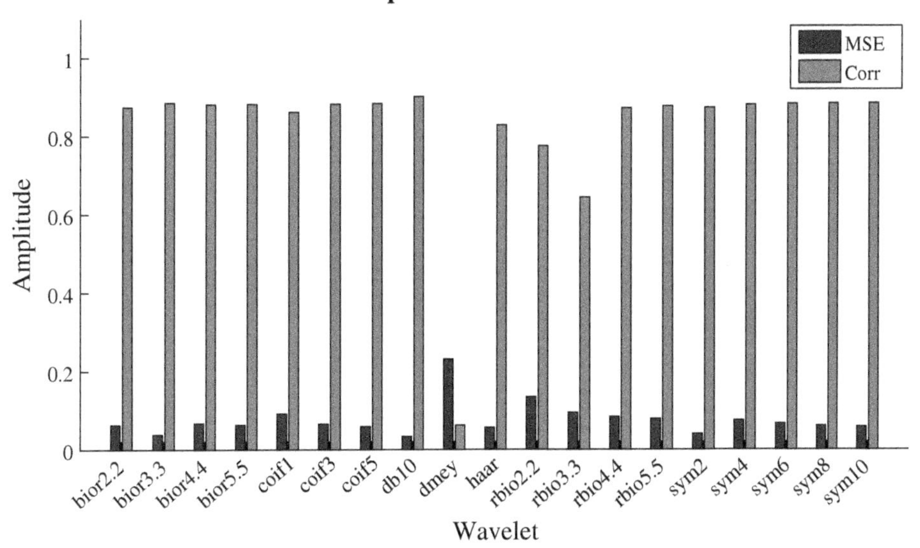

(b) Evaluation of approximation component at level-9 using experiment 1 dataset

Figure 5.24 Performance evaluation of DWT: (a) DB10 wavelet; (b) various types of mother wavelets.

Table 5.7 Evaluation of a DWT approximation component and an FIR LPF output signal in comparison to respiration belt signal.

Method		Parameter order	Others	Experiment 1 dataset		Experiment 2 dataset		Experiment 3 dataset	
				MSE	Corr	MSE	Corr	MSE	Corr
LPF FIR	Kaiser window	600	fc = 1 Hz; $\beta = 6.5$	0.1501	0.5482	0.1050	0.7940	0.0421	0.8566
LPF FIR	Gaussian window	600	fc = 1 Hz; $\alpha = 2.5$	0.1485	0.5509	0.1038	0.7956	0.0421	0.8570
LPF FIR	Least-square filter	600	fc = 1 Hz	0.1296	0.5825	0.0931	0.8139	0.0419	0.8617
LPF FIR	Blackman window	600	fc = 1 Hz	0.1536	0.5420	0.1081	0.7873	0.0422	0.8555
LPF FIR	Bartlett window	600	fc = 1 Hz	0.1457	0.5556	0.1017	0.7997	0.0420	0.8578
LPF FIR	Hamming window	600	fc = 1 Hz	0.1467	0.5541	0.1024	0.7986	0.0420	0.8575
LPF FIR	Hanning window	600	fc = 1 Hz	0.1495	0.5491	0.1046	0.7948	0.0421	0.8567
LPF FIR	Blackman-Harris window	600	fc = 1 Hz	0.1576	0.5352	0.1119	0.7795	0.0424	0.8542
LPF FIR	Chebyshev window	600	fc = 1 Hz	0.1593	0.5061	0.1711	0.7350	0.0449	0.8482
Wavelet	Daubechies (Db10)	Level 9 approximation		0.0335	0.9017	0.0725	0.8834	0.0408	0.9136

Table 5.8 Performance evaluation on five additional subjects with different motion artefacts.

		Dataset 1				Dataset 2			
		DWT		LPF-Kaiser		DWT		LPF-Kaiser	
Subject	Type of artefact	MSE	Corr	MSE	Corr	MSE	Corr	MSE	Corr
1	Small body movement	0.0682	0.8791	0.0724	0.8704	0.0870	0.7420	0.0946	0.7223
2	Both hand swing	0.0563	0.8781	0.1487	0.6930	0.0233	0.9550	0.1009	0.8690
3	Right arm swing	0.0334	0.9258	0.0690	0.8121	0.0232	0.9136	0.0378	0.8629
4	Left hand swing	0.0414	0.9140	0.0603	0.8545	0.0354	0.9549	0.0593	0.8923
5	Head motion	0.0502	0.8912	0.0596	0.8829	0.0419	0.9177	0.0503	0.9079

[breathing rate at 0.2441 Hz or 14.6 breaths/min as shown in Figure 5.23(b) deduced from a multi-level DWT analysis], the breathing patterns have also been recovered and the temporal functions have a high degree of correlation to the reference respiration belt, which is the main aim of this book. Using the reconstructed respiration patterns, we can further derive the inhalation to exhalation time ratio from the local maxima and minima of the curve as well as other parameters, i.e. respiration disorder [240, 393].

Another trial was conducted with the subject in a supine position and the results are shown in Figure 5.23(a). The subject was asked to breathe normally and create some motion artefacts with movements of the right arm occasionally within a time span of 6 minutes. There were three jerking events that corrupted the breathing signals where the detailed components from DWT highlighted the occurrence of such disturbance, as shown in Figure 5.23(a). The decomposition results from DWT recover the breathing signal as shown in the figure and the spectrum analysis for all [approximation component, LPF output signal, and raw data (I signal)] shows the same peak at 0.1659 Hz, which corresponds to 9.954 breaths/min. Trials with two different orientations were performed to further establish the feasibility and robustness of Doppler radar in detecting and monitoring breathing in various types of applications, ranging from sleep monitoring to diagnostics and rehabilitation.

Additionally, two more datasets were collected from five other participants, demonstrating specific motion artefacts while breathing normally and the results are shown in Table 5.8. In these experiments, we evaluated the performance between the DWT and the LPF [100] methods in comparison with the reference respiration strap using the mean squared error and Pearson correlation coefficient. Several different motions were evaluated, as depicted in Table 5.8. Overall, the DWT method performs better than the LPF in terms of MSE and correlation coefficient as well as the reconstructed respiration patterns.

5.7.3 Summary

We investigated Doppler radar-based measurements and monitoring of breathing subject to certain disturbances (motion artefacts), for two different patient postures common in a multitude of applications – seated and supine. We proposed to analyse the raw signals using multi-level DWT to filter out the motion artefacts and recover the desired breathing signal in order to derive the respiration rate as well as the breathing patterns. A higher degree of correlation was observed between the corresponding belt strap measurements and the approximation signal from the multi-level DWT analysis. It was shown that the DWT analysis is capable of filtering out the motion due to limb and small body movements effectively, as discussed in the results section, and we accomplished this with no additional complexity to the receiver architecture.

5.8 Separation of Doppler Radar-based Respiratory Signatures

Respiration detection using microwave Doppler radar has attracted significant interest primarily due to its unobtrusive form of measurement. With less preparation in comparison to attaching physical sensors on the body or wearing special clothing, Doppler radar for respiration detection and monitoring is particularly useful for long-term monitoring (LTM) applications such as sleep studies (i.e. sleep apnoea, SIDS). However, motion artefacts and interference from multiple sources limit the widespread use and the scope of potential applications of this technique.

Most of the reported results focused on obtaining non-contact respiration measurements for a single subject. In many practical applications, the received signal can easily be adversely affected when there is more than one subject present in the immediate neighbourhood, especially in a home-based, long-term monitoring application where the subjects share a bed with the partner. Thus, it is necessary to take this case into account for the Doppler radar system to be able to consistently measure the relevant respiratory signatures when subjected to multiple competing signals. In certain earlier work of source separation in Doppler radar signal processing, Vergara *et al.* [364] demonstrated the use of the real analytical constant modulus algorithm (RACMA) to separate certain human body motions and the respiratory signal from the received signals. The results reported in the book used the wider frequency range of the motion artefacts compared to the targeted normal breathing frequency to clearly identify two distinct sources present when the experiment was conducted under normal breathing conditions.

Furthermore, Boric–Lubecke *et al.* [47] demonstrated the use of single and multiple antenna systems (SIMO /MIMO) in sensing multiple subjects using Doppler radar. Here, the experimental results indicated that it is possible to separate respiratory sources through multiple antenna configuration schemes albeit the experiment considered only the continuous breathing of subjects and the signal analysis was entirely based on the fast Fourier transform (FFT). This is unlikely in the case of a number of potential applications where abnormal breathing patterns, for instance, apnoea symptoms, are present and hence FFT analysis is no longer suitable to represent the respiration signal state [197, 198]. Therefore, it is also vital to analyse the subject's respiration patterns along with the respiration rate [393] in the presence of possible abnormal breathing patterns.

5.8.1 Respiration sensing using Doppler radar

Extending the theory from Section 5.2, in a multiple subject (N) environment, the modulated signal of each person due to respiration is R_i, where $i = 1, 2, ..., N$ refers to each subject and can be expressed as

$$
\begin{aligned}
R_1(t) &\approx \cos\left(2\pi f_0 t - \frac{4\pi d_0}{\lambda} - \frac{4\pi x_1(t)}{\lambda} + \phi\left(t - \frac{2d_0}{c}\right)\right), \\
R_2(t) &\approx \cos\left(2\pi f_0 t - \frac{4\pi d_0}{\lambda} - \frac{4\pi x_2(t)}{\lambda} + \phi\left(t - \frac{2d_0}{c}\right)\right), \\
&\vdots \\
R_N(t) &\approx \cos\left(2\pi f_0 t - \frac{4\pi d_0}{\lambda} - \frac{4\pi x_N(t)}{\lambda} + \phi\left(t - \frac{2d_0}{c}\right)\right).
\end{aligned} \tag{5.40}
$$

Considering that the reflected signal from each target is instantaneously linearly mixed, for instance in the case of $N = M = 2$ where N is the number of subjects and M is the number of transceivers, the received signal (observation) can be represented by

$$
\begin{bmatrix} Receiver_1(t) \\ Receiver_2(t) \end{bmatrix} = \begin{bmatrix} a_{11} & a_{12} \\ a_{21} & a_{22} \end{bmatrix} \begin{bmatrix} R_1(t) \\ R_2(t) \end{bmatrix}, \tag{5.41}
$$

where a_{ij} are the mixing parameters. In general, for $N = M = X$, where $X = 2, 3, ..., N$, equation (5.41) can be represented as follows:

$$
\begin{bmatrix} Receiver_1(t) \\ Receiver_2(t) \\ \vdots \\ Receiver_M(t) \end{bmatrix} = \begin{bmatrix} a_{11} & a_{12} & \cdots & a_{1N} \\ a_{21} & a_{22} & \cdots & a_{2N} \\ \vdots & \vdots & \cdots & \vdots \\ a_{M1} & a_{M2} & \cdots & a_{MN} \end{bmatrix} \begin{bmatrix} R_1(t) \\ R_2(t) \\ \vdots \\ R_N(t) \end{bmatrix}. \tag{5.42}
$$

In quadrature receiver architecture generally, equation (5.14) can be structured as

$$
\begin{bmatrix} I_1 \\ I_2 \\ \vdots \\ I_M \end{bmatrix} = \begin{bmatrix} a_{11} & a_{12} & \cdots & a_{1N} \\ a_{21} & a_{22} & \cdots & a_{2N} \\ \vdots & \vdots & \cdots & \vdots \\ a_{M1} & a_{M2} & \cdots & a_{MN} \end{bmatrix} \begin{bmatrix} \cos(\theta + Cx_1(t) + \Theta) \\ \cos(\theta + Cx_2(t) + \Theta) \\ \vdots \\ \cos(\theta + Cx_N(t) + \Theta) \end{bmatrix} \tag{5.43}
$$

while equation (5.15) is represented as

$$
\begin{bmatrix} Q_1 \\ Q_2 \\ \vdots \\ Q_M \end{bmatrix} = \begin{bmatrix} a_{11} & a_{12} & \cdots & a_{1N} \\ a_{21} & a_{22} & \cdots & a_{2N} \\ \vdots & \vdots & \cdots & \vdots \\ a_{M1} & a_{M2} & \cdots & a_{MN} \end{bmatrix} \begin{bmatrix} \sin(\theta + Cx_1(t) + \Theta) \\ \sin(\theta + Cx_2(t) + \Theta) \\ \vdots \\ \sin(\theta + Cx_N(t) + \Theta) \end{bmatrix}, \tag{5.44}
$$

where $\theta = \frac{4\pi d_0}{\lambda}$, $C = \frac{4\pi}{\lambda}$ and $\Theta = \Delta\phi(t)$.

5.8.2 Signal processing source separation (ICA)

Independent component analysis (ICA) is a statistical method that performs the transformation of a multi-dimensional random vector observation into sources that are statistically as independent from each other as possible [151] and generally uses techniques involving higher-order statistics [155]. The different implementations of ICA can be found in the literature [34, 76, 152] but we will be only focusing on the fastICA algorithm [151] in this application.

In essence, independent component analysis assumes that in a set of m-dimensional measured time series, a vector denoted as $x(t) = [x_1(t), x_2(t), ..., x_m(t)]^T$ is a linear combination of n-dimensional source vectors whose components are assumed to be statistically independent and given as $s(t) = [s_1(t), s_2(t), ..., s_n(t)]^T]$. Further, it assumes that the dimension of x and s are equal, for instance $m = n$ [151]. ICA is often represented as

$$x(t) = As(t), \tag{5.45}$$

where A is a full rank $m \times n$ mixing matrix [155]. Then, a separating or de-mixing matrix W must be estimated under certain assumptions and constraints [151] in order to extract each independent source signal from the observations such that

$$s(t) = Wx(t). \tag{5.46}$$

FastICA is based on a fixed point iteration scheme that attempts to separate the underlying sources from a given set of mixed measurements (observations) by finding their maximum of the non-Gaussianity, $w^T x$, as measured in equation (5.47). Here, the measure of non-Gaussian, negentropy $J(y)$ is based on the information theoretic quantity of differential entropy, which can be given as $J(y) = H(y_{\text{Gauss}} - H(y))$. Here $H(.)$ is the differential entropy and y_{Gauss} is a Gaussian random variable with the same covariance matrix as output signal y. In FastICA, the approximation of the negentropy was referred to as

$$J(y) \approx \rho[EG(y) - EG(v)]^2, \tag{5.47}$$

where ρ is a positive constant, v is a Gaussian variable with zero mean and unit variance and $G(.)$ is any non-quadratic function as typically suggested in Hyvarinen [151].

The basic form of the fastICA algorithm (one unit) can be denoted as follows:

Algorithm 1 FastICA

1: **procedure** FASTICA
2: Data centring
3: Whitening
4: Initialise weight vector, w,
5: *loop*:
6: $w^+ = E\{xg(w^T x)\} - E\{g'(w^T x)\}$,
7: $w = \frac{w^+}{\|w^+\|}$,
8: goto loop if not converged.

where g is the derivative of the non-quadratic function G, for instance:

$$G_1(u) = \frac{1}{a_1}\text{log.cosh}(a_1 u),$$
$$g_1(u) = \tanh(a_1 u),$$
$$g_1'(u) = a(1 - \tanh^2(a_1 u)). \tag{5.48}$$

Here, $1 \leq a \leq 2$ is a suitable constant, normally taken as $a = 1$. To estimate several independent components (ICs), the basic form of one unit of FastICA needs to be extended using several units with weight vectors $w_i, i = 1, 2..., n$. The outputs $w_1^T x, ..., w_n^T$ need to be decorrelated to prevent vectors from converging to the same maxima after every iteration [153]. One method of achieving decorrelation is through Gram-Schmidt-like decorrelation, where estimation of IC is performed one by one. When p independent components or p vectors $w_1, ..., w_p$ were estimated, one unit of FastICA is performed for w_{p+1} and after each iteration, the projections of $w_{p+1}^T w_j w_j, j = 1, ..., p$ of the previously estimated p vectors are subtracted from w_{p+1} and renormalised. The entire process can be represented as

$$\begin{aligned} &1.\ \text{Let} \quad w_{p+1} = w_{p+1} - \sum_{j=1}^{p} w_{p+1}^T w_j w_j, \\ &2.\ \text{Let} \quad w_{p+1} = w_{p+1} / \sqrt{w_{p+1}^T w_{p+1}}. \end{aligned} \tag{5.49}$$

Another approach to achieve decorrelation is through a symmetric decorrelation where no vectors are "priviliged" over others [153, 165]. It can be implemented via a classical method involving matrix square roots as follows [153]:

1. Let $W = (WW^T)^{-0.5} W$,

where W is the matrix of $w_1, ..., w_n$ of the vectors. The inverse square root of $(WW^T)^{-0.5}$ is computed from the eigenvalue decomposition WW^T [165]. A more simplistic alternative is given in the form of the following iterative algorithm [151],

1. Let $W = W/\sqrt{\| WW^T \|}$.
2. Let $W = 1.5W - 0.5WW^T W$.
3. Repeat 2 until it converges.

5.8.3 Experiment protocol for real data sensing

In this experiment, two 2.4 GHz Doppler radar modules [126] were used to capture the respiration signals from two subjects (in each experiment), as shown in Figure 5.25(a). Each of the systems transmitted a continuous wave (CW) of 2.4 GHz and was attached to two patch antennae (transmitter and receiver) connected to a data acquisition module (DAQ: NI-USB6009). The received signals were then further processed in a MATLAB environment. In each experiment, the subjects were positioned 1 metre away from the antennae, where the antennae were aligned to focus on the abdomen of the subjects. Respiration signals from the subjects were collected with normal clothing and in a seated position in the laboratory. An external respiration strap (MLT1132 Piezo Respiratory Belt Transducer) attached to PowerLab (ADInstruments) was used as a reference signal to evaluate the performance results of the source separation technique on the measurements obtained from the Doppler radar.

Two sets of experiments were then carried out to evaluate the separation technique between two breathing signals captured from the Doppler radar measurements. In the first experiment, two subjects were breathing normally but at different rates, while in the second experiment, one of the subjects was breathing normally while the other subject role-played apnoea by stopping breathing for a certain duration multiple times. The last

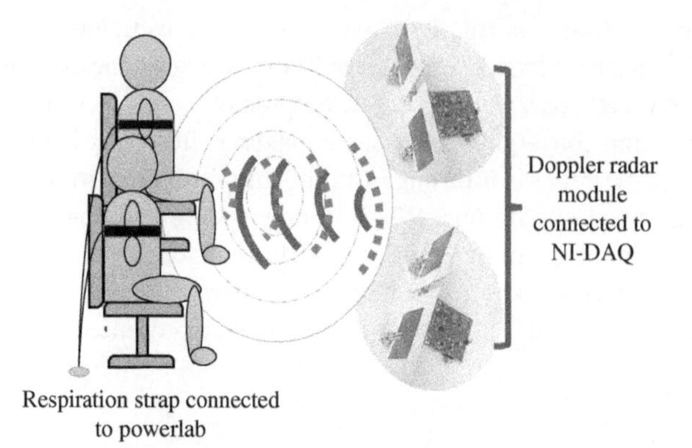

(a) Experimental setup for the data acquisition of two subjects

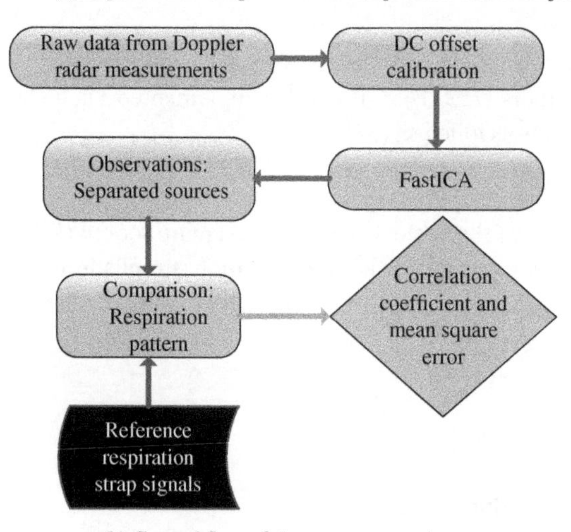

(b) General flow of the source separation process

Experiment	Person 1	Person 2
1 Two subjects	Normal	Normal (faster pace)
2 Two subjects	Normal + apnoea	Normal
3 One subject	Normal + arm swinging	Nil

(c) Experimental protocol for different modes of breathing

Figure 5.25 Experiment setup, signal processing flow and experiment protocol.
Source: Yee Siong Lee.

experimental trial was performed on one subject under the influence of motion artefacts (arm swing). The summary of the experiments is shown in Figure 5.25(c). The general flow of the source separation process is shown in Figure 5.25(b).

5.8.4 Two simulated respiratory sources

In order to commence with the basic idea of source separation, two sources mimicking respiration were considered in this simulation. As depicted in Figure 5.26, two respiration

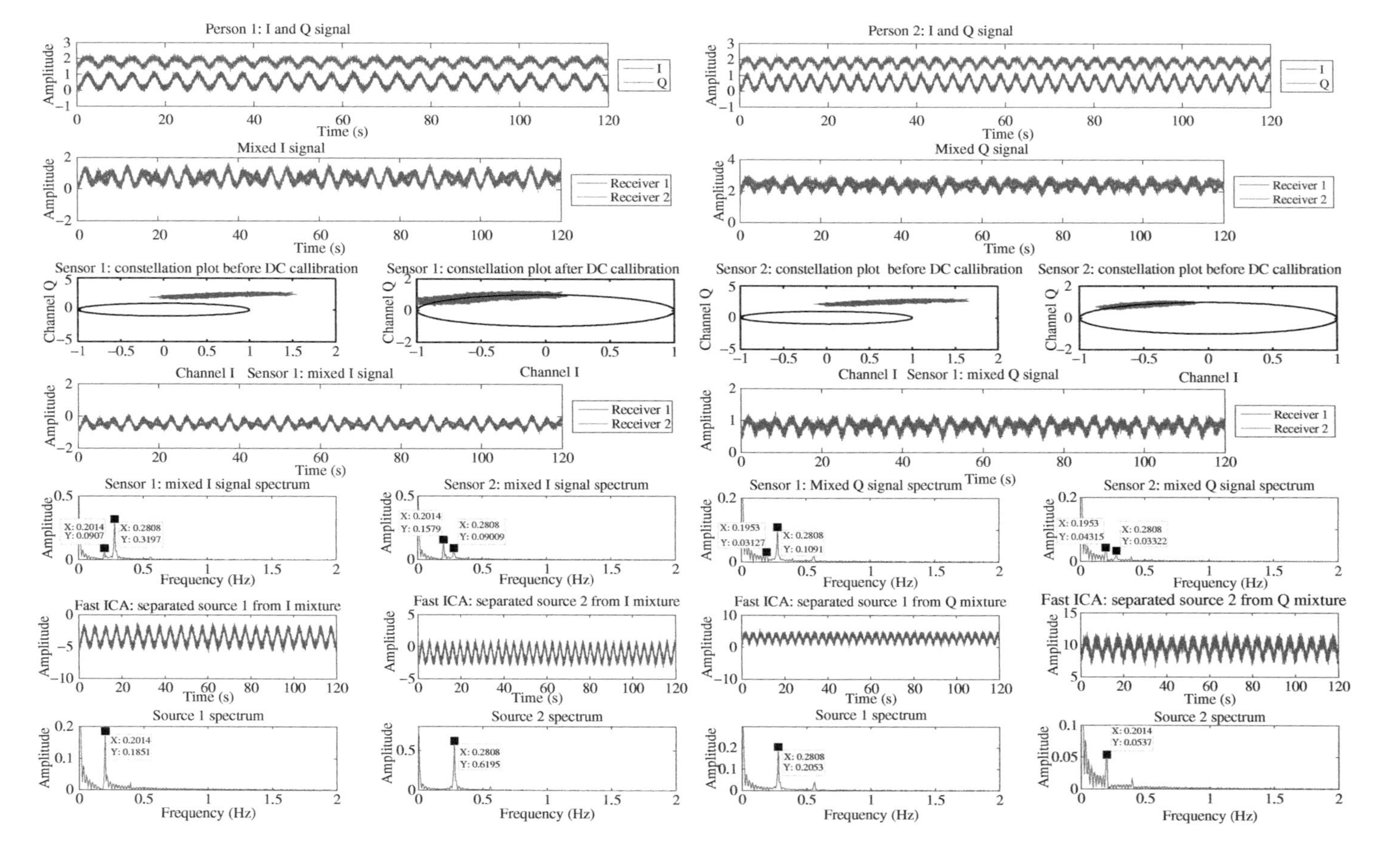

Figure 5.26 Two simulated sources with normal breathing captured from Doppler radar modules. Figure (from top row to bottom row): baseband signal of two respiration signals; randomly mixed baseband signals; constellation plot before and after DC offset calibration; mixture of sources after DC calibration; frequency spectrum of the mixture; independent sources; frequency spectrum of the independent sources.

baseband signals (with a DC offset and Gaussian noise) at 0.20 Hz and 0.28 Hz, respectively, were simulated using equations (5.14) and (5.15). Then, the baseband signals [in-phase (I) and quadrature-phase (Q)] were randomly mixed (see equations 5.43 and 5.44) and re-calibrated to eliminate the DC offset. From the constellation plot, it was evident that DC offsets have been corrected. In this particular step, we have used the curve-fitting technique [348] to fit the I and Q data to a circle, as shown in Figure 5.26. Two different frequency spectra were clearly observed from the spectrum plot denoting the respective baseband respiration frequencies that were simulated. The main aim of this simulation was to evaluate the source separation technique in separating the simulated Doppler radar combined baseband respiration signals into its relevant independent sources while preserving the frequency and the breathing pattern information.

In this simulation, a fast fixed-point algorithm for an independent component analysis (fastICA) was used. From the results shown in Figure 5.26, the mixed signals were successfully separated where the approximated spectral frequencies were 0.2014 Hz and 0.2808 Hz, respectively. This was achieved while preserving similarity of the baseband signal pattern to the simulated respiration signals. From the simulation, it was observed that by using the in-phase signals from two baseband signals, the spectral frequency and the patterns of the baseband gave a better representation rather than using the quadrature-phase signals.

5.8.5 Experiment involving real subjects

Simulation results indicated promising results in separating the mixed signals from the respective I/Q channel into its respective independent component of a baseband respiration signal. Thus, we have implemented the corresponding algorithm with the real respiration measurement data from the Doppler radar. For these particular experiments, two subjects were located in front of two Doppler radar transceivers while breathing according to the conditions listed in Figure 5.25(c). In the first experiment, two subjects were asked to breathe normally at different rates while in the second experiment, one of the subjects was asked to hold their breath (role-playing central apnoea) while the other subject was continually breathing in a normal mode.

Figure 5.27(a) depicts a compilation of the results observed from Experiment 1. The raw mixed I signals were pre-processed for DC offset calibration before the separation process. From one of the spectrum plots in the calibrated mixed I signals, there were two dominant breathing frequencies of 0.3052 Hz and 0.4425 Hz. The expected outcome of this experiment would be two respiration signals from two different subjects in terms of the respiration rates and patterns. For this purpose, we have used the fastICA algorithm to separate the mixed I signals into its independent components where the separation results were then compared to the independent measurements of the respiration strap. The results are shown in Table 5.9 in terms of the mean square error (MSE) and correlation coefficient (Corr). Under normal breathing conditions, the respiration rate can be estimated using the fast Fourier transform (FFT) [127, 198] and, as shown in Figure 5.27(a), the breathing rates estimated from the independent respiration strap were similar to the breathing frequency of each independent source derived from fastICA (using the measurements acquired from the Doppler radar).

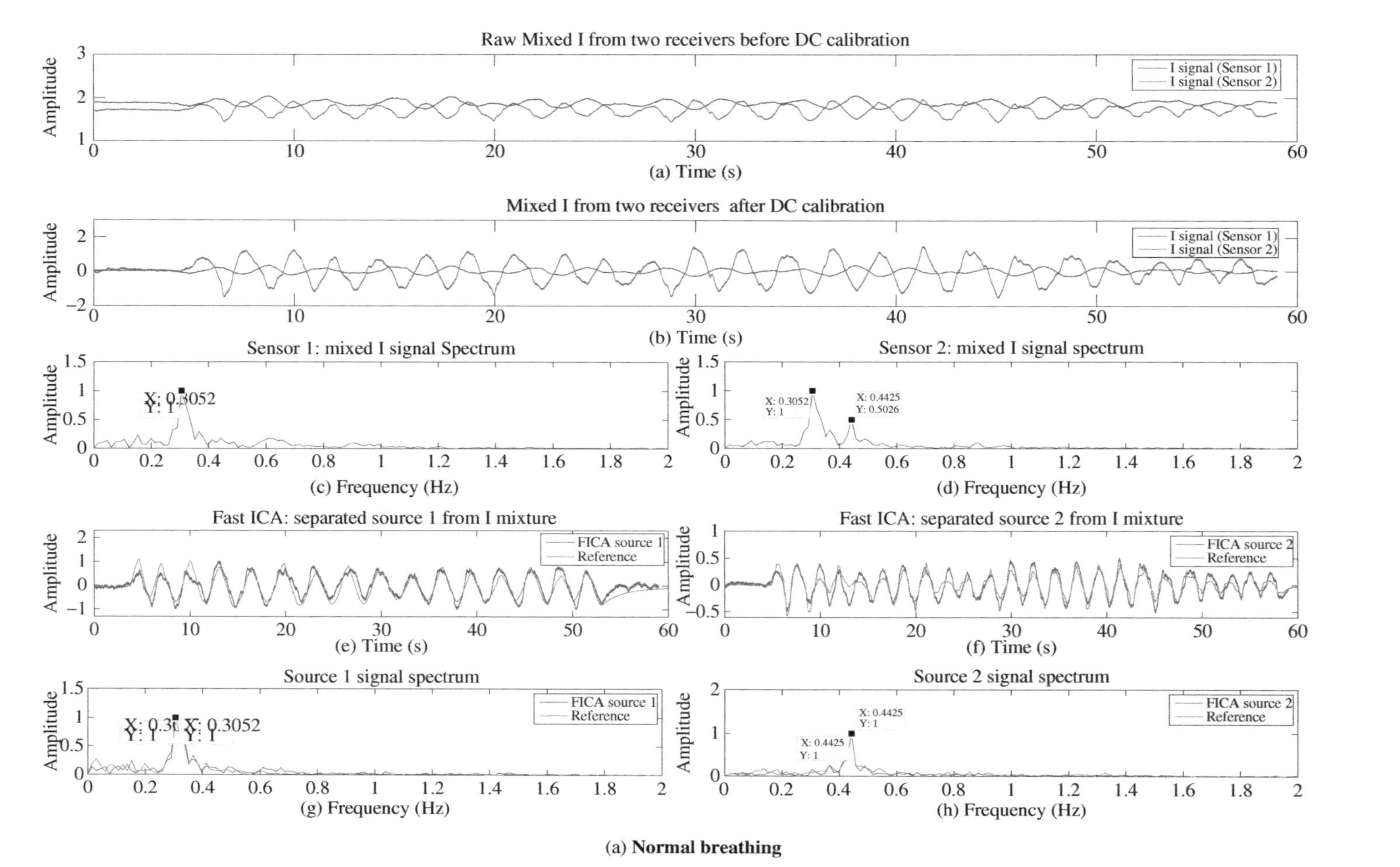

Figure 5.27 Two source breathing signals captured from Doppler radar modules (top to bottom): (a) raw data of mixed in-phase (I) signal; (b) DC offset calibrated mixed I signal; (c), (d) spectrum of each mixed I channel signal; (e, f) separated breathing source for each subject; (g, h) spectrum of each independent breathing component.

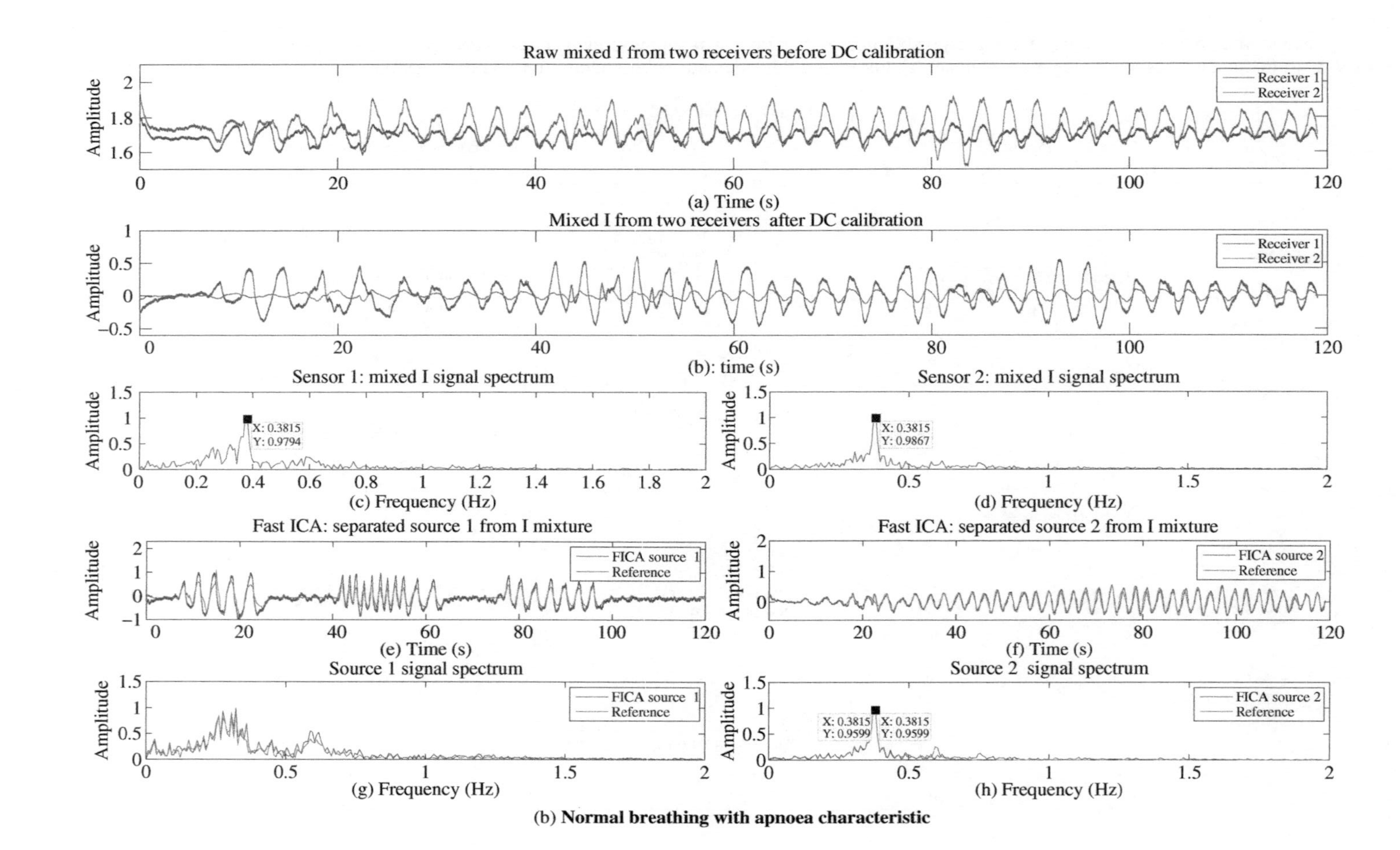

Figure 5.27 *(Continued)*

Table 5.9 Performance evaluation on FastICA using different non-linearity *g* function with reference respiration strap signal patterns.

(a) Experiment 1

	Source 1		Source 2		
g function	**MSE**	**Corr**	**MSE**	**Corr**	**Time (s)**
tanh	**0.2055**	**0.8765**	**0.0474**	**0.8811**	**0.0536**
power	0.2098	0.8765	0.0838	0.8819	0.0575
gauss	0.2068	0.8764	0.0473	0.8809	1.3024
skew	0.2207	0.7959	0.0372	0.7737	1.2907

(b) Experiment 2

	Source 1		Source 2		
g function	**MSE**	**Corr**	**MSE**	**Corr**	**Time (s)**
tanh	**0.1670**	**0.9080**	**0.1280**	**0.9543**	**0.0766**
power	0.1717	0.9071	0.1281	0.9543	0.1183
gauss	0.1713	0.9072	0.1279	0.9543	0.0926
skew	0.1766	0.9057	0.1269	0.9520	1.7229

As for the second experiment, the results of the source separation are shown in Figure 5.27(b). In this particular experiment, one of the subjects was asked to hold his breath multiple times in 2 minutes of recording to mimic the condition of central sleep apnoea [68]. This experiment was specifically designed to evaluate fastICA source separation capabilities in dealing with the mixed signals of normal breathing and abnormal breathing patterns. As shown in Figure 5.27(b), (see Figure 5.27b, part e), the results show the occurrence of multiple cessation of breathing in the patterns that correspond to the central apnoea type of breathing. The derived independent component for this is highly correlated with the independent measurement of the respiration strap. As for the second subject, a normal breathing pattern was observed (see Figure 5.27b, part f) as expected. From the spectral analysis (see Figure 5.27b, part h), the normal breathing source was successfully separated from the mixed sources where the peak of 0.3815 Hz was detected in both the reference respiration strap and the separated source. As for the spectral analysis shown in Figure 5.27(b), part (g), the FFT-based evaluation for an abnormal respiration pattern is not a good measure to estimate the respiration frequency. A technique such as time-frequency analysis is needed to cater for such respiration dynamics [198].

We have also performed the separation of sources using fastICA with a different type of non-linearities, i.e. g functions (refer to Table 5.10), where the performance of each function was shown in Table 5.9 and by comparing each independent source with the reference respiration strap. Considering data from four different subjects in two different experiments, *tanh* as the g function performs better (in terms of MSE) in separating the sources from the Doppler radar-based measurements compared to other common non-linear functions considered. A reasonable correlation with shorter processing time was observed as shown in Figure 5.28.

Table 5.10 Non-linearity of the g function.

Non-linearity	$g(u)$
tanh	$g(u) = \tanh(au)$
power	$g(u) = u^3$
gauss	$g(u) = u * \exp(-au^2/2)$
skew	$g(u) = u^2$

Experiment 1: Evaluation on FastICA using different non-linearity g function
Source 1: MSE
Source 1: Corr
Source 2: MSE
Source 2: Corr
Execution time (s)
Amplitude
1.4
1.2
1
0.8
0.6
0.4
0.2
0
tanh
power(u3)
gauss
skew
Non-linearity g function

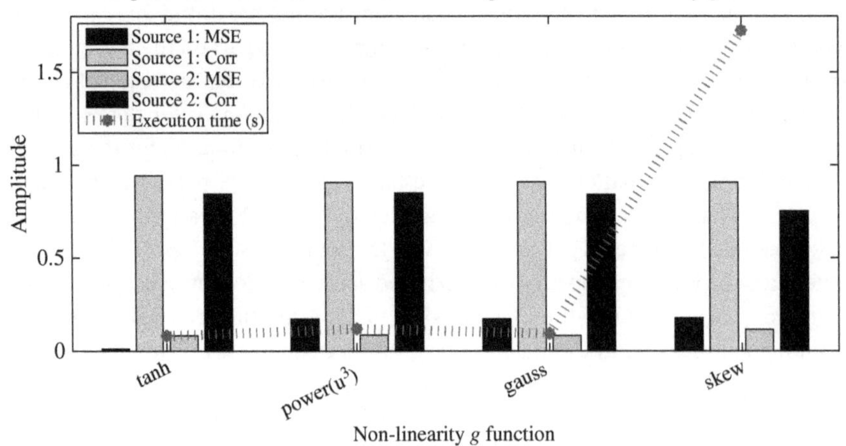

Figure 5.28 Pictorial representation of the performance evaluation on FastICA using different non-linearity functions with reference respiration strap signal patterns.

Table 5.11 Performance evaluation on different BSS algorithms separated sources with reference respiration strap signal patterns for Experiment 1.

(a) Experiment 1

	Source 1		Source 2		
Algorithm	**MSE**	**Corr**	**MSE**	**Corr**	**Time (s)**
EFICA	0.2055	0.8765	0.0472	0.8832	0.2118
WASOBI	0.1566	0.7683	0.0437	0.8037	0.3643
COMBI	0.2053	0.8765	0.0472	0.8832	0.2097
MCOMBI	0.2053	0.8765	0.0472	0.8832	0.1700
FCOMBI	0.2053	0.8765	0.0472	0.8832	0.1629
BEFICA	0.1922	0.8751	0.0469	0.8837	0.2841
BARBI	0.2332	0.8434	0.0473	0.8745	0.1309

(b) Experiment 2

	Source 1		Source 2		
Algorithm	**MSE**	**Corr**	**MSE**	**Corr**	**Time (s)**
EFICA	0.1752	0.9063	0.1264	0.9514	0.1044
WASOBI	0.1797	0.9043	0.1253	0.9502	0.0456
COMBI	0.1752	0.9063	0.1264	0.9514	0.1015
MCOMBI	0.1752	0.9063	0.1264	0.9515	0.1228
FCOMBI	0.1753	0.9062	0.1265	0.9515	0.1763
BEFICA	0.1783	0.9050	0.1286	0.9535	0.1745
BARBI	0.1671	0.7314	0.1198	0.9539	0.0432

We have also evaluated the performance of various BSS algorithms [283] on all the datasets and the results are given in Table 5.11. The tested algorithms include EFICA [180], WASOBI [352], COMBI [353], MCOMBI [354], FCOMBI [120], BEFICA [180] and BARBI [351]. From the results, most of the BSS algorithms are capable of performing the separation with a low MSE and good correlation coefficient but fastICA with the *tanh function* performs better; it is more suitable for real-time applications, particularly due to a shorter processing time.

5.8.6 Separation of hand motion

In addition to the decomposition of signals, we have also explored the use of fastICA in the removal of motion artefacts. In particular, we investigated separating the respiratory signal from a typical interfering signal, i.e. a subject's right arm movement as a motion artefact. The performance of the separation task is shown in the Figure 5.29. The separated hand motion from the mixture of signals is clearly observed in Figure 5.29(d) as marked, while the

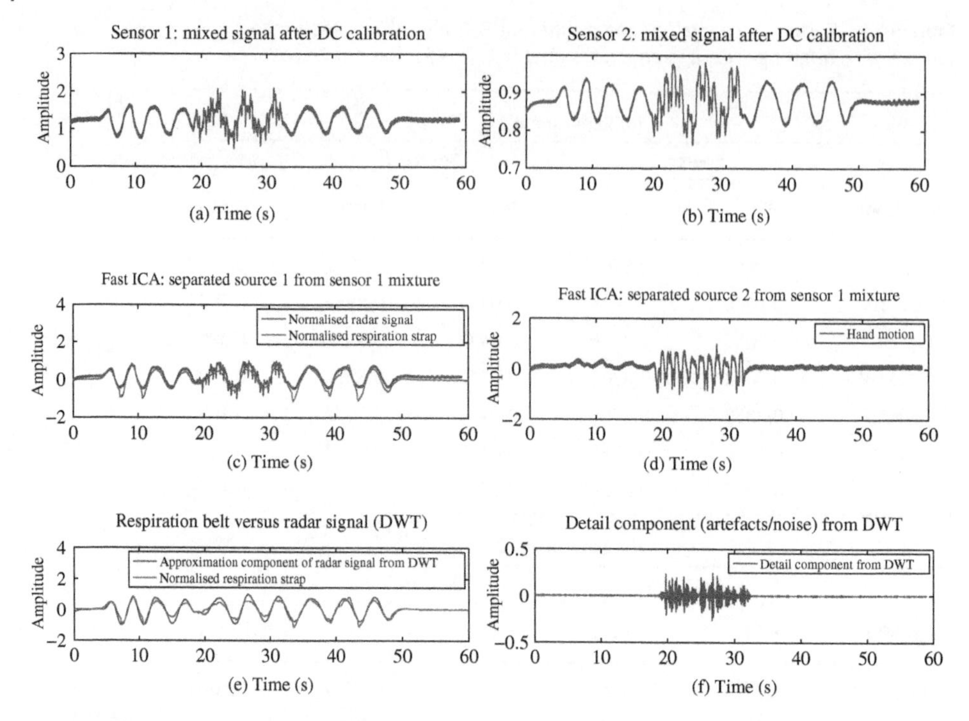

Figure 5.29 A mixture of breathing signals with hand motion (swinging of the arm) from one subject: (a) and (b) are the mixtures of the signal from two antennas; (c) and (d) are the separated breathing source/hand motion signals (using FastICA); (e) and (f) are the filtered breathing source (DWT-approximation component) and the filtered hand motion/noise signal source (DWT-detailed component).

filtered respiration signal pattern is shown in Figure 5.29(c). Although the motion due to the swinging of the arm is separated from the mixture of the signals, the separated respiration signal is not entirely free from other noise inputs (for instance, possible artefacts due to the slight motion of the body during the swinging process), as shown in Figure 5.29(c). Therefore, even for the case of respiration detection involving a single subject, the use of a multiple antenna configuration would significantly improve the coverage, detection and the characterisation capabilities of human motion in general. For instance, in characterising the capabilities of human motion, multiple antennas can be used to identify the motion from various body parts, such as leg, arm or head, in correspondence to the placement of the antenna. In particular sleep studies where periodic limb movements are of interest, the source separation can be used to analyse the artefacts.

Typically, DWT uses a dyadic grid [16] where the dyadic wavelet transform of signal $s(t)$ is given as

$$DWT\ s(m,n) = 2^{-m/2} \int_{-\infty}^{\infty} s(t)\varphi * \left(\frac{t-2^m n}{2^m}\right) dt, \tag{5.50}$$

where m and n are scale and time shift parameters, respectively, * is the complex conjugate and $\varphi(t)$ is the given basis function (mother wavelet).

In general, the output of the low-pass filter is known as the wavelet approximation (scaling) coefficient cA_m and the output of the high-pass filter is called the wavelet detail coefficient cD_m [54, 367]. The approximation and the detail coefficient at the mth level can be denoted as equations (5.51) and (5.52), respectively:

$$cA_m[n] = \sum_{-\infty}^{\infty} l_d[k]cA_{m-1}[2n-k], \tag{5.51}$$

$$cD_m[n] = \sum_{-\infty}^{\infty} h_d[k]cA_{m-1}[2n-k]. \tag{5.52}$$

The taps of the high-pass filter h_d and the low-pass filter l_d are derived from the scaling and wavelet function of a chosen mother wavelet family, i.e. Daubechies, Haar or Coiflet.

As shown in Figure 5.29(e), we used the discrete wavelet transform (DWT) to filter the artefacts/noise from the signal mixture, where the reconstructed respiration signal (approximation component from the DWT) is highly correlated and smoother than the results obtained from fastICA. The type of wavelet used in this experiment trial is Daubechies (Db10) with 10 levels of decomposition. The motion artefacts/noise can be obtained from the detail components as shown in Figure 5.29(f). The DWT technique is particularly useful in filtering the artefacts/noises from the mixture of signals due to its nature. From our observations, this is not the case in separating multiple respiration sources due to multiple subjects.

Indeed the DWT technique can be useful and yield a better result compared to the blind source separation technique as demonstrated in Vergara *et al.* [364] and, more importantly, it can be implemented with only a single antenna configuration. Nevertheless, using DWT solely will not improve the respiration detection coverage, especially when more than one source of respiration exists.

5.8.7 Conclusion

In this book, we demonstrated the feasibility of Doppler radar as a sensing mechanism for respiration detection and monitoring. In particular, we consider the use of Doppler radar in capturing various dynamics of respiration patterns ranging from common types to patterns characteristic of certain respiration disorders. Indeed the experiments were performed by professional role playing of known breathing patterns, yet the results are encouraging as an alternative to the standard respiratory belt measurements. Results also show that breathing is not always periodic and purely spectral methods are not adequate and frequency as a function of time is not the only essential matrix that can be used to extract useful information from the radar signal. Thus, the CWT is more suited for the detailed analysis of breathing patterns.

Doppler radar based techniques were considered for estimating the tidal volume; deriving the relationship between Doppler radar signal and the corresponding flow of air in and out due to inhalation and exhalation. When the respiration signal and artefacts are of particularly interest, the discrete wavelet transform can be used at an effective tool in reconstructing the desired breathing signal as well as in isolating the artefacts signatures (i.e. jerking, arm's swing and body movement). The reconstructed respiration patterns from the approximated components of DWT are highly correlated with the information captured through the respiration strap that is used and the reference. On the other hand, the detailed components of DWT are deemed as motion artefacts signatures.

The respiration detection of multiple subjects simultaneously using Doppler radar can be achieved with the use of the multiple Doppler radar systems. Capturing multiple respiratory sources is useful particularly for long term home-based monitoring applications such as monitoring sleep apnoea in sleep studies, where typically the interested patient is sleeping with a partner in the same bed. For this purpose, we used the fastICA algorithm in separating the mixed Doppler radar measurements of two people under two different scenarios involving either normal breathing only or abnormal breathing (mixture of normal breathing and apnoea), as discussed in the experiment protocol.

Future work would be extended to involve real patient experiments (clinical trial) and investigation of the use of wavelet coefficients from CWT for feature extraction (i.e. energy, entropy, frequency distribution, power etc.) along with patterns recognition techniques to characterise breathing disorders and suppressing the motion artefacts from the measurements captured in this non-contact sensing environment.

6

Appendix

6.1 Static Estimators

6.1.1 Least-squares estimation

LSE is to estimate the parameters by minimising the squared discrepancies between the observed data and their expected or predicted values. Let $X_i \in \mathbf{R}^n$ indicate an independent dataset that is modelled to generate an observable dataset Y_i such that

$$Y_i = f(X_i) + v_i,$$

where $f : \mathbf{R}^k \rightarrow \mathbf{R}^n$ is the regression function and v_i is the noise.

Linear regression

This is when the function f is linear in the parameters – not necessarily in the explanatory variables. Assuming x_i indicates either the explanatory variable or a function of the explanatory variable

$$y = Xb + v,$$

where

$$y = \begin{bmatrix} y_1 \\ \vdots \\ y_n \end{bmatrix} \in \mathbf{R}^n \; X = \begin{bmatrix} 1 & x_2 & \cdots & x_k \\ \vdots & \vdots & \vdots & \vdots \\ 1 & x_2 & \cdots & x_k \end{bmatrix} \in \mathbf{R}^{n \times k} \; b = \begin{bmatrix} b_1 \\ \vdots \\ b_k \end{bmatrix} \in \mathbf{R}^k, v = \begin{bmatrix} v_1 \\ \vdots \\ v_n \end{bmatrix} \in \mathbf{R}^n.$$

Solving for the parameters,

$$b = \left(X^\top X\right)^{-1} X^\top y, \det\left(X^\top X\right) \neq 0. \tag{6.1}$$

Therefore X should have a rank of k and this implies $n \geq k$, i.e. there needs to be at least the same number of measurements as the number of parameters.

Non-linear regression

When f is a non-linear function of the parameters b, then $X = \partial f / \partial b$.

6.1.2 Maximum likelihood estimation

This is the same as least-squares when the noise is normally distributed with equal variances.

Human Motion Capture and Identification for Assistive Systems Design in Rehabilitation, First Edition.
Pubudu N. Pathirana, Saiyi Li, Yee Siong Lee and Trieu Pham.

6.2 Model-based Estimators

6.2.1 Kalman filter (KF)

The Kalman filter is one of the most popular filters using Bayesian estimation techniques to provide tracking and estimation for both linear and non-linear dynamic systems with stochastic normal or Gaussian distributed noise [231]. This filter was introduced with the purpose of addressing the limitations of other filters when they were solving the Wiener problems [163]. Nowadays, the Kalman filter has been extended to various versions, ranging from optimal, also called standard, to extended and unscented versions for the data from both linear and non-linear systems. Since our system is a linear system with position, velocity and acceleration only, the optimal Kalman filter was implemented and applied for data fusion.

To apply an optimal Kalman filter in our system, the first step is to describe our system as

$$X_{t+1} = AX_t + Bu_t + w_t, \tag{6.2}$$

$$Y_t = CX_t + v_t. \tag{6.3}$$

Here, X_t is the state of the system at time t, Y_t is the observation of the system at time t, A is the state transition matrix, C is the observation matrix, $w_t\ N(0, S_w)$ is the process noise and $v_t\ N(0, S_v)$ the measurement noise. Moreover, $S_w = E(w_t(w_t)^\top)$ corresponds to the process noise covariance and $S_v = E(v_t(v_t)^\top)$ is the measurement noise covariance.

After the system is defined, the states in the system will be estimated with the observations collected from one or more sensors by the following forward process equations:

$$\hat{X}_{t+1|t} = A\hat{X}_{t|t}, \tag{6.4}$$

$$P_{t|t} = AP_{t|t}A^\top + S_w, \tag{6.5}$$

$$K_{t+1} = P_{t+1|t}C^\top(CP_{t+1|t}C^\top + S_v)^{-1}, \tag{6.6}$$

$$\hat{X}_{t+1|t+1} = \hat{X}_{t+1|t} + K_{t+1}(y_{t+1} - C\hat{X}_{t+1|t}), \tag{6.7}$$

$$P_{t+1|t+1} = P_{t+1|t} - K_{t+1}CP_{t+1|t}, \tag{6.8}$$

where P is the state estimate covariance matrix, K is the Kalman gain matrix and $\hat{X}$ is the estimated states of the system.

Sometimes, we are also able to optimise S_w and S_v with a backward process, defined by

$$L_t = P_{t|t}A^\top P^{-1}_{t+1|t}, \tag{6.9}$$

$$\hat{x}_{t|T} = \hat{x}_{t|t} + L_t(x_{t+1|T} - \hat{x}_{t+1|T}), \tag{6.10}$$

$$P_{t|T} = P_{t|t} + L_t(P_{t+1|T} - P_{t+1|t})L_t^\top. \tag{6.11}$$

The EM algorithm was then used to optimise this forward-backward process by computing the maximum log-likelihood,

$$\max_{S_w, S_v} \log \mathcal{L}(S_w, S_v) = \max_{S_w, S_v} \log P(y_{0:K}; S_w, S_v). \tag{6.12}$$

The probability of y_{t+1} was computed from observations $y_0, \ldots, y_t$:

$$P\left(y_{t+1}|y_{0:t}\right) = \frac{1}{(2\pi)^{d/2}|\Sigma_{y_{t+1|t}}|^{1/2}} e^{-\frac{1}{2}(y_{t+1}-\hat{y}_{t+1|t})^\top \Sigma_{y_{t+1|t}} (y_{t+1}-\hat{y}_{t+1|t})}, \tag{6.13}$$

where

$$\hat{y}_{t+1|t} = C\hat{x}_{t+1|t},$$
$$\Sigma_{y_{t+1|t}} = CP_{t+1|t}C^{\top} + \Sigma_t. \tag{6.14}$$

Using Bayes' rule, we then can write the likelihood of $y_{0:K}$ as

$$P\left(y_0, \ldots, y_T\right) = P\left(y_0\right)\prod_{t=1}^{T} P\left(y_t|y_{0:t-1}\right)$$
$$= \prod_{t=1}^{t} \frac{1}{(2\pi)^{d/2}|\Sigma_{y_{t+1|t}}|^{1/2}} e^{-\frac{1}{2}\left(y_{t+1}-\hat{y}_{t+1|t}\right)^{\top}\Sigma^{-1}_{y_{t+1|t}}\left(y_{t+1}-\hat{y}_{t+1|t}\right)}. \tag{6.15}$$

The log-likelihood can then be computed as

$$\log \mathcal{L} = \sum_t \log P\left(y_{t+1}|y_{0:t}\right).$$

The optimal S_w and S_v are computed as

$$S_w = \frac{1}{T}\sum_{t=0}^{T-1}\left[\left(\hat{x}_{t+1|T} - A\hat{x}_{t|T}\right)\left(\hat{x}_{t+1|T} - A\hat{x}_{t|T}\right)^{\top}\right.$$
$$\left. + A_t P_{t|T} A_t^{\top} + P_{t+1|T} - P_{t+1|T}L_t^{\top}A^{\top} - AL_t P_{t+1|T}\right], \tag{6.16}$$

$$S_v = \frac{1}{T+1}\sum_{t=0}^{T}\left(y_t - C\hat{x}_{t|T}\right)\left(y_t - C\hat{x}_{t|T}\right)^{\top} + CP_{t|T}C^{\top}. \tag{6.17}$$

Although the optimal Kalman filter is pervasively applied in various fields, one of the limitations of the Kalman filter is obvious, which is that the noise in the system is assumed to be Gaussian-distributed. However, in the real world, it is not always true. Therefore, if the noise is out of the range of the Gaussian distribution pre-assumed in the system equations the Kalman filter is unlikely to perform as expected.

6.3 Particle Filter

Similar to the Kalman filter, the particle filter, also called the sequential Monte Carlo method, is also widely used for state estimation for both linear and non-linear dynamic systems. The only differences between the particle filter and the Kalman filter is that the noise in the former does not necessarily belong to a Gaussian distribution, which extends the fields where a particle filter can be applied. In a particle filter, the range of noise is controlled by the probability density function. When this function is a Gaussian function, the particle filter becomes a standard Kalman filter. The application of the particle filter in our linear system is as follows:

1. *Building system*: Similar to a Kalman filter, before applying the particle filter, the system should be defined as follows:

$$X_{t+1} = AX_t + Bw_t, \tag{6.18}$$
$$Y_t = CX_t + v_t. \tag{6.19}$$

Here, X_t is the system state at time t, Y_t is the system observation at time t, A is the state transition matrix C is the observation matrix, w_t and v_t are system input and measurement noise with covariances of $S_w = E(w_t(w_t)^{\top})$ and $S_v = E(v_t(v_t)^{\top})$, respectively. Unlike the Kalman filter, w_t and v_t are not necessary to be Gaussian distributed and w_t also includes certain system noise.

2. *Establishing probability density function (PDF) of noise*: Since in the previous step, the probability density function of the state vector, as well as each particle for a state vector, can be denoted as $p(\cdot)$. Given the past observations $Y_{1:t-1}$, the posterior probability distribution of the state vector and particles in our linear system with discrete time can be computed as

$$p(X_t|Y_{1:t}) = \frac{p(Y_t|X_t)p(X_t|Y_{1:t-1})}{p(Y_{t|1:t})}, \tag{6.20}$$

where $p(Y_t|X_t)$ can also be written as $p(Y_t - CX_t)$, and it will be used for weight update in the following steps. The probability density function in a particle filter can be defined according to the known noise distribution in the particular system. If it is defined as a Gaussian distribution function, then the optimal solution to this filter is the Kalman filter [134].

3. *Particle initialisation*: Generate particle-weight pairs, $\{P_i^t, W_i^t\}, i = 1, 2, \dots Np, t = 1, 2, \dots, T$, for each state vector $X_t, t = 1, 2, \dots, T$ of the system. Here, Np is the total number of particles (samples) for each state vector. Additionally, the weight for each particle will be initialized to $1/Np$.

4. *Particle and weight updates*: Given the particles at time t, the posterior particles can be updated as

$$P_i^{t+1} = AP_i^t + w_t. \tag{6.21}$$

At the same time, the associated weight for this particle will be updated as:

$$\begin{aligned} W_i^{t+1} &= W_i^t p(Y_{t+1}|P_i^{t+1}) \\ &= W_i^t p(Y_{t+1} - CP_{i+1}^t). \end{aligned} \tag{6.22}$$

After updating all the particles from time t to $t+1$, the weights at $t+1$ will be normalised to

$$W_i^{t+1} = \frac{W_i^{t+1}}{\sum_{i=1}^{Np} W_i^{t+1}}. \tag{6.23}$$

5. *Computing estimated state*: With the updated particles and the normalised weights, the estimated state at time $t+1$ can be computed as

$$X_{t+1} = \sum_{i=1}^{Np} P_i^{t+1} W_i^{t+1}. \tag{6.24}$$

6. *Sampling importance re-sampling (SIR)*: From equation (6.21), we can see that when the particles are updated, some noise will be involved. In other words, all the particles will not be updated to an expected position from time t to $t+1$. Some of them, being close to the position where the estimated state will be, will have a heavier weight, while those drifting away from this position may have a lighter weight. The former will contribute to the more accurate estimation while the latter does not. Therefore, in order to avoid the

degeneracy of the filter, particles that are too far away (with lighter weight) should be dropped. For this purpose, a threshold N_{th} should be set and when the effective number of particles

$$N_{eff} = \frac{1}{\sum_{i=1}^{Np} W_i^{t+1}} < N_{th},$$

the re-sampling process will be activated.

7. *New loop*: Set $t = t + 1$ and start the new loop from step 4.

Although a particle filter can define the probability density function according to the system, one of its disadvantages should be taken into account, which is the efficiency. In step 4, all the particles should be updated. However, the number of particles will increase with the dimension of state [310]. If there are a large number of particles, this step will take a considerably long time. In this case, it might not be quite suitable for online tracking.

6.3.1 Robust filtering with linear measurements

Consider non-linear uncertain systems of the form,

$$\begin{aligned} &\dot{x} = A(x, u) + Dw, z = Kx, \\ &y = Cx + n, \end{aligned} \tag{6.25}$$

defined on $[0, T]$ with $x(t) \in \mathbb{R}^n$ denoting the state of the system and $y(t) \in \mathbb{R}^l$ the measurements vector. Further, $z(t), u(t), w(t)$ denote the *uncertainty output* and the *uncertainty inputs*, respectively.

Assumption 2.

$$\begin{aligned} &\left(x(0) - x_0\right)^{\mathsf{T}} N \left(x(0) - x_0\right) \\ &\quad + \frac{1}{2}\int_0^T \left[w(t)^{\mathsf{T}} Qw(t) + n(t)^{\mathsf{T}} Rn(t)\right] dt \\ &\leq d + \frac{1}{2}\int_0^T z(t)^{\mathsf{T}} z(t). \end{aligned} \tag{6.26}$$

Introduce the following Riccati differential equation (RDE):

$$\begin{aligned} &\dot{S} + \nabla_x A(\tilde{x}, u)^{\mathsf{T}} S + S\nabla_x A(\tilde{x}, u) + SDQ^{-1}D^{\mathsf{T}} S \\ &\quad -C^{\mathsf{T}} RC + K^{\mathsf{T}} K = 0, S(0) = N. \end{aligned} \tag{6.27}$$

Then the state propagation is given by

$$\begin{aligned} &\dot{\tilde{x}}(t) = A(\tilde{x}(t), u^0) \\ &\quad + S^{-1}(t)\left[C^{\mathsf{T}} R\left[y_c(t) - C\tilde{x}(t)\right] + K^{\mathsf{T}} K\right], \tilde{x}(0) = x_0 \end{aligned} \tag{6.28}$$

The reference frame is oriented with the following assumptions:

1. Accelerations apart from gravity are small as no jerky movements are applicable in assessing movement disorders [249, 356, 405].
2. The reference frame is such that the direction of the magnetic field is perpendicular to the X axis.

Remark 2. *Notice here that there is a significant component of the earth's magnetic field in the Z direction in Australia and this cannot be neglected, unlike in the case of locations close to the equator.*

Robustness of the estimation

The approximate solution for the set of estimated states for the robust set valued state estimation is

$$\chi_s = \left\{ x \in \mathbf{R}^n : \frac{1}{2}(x - \tilde{x}(s))^\top X(s)\,(x - \tilde{x}(s)) \leq d - \phi(s) \right\}, \tag{6.29}$$

where

$$\phi(t) \triangleq \int_0^t \left[\frac{1}{2}(y - C\tilde{x})^\top R\,(y - C\tilde{x}) - \tilde{x}^\top K^\top K\tilde{x} \right] d\tau.$$

Therefore, the centroid of the ellipsoidal set is taken as the estimated state. Let Φ and Θ denote the diagonalising and the resulting diagonal matrix, respectively, while a_i and a_j denote the spectral densities of $1/\sqrt{d - \phi(s)\Theta}$ and $\sqrt{d - \phi(s)}\ \Theta^{-1}$, respectively. Taking $\delta_+ = \left[0 \cdots a_i \cdots\right]^\top \in \mathbf{R}^n$ and $\delta_- = \left[0 \cdots a_j \cdots\right]^\top \in \mathbf{R}^n$ and noticing $\Phi^\top X(s)\Phi = \Theta$, $\hat{x}_+ = x(s) + \Phi\delta_+$ and $\hat{x}_- = x(s) + \Phi\delta_-$ indicate the major axis and the minor axis of the set values state estimation. This provides a measure of the estimation bounds.

6.3.2 Constrained optimisation

Consider the following optimization problems:

$$\mathrm{OP}_1 : \min\ F(x) \text{subjected to } x \in \Omega,$$
$$\mathrm{OP}_2 : \min\ G(x) \text{subjected to } x \in \Omega,$$
$$\mathrm{OP}_3 : \min\ G(x) \text{subjected to } x \in \Lambda.$$

Lemma 2. *Consider the problem OP_3. Assuming that x^* is an optimal point of the problem and two points $z^1 \in \Omega_2$, $z^2 \in \Lambda$. If there exists a real number $\lambda \in (0;1)$ such that $x^* = \lambda z^1 + (1 - \lambda)z^2$ then z^1 and z^2 are also optimal points.*

Proof: Since x^* is an optimal point of the problem and $z^1 \in \Lambda$, $z^2 \in \Lambda$, we have $G(x^*) \leq G(z^1)$ and $G(x^*) \leq G(z^2)$. If z^1 is not an optimal point of the problem $(OP)_3$ then $G(x^*) < G(z^1)$. By the linearity of functional $G(x)$, we have

$$\begin{aligned} G(x^*) &= G(\lambda z^1 + (1 - \lambda)z^2) \\ &= \lambda G(z^1) + (1 - \lambda)G(z^2) \\ &< \lambda G(x^*) + (1 - \lambda)G(x^*) = G(x^*). \end{aligned} \tag{6.30}$$

This is a contradiction. Thus, z^1 is an optimal point. Similarly, we also prove that z^2 is an optimal point. □

Lemma 3. *Problem* OP_3 *has at least an optimal point, which belongs to* Ω.

Proof: From lemma 2, we can imply that problem OP_3 has an optimal point x^* which belongs to the boundary of Λ. This means that $x^* \in \partial\Lambda$. Note that $\Lambda_i \cap \Lambda_j = \emptyset$, $\lambda_i \cap \Omega = \emptyset$, $i \neq j, i, j = 4 \ldots 8$ and

$$\partial\Lambda = \Omega \cup \left(\bigcup_{i=4}^{8} \Lambda_i \right).$$

Therefore, if $x^* \notin \Omega$ then there is an index $i \in \{4, 5, \cdots, 8\}$ such that $x^* \in \Lambda_i$. Without loss of generality, we assume that $x^* \in \Lambda_4$. By using lemma 2, we can imply that one of three points A_2, A_3 and A_4 must be an optimal point. This means that problem OP_3 has at least an optimal point belonging to Ω. □

Now let us recall the definition of *equivalence of optimisation problems* as given in Boyd and Vandenberghe [50] as follows.

Definition. Two optimisation problems are equivalent if from a solution of one, a solution of the other is readily found and vice versa.

Lemma 4. *The problems are equivalent.*

Proof: By expanding functional $F(x)$ and using the constraint $x_4^2 + x_5^2 + x_6^2 + x_7^2 = 1$, we can see that $OP_1 = 1 + P^2 + Q^2 + R^2 + S^2 + OP_2$. This implies that if x^* is an optimal point of problem OP_1 then it is also an optimal point of problem OP_2 and vice versa. Therefore, problem OP_1 is equivalent to problem OP_2. On the other hand, it is easy to see that if $x^* \in \Omega$ is an optimal point of problem OP_3 then it also is an optimal point of problem OP_2. Note that in the case $x^* \notin \Omega$ then we can use lemma 3 to find another optimal point $x^{**} \in \Omega$. Certainly, this point x^{**} is an optimal point of problem OP_2. The rest of the proof is to prove the converse. This means that if $x^* \in \Omega$ is an optimal point of problem OP_2 then we must prove that it also is an optimal point of problem OP_3. If we assume that x^* is not an optimal point of problem OP_3 then there is another point $z^1 \in (\Lambda \setminus \Omega)$ such that $G(z^1) < G(x^*)$. By lemma 3, there exists $z^2 \in \Omega$ such that $G(z^2) = G(z^1)$. This implies that $G(z^2) < G(x^*)$. This is in contradiction to the fact that x^* is an optimal point of problem OP_2. Therefore, x^* must be an optimal point of problem OP_3. The proof of lemma 4 is completed. □

Bibliography

1 http://foodcoachinstitute.com/resources/graphics-pack/. Accessed: 2015-05-22.
2 Elbow stretches. https://www.physioadvisor.com.au/exercises/flexibility-joints/elbow/ Accessed: 1 March 2021.
3 Jointtype enumeration. https://msdn.microsoft.com/en-us/library/microsoft.kinect.jointtype.aspx. Accessed: 2015-05-22.
4 Kinect for windows features. https://www.microsoft.com/en-us/kinectforwindows/meetkinect/features.aspx. Accessed: 2015-05-22.
5 Rokoko smartsuit pro. https://www.rokoko.com/en?gclid=CjwKCAjw_47YBRBxEiwAYuKdwynSSU8rlweQF-R8QBmqMjHq_aEzrUOffS7E0CYpc2pRCkeV9LAZ-xoCrT4QAvD_BwE. Accessed: 2018-05-20.
6 Tracking users with kinect skeletal tracking. https://msdn.microsoft.com/en-us/library/jj131025.aspx. Accessed: 2015-05-22.
7 Vicon. https://www.vicon.com. Accessed: 2018-05-20.
8 Welcome to poeticon++. Accessed: 12 July 2013.
9 World report on disability. Accessed: 31-03-2015.
10 National institute on disability and rehabilitation research; notice of final funding priorities for fiscal years 1998 to 1999 for certain centers. *Federal Register*, 63(113):4, 1998.
11 Fourth International Dystonia Symposium. *Movement Disorders*, 17(5):1115–1142, 2002.
12 Sixteenth Annual Symposia on Etiology, Pathogenesis, and Treatment of Parkinson's Disease and Other Movement Disorders: Program. *Movement Disorders*, 17(5):1103–1114, 2002.
13 Hervé Abdi. The method of least squares. *Encyclopedia of Measurement and Statistics.* CA, USA: Thousand Oaks, 2007.
14 A. Abushakra, M. Faezipour, and A. Abumunshar. Efficient frequency-based classification of respiratory movements. In *Electro/Information Technology (EIT), 2012 IEEE International Conference on*, pages 1–5, 2012.
15 R. J. Adams, M. D. Lichter, E. T. Krepkovich, A. Ellington, M. White, and P. T. Diamond. Assessing upper extremity motor function in practice of virtual activities of daily living. *Neural Systems and Rehabilitation Engineering, IEEE Transactions on*, 23(2):287–296, 2015.
16 Paul S Addison. Wavelet transforms and the ecg: a review. *Physiological Measurement*, 26(5):R155, 2005.
17 B.J. Aehlert and R. Vroman. Paramedic Practice Today: Above and Beyond. Number volume 2 in *Paramedic Practice Today: Above and Beyond*. Jones & Bartlett Learning, LLC, 2011.

Human Motion Capture and Identification for Assistive Systems Design in Rehabilitation, First Edition.
Pubudu N. Pathirana, Saiyi Li, Yee Siong Lee and Trieu Pham.

18 Shabnam Agarwal, Garry T. Allison, and Kevin P. Singer. Validation of the Spin-T goniometer, a cervical range of motion device. *Journal of Manipulative and Physiological Therapeutics*, 28:604–609, 2005.

19 M. Ahmad and Seong-Whan Lee. Recognizing human actions based on silhouette energy image and global motion description. In *Automatic Face Gesture Recognition, 2008. FG '08. 8th IEEE International Conference on*, pages 1–6, 2008.

20 Norhafizan Ahmad, Raja Ariffin Raja Ghazilla, Nazirah M Khairi, and Vijayabaskar Kasi. Reviews on various inertial measurement unit (IMU) sensor applications. *International Journal of Signal Processing Systems*, 1(2):256–262, 2013.

21 T. Al-ani, C.K. Karmakar, A.H. Khandoker, and M. Palaniswami. Automatic recognition of obstructive sleep apnoea syndrome using power spectral analysis of electrocardiogram and hidden markov models. In *Intelligent Sensors, Sensor Networks and Information Processing, 2008. ISSNIP 2008. International Conference on*, pages 285–290, Dec. 2008.

22 Alberto Albanese, Francesca Del Sorbo, Cynthia Comella, H. A. Jinnah, Jonathan W. Mink, Bart Post, Marie Vidailhet, Jens Volkmann, Thomas T. Warner, Albert F. G. Leentjens, Pablo Martinez-Martin, Glenn T. Stebbins, Christopher G. Goetz, and Anette Schrag. Dystonia rating scales: Critique and recommendations. *Movement Disorders*, 28(7):874–883, 2013.

23 D Alciatore and C. Ng. Determining Manipulator Workspace Boundaries using the Monte Carlo Method and Least Squares Segmentation. In *Proceedings of the ASME Biennial Technical Conference, Robotics, Kinematics, Dynamics and Controls*, pages 141–146, 1974.

24 M. Alt Murphy, K. S. Sunnerhagen, B. Johnels, and C. Willen. Three-dimensional kinematic motion analysis of a daily activity drinking from a glass: a pilot study. *J Neuroeng Rehabil*, 3:18, 2006.

25 M. Alt Murphy, C. Willen, and K. S. Sunnerhagen. Responsiveness of upper extremity kinematic measures and clinical improvement during the first three months after stroke. *Neurorehabilitation and Neural Repair*, 27(9):844–53, 2013.

26 Oliver Amft, Holger Junker, and Gerhard Troster. Detection of eating and drinking arm gestures using inertial body-worn sensors. In *Wearable Computers, 2005. Proceedings. Ninth IEEE International Symposium on*, pages 160–163. IEEE, 2005.

27 B.D.O Anderson and J.B. Moore. *Optimal Filtering*. Prentice Hall, Englewood Cliffs, N.J., 1979.

28 Stylianos Asteriadis, Anargyros Chatzitofis, Dimitrios Zarpalas, Dimitrios S. Alexiadis, and Petros Daras. Estimating human motion from multiple kinect sensors. In *Proceedings of the 6th International Conference on Computer Vision/Computer Graphics Collaboration Techniques and Applications*, MIRAGE '13, pages 3:1–3:6, New York, NY, USA, 2013. ACM.

29 F. Auger, M. Hilairet, J.M. Guerrero, E. Monmasson, T. Orlowska-Kowalska, and S. Katsura. Industrial applications of the Kalman filter: A review. *Industrial Electronics, IEEE Transactions on*, 60(12):5458–5471, Dec. 2013.

30 Ameet Bakhai. The burden of coronary, cerebrovascular and peripheral arterial disease. *Journal of Pharmacoeconomics*, 22(4):11–18, 2004.

31 S. Balasubramanian, A. Melendez-Calderon, and E. Burdet. A robust and sensitive metric for quantifying movement smoothness. *IEEE Transactions on Biomedical Engineering*, 59(8):2126–36, 2012.

32 M.S. Barreiro, A.F. Frere, N.E.M. Theodorio, and F.C. Amate. Goniometer based to computer. In *Engineering in Medicine and Biology Society, 2003. Proceedings of the 25th Annual International Conference of the IEEE*, volume 4, pages 3290–3293, Sept 2003.

33 L. E. Baum. An Inequality and Associated Maximization Technique in Statistical Estimation for Probabilistic Functions of a Markov Process. *Inequalities*, 3, 1972.

34 Anthony J Bell and Terrence J Sejnowski. An information-maximization approach to blind separation and blind deconvolution. *Neural Computation*, 7(6):1129–1159, 1995.

35 G.D. Bergland. A guided tour of the fast fourier transform. *Spectrum, IEEE*, 6(7):41–52, 1969.

36 Nikolaĭ Aleksandrovich Bernshteĭn. *The Co-ordination and Regulation of Movements*. Pergamon Press, 1967.

37 James C. Bezdek. *Pattern Recognition with Fuzzy Objective Function Algorithms*. Kluwer Academic Publishers, 1981.

38 James Biggs, Ken Horch, and Francis J. Clark. Extrinsic muscles of the hand signal fingertip location more precisely than they signal the angles of individual finger joints. *Experimental Brain Research*, 125:221–230, 1999.

39 Martin Bilodeau, Douglas A. Keen, Patrick J. Sweeney, Robert W. Shields, and Roger M. Enoka. Strength training can improve steadiness in persons with essential tremor. *Muscle & Nerve*, 23(5):771–778, 2000.

40 N. Birsan, D. Munteanu, G. Iubu, and T. Niculescu. Time-frequency analysis in Doppler radar for noncontact cardiopulmonary monitoring. In *E-Health and Bio-engineering Conference (EHB), 2011*, pages 1–4, 2011.

41 Adrian N. Bishop, Bariş Fidan, Brian D.O. Anderson, Kutluyil Doğançay, and Pubudu N. Pathirana. Optimality analysis of sensor-target localization geometries. *Automatica*, 46(3):479–492, 2010.

42 A.N. Bishop, A.V. Savkin, and P.N. Pathirana. Vision-based target tracking and surveillance with robust set-valued state estimation. *Signal Processing Letters, IEEE*, 17(3):289–292, 2010.

43 Harold D Black. Early development of transit, the navy navigation satellite system. *Journal of Guidance, Control, and Dynamics*, 13(4):577–585, 1990.

44 A.P.L. Bo, M. Hayashibe, and P. Poignet. Joint angle estimation in rehabilitation with inertial sensors and its integration with kinect. In *Engineering in Medicine and Biology Society,EMBC, 2011 Annual International Conference of the IEEE*, pages 3479–3483, Aug. 2011.

45 Barry P.M. D. Boden, G. Scott M. D. Dean, John A. Jr M. D. Feagin, and William E. Jr M. D. PhD Garrett. Mechanisms of anterior cruciate ligament injury. *Orthopedics*, 23(6):573–578, 2000.

46 M. Borghetti, E. Sardini, and M. Serpelloni. Sensorized glove for measuring hand finger flexion for rehabilitation purposes. *Instrumentation and Measurement, IEEE Transactions on*, 62(12):3308–3314, Dec. 2013.

47 O. Boric-Lubecke, V.M. Lubecke, A. Host-Madsen, D. Samardzija, and K. Cheung. Doppler radar sensing of multiple subjects in single and multiple antenna systems. In *Telecommunications in Modern Satellite, Cable and Broadcasting Services, 2005. 7th International Conference on*, volume 1, pages 7–11, vol. 1, 2005.

48 Olga Boric-Lubecke, Victor M Lubecke, Anders Host-Madsen, Dragan Samardzija, and Ken Cheung. Doppler radar sensing of multiple subjects in single and multiple antenna systems. In *Telecommunications in Modern Satellite, Cable and Broadcasting Services, 2005. 7th International Conference on*, volume 1, pages 7–11. IEEE, 2005.

49 Olga Boric-Lubecke, Victor M Lubecke, Isar Mostafanezhad, Byung-Kwon Park, Wansuree Massagram, and Branka Jokanovic. Doppler radar architectures and signal processing for heart rate extraction. *Mikrotalasna Revija, Decembar*, 2009.

50 Stephen Boyd and Lieven Vandenberghe. *Convex Optimization*. Cambridge University Press, New York, NY, USA, 2004.

51 J. Boyle, N. Bidargaddi, A. Sarela, and M. Karunanithi. Automatic detection of respiration rate from ambulatory single-lead ECG. *Information Technology in Biomedicine, IEEE Transactions on*, 13(6):890–896, 2009.

52 Paul W. Brand and Anne Hollister. *Clinical Mechanics of the Hand*. Mosby Year Book, 1993.

53 Timo Breuer, Christoph Bodensteiner, and Michael Arens. Low-cost commodity depth sensor comparison and accuracy analysis. vol. 9250, pages 92500G–92500G-10, 2014. 10.1117/12.2067155.

54 Lori Mann Bruce, Cliff H Koger, and Jiang Li. Dimensionality reduction of hyperspectral data using discrete wavelet transform feature extraction. *Geoscience and Remote Sensing, IEEE Transactions on*, 40(10):2331–2338, 2002.

55 Anne Bruton, Bridget Ellis, and Jonathan Goddard. Comparison of visual estimation and goniometry for assessment of metacarpophalangeal joint angle. *Physiotherapy*, 85:201–208, 1999.

56 Chris D. Bryce and April D. Armstrong. Anatomy and biomechanics of the elbow. *Orthopedic Clinics*, 39(2):141–154.

57 G. C. Burdea. Virtual rehabilitation - benefits and challenges. *Methods Archive*, 42(5):519–523, 2003.

58 Robert E. Burke, Stanley Fahn, C. David Marsden, Susan B. Bressman, Carol Moskowitz, and Joseph Friedman. Validity and reliability of a rating scale for the primary torsion dystonias. *Neurology*, 35(1):73, 1985.

59 Terry Caelli, Andrew McCabe, and Gordon Binsted. On learning the shape of complex actions. In *Visual Form 2001*, pages 24–39. Springer, 2001.

60 Q Cai and J.K Aggawal. Tracking human motion using multiple cameras. *International Conference or Patterns Recognition(ICPR)*, pages 68–72, 1996.

61 Jorge Cancela, Matteo Pastorino, Alexandros Tzallas, Markos Tsipouras, Giorgios Rigas, Maria Arredondo, and Dimitrios Fotiadis. Wearability assessment of a wearable system for parkinson's disease remote monitoring based on a body area network of sensors. *Sensors*, 14(9):17235–17255, 2014.

62 A. Cangelosi, G. Metta, G. Sagerer, S. Nolfi, C. Nehaniv, K. Fischer, Jun Tani, T. Belpaeme, G. Sandini, F. Nori, L. Fadiga, B. Wrede, K. Rohlfing, E. Tuci, K. Dautenhahn, J. Saunders, and A. Zeschel. Integration of action and language knowledge: A roadmap for developmental robotics. *Autonomous Mental Development, IEEE Transactions on*, 2(3):167–195, 2010.

63 Bruce M Cappo and David S Holmes. The utility of prolonged respiratory exhalation for reducing physiological and psychological arousal in non-threatening and threatening situations. *Journal of Psychosomatic Research*, 28(4):265–273, 1984.

64 MA Carey, DE Laird, KA Murray, and JR Stevenson. Reliability, validity, and clinical usability of a digital goniometer. *Work*, 36:55–66, 2010.

65 Marco Ceccarelli. A formulation for the workspace boundary of general N-Revolute Manipulators. *Mechanism and Machine Theory*, 31(5):637–646, 1996.

66 Allal Chabli, Martin Michaud, and Jacques Montplaisir. Periodic arm movements in patients with the restless legs syndrome. *European Neurology*, 44(3):133–138, 2000.

67 Chien-Yen Chang, Belinda Lange, Mi Zhang, Sebastian Koenig, Phil Requejo, Noom Somboon, Alexander A Sawchuk, and Albert A Rizzo. Towards pervasive physical rehabilitation using microsoft kinect. In *Pervasive Computing Technologies for Healthcare (PervasiveHealth), 2012 6th International Conference on*, pages 159–162. IEEE, 2012.

68 Sudhansu Chokroverty. *Sleep Disorders Medicine*. Elsevier Inc, 2009.

69 Selina Chu, Eamonn J Keogh, David Hart, Michael J Pazzani, et al. Iterative deepening dynamic time warping for time series. In *SDM*, 2002.

70 Chin-Seng Chua, Haiying Guan, and Yeong-Khing Ho. Model-based 3D hand posture estimation from a single 2D image. *Image and Vision Computing*, 20(3):191–202, March 2002.

71 Liu Chun-Lin. A tutorial of the wavelet transform. *NTUEE*, Taiwan, 2010.

72 Pamela G Clark, Stephen J Dawson, Cynthia Scheideman-Miller, and Micha L Post. Telerehab: Stroke teletherapy and management using two-way interactive video. *Journal of Neurologic Physical Therapy*, 26(2):87–93, 2002.

73 D E Cleveland. Diagrams for showing limitation of movements through joints, as used by the board of pensions commissioners for canada. *Canadian Medical Association Journal*, 8(12):1070–1076, dec 1918.

74 Frank W. Jobe, Clive E. Brewster, and Diane Radovich Moynes. Rehabilitation for anterior cruciate reconstruction. *Journal of Orthopaedic & Sports Physical Therapy*, 5(3):121–126, 1983.

75 C. A. Coburn and D. R. Peddle. A low-cost field and laboratory goniometer system for estimating hyperspectral bidirectional reflectance. *Canadian Journal of Remote Sensing*, 32:244–253, 2006.

76 Pierre Comon. Independent component analysis, a new concept? *Signal Processing*, 36(3):287–314, 1994.

77 D James Cooper and Michael D Buist. Vitalness of vital signs, and medical emergency teams. *Medical Journal of Australia*, 188(11):630, 2008.

78 F. Cordella, F. Di Corato, L. Zollo, B. Siciliano, and P. van der Smagt. Patient performance evaluation using kinect and monte carlo-based finger tracking. In *Biomedical Robotics and Biomechatronics (BioRob), 2012 4th IEEE RAS & EMBS International Conference on*, pages 1967–1972.

79 H.S. Corning. *Mosby's PDQ for Respiratory Care - Revised Reprint*. Elsevier Health Sciences, 2012.

80 Michelle A Cretikos, Rinaldo Bellomo, Ken Hillman, Jack Chen, Simon Finfer, and Arthas Flabouris. Respiratory rate: the neglected vital sign. *Medical Journal of Australia*, 188(11):657, 2008.

81 D'Alonzo Gilbert E. Criner, Gerard J. *Critical Care Study Guide*. Springer, 2002.

82 E G Cruz and D G Kamper. Kinematics of point-to-point finger movements. *Experimental Brain Research*, 174(1):29–34, September 2006.

83 Leandro Cruz, Djalma Lucio, and Luiz Velho. Kinect and rgbd images: Challenges and applications. *2012 25th SIBGRAPI Conference on Graphics, Patterns and Images Tutorials*, 0:36–49, 2012.

84 Sara Cuccurullo. Physical Medicine and Rehabilitation Board Review. 2004.

85 Jean-François Daneault, Benoit Carignan, Abbas F Sadikot, Michel Panisset, and Christian Duval. Drug-induced dyskinesia in Parkinson's disease. should success in clinical management be a function of improvement of motor repertoire rather than amplitude of dyskinesia? *BMC Medicine*, 11(1):76, 2013.

86 N.H. Dardas and Nicolas D. Georganas. Real-time hand gesture detection and recognition using bag-of-features and support vector machine techniques. *Instrumentation and Measurement, IEEE Transactions on*, 60(11):3592–3607, Nov. 2011.

87 W G Darling, K J Cole, and G F Miller. Coordination of index finger movements. *Journal of Biomechanics*, 27(4):479–91, April 1994.

88 Ingrid Daubechies. Orthonormal bases of compactly supported wavelets. *Communications on Pure and Applied Mathematics*, 41(7):909–996, 1988.

89 Charles Anthony Davie. A review of Parkinson's disease. *British Medical Bulletin*, 86(1):109–127, 2008.

90 Lonneke ML de Lau and Monique MB Breteler. Epidemiology of Parkinson's disease. *The Lancet Neurology*, 5(6):525–535, 2006.

91 Ana de los Reyes-Guzmán, Iris Dimbwadyo-Terrer, Fernando Trincado-Alonso, Félix Monasterio-Huelin, Diego Torricelli, and Angel Gil-Agudo. Quantitative assessment based on kinematic measures of functional impairments during upper extremity movements: A review. *Clinical Biomechanics*, 29(7):719–727, 2014.

92 Domitilla Del Vecchio, Richard M Murray, and Pietro Perona.Decomposition of human motion into dynamics-based primitives with application to drawing tasks. *Automatica*, 39(12):2085–2098, 2003.

93 C. B. Delaplain, C. E. Lindborg, S. A. Norton, and J. E. Hastings. Tripler pioneers telemedicine across the pacific. *Hawaii Medical Journal*, 52(12):338–339, 1993.

94 Johanne Desrosiers, Réjean Hébert, Elisabeth Dutil, Gina Bravo, and Louisette Mercier. Validity of the tempa: A measurement instrument for upper extremity performance. *The Occupational Therapy Journal of Research*, 14(4):267–281, 1994.

95 Gail D. Deyle, Nancy E. Henderson, Robert L. Matekel, Michael G. Ryder, Matthew B. Garber, and Stephen C. Allison. Effectiveness of manual physical therapy and exercise in osteoarthritis of the knee a randomized, controlled trial. *Annals of Internal Medicine*, 132(3):173–181, 2000.

96 Manfredo Perdigao Do Carmo and Manfredo Perdigao Do Carmo. *Differential Geometry of Curves and Surfaces*, volume 2. Prentice-Hall Englewood Cliffs, 1976.

97 M. Donno, Elia Palange, F. Di Nicola, G. Bucci, and F. Ciancetta. A new flexible optical fiber goniometer for dynamic angular measurements: Application to human joint movement monitoring. *Instrumentation and Measurement, IEEE Transactions on*, 57(8):1614–1620, 2008.

98 Massimiliano Donno, Elia Palange, Fabio Di Nicola, Giovanni Bucci, and Fabrizio Ciancetta. A new flexible optical fiber goniometer for dynamic angular, measurements: Application to human joint movement monitoring. *IEEE Transactions on Instrumentation and Measurement*, 57:1614–1620, 2008.

99 Adam S. Dowrick, Belinda J. Gabbe, Owen D. Williamson, and Peter A. Cameron. Outcome instruments for the assessment of the upper extremity following trauma: A review. *Injury*, 36(4):468–476, 2005.

100 Amy Diane Droitcour. Non-contact measurement of heart and respiration rates with a single-chip microwave Doppler radar. 2006.

101 Guanglong Du, Ping Zhang, and Di Li. Human-manipulator interface based on multisensory process via Kalman filters. *Industrial Electronics, IEEE Transactions on*, 61(10):5411–5418, Oct 2014.

102 William K Durfee, Samantha A Weinstein, Ela Bhatt, Ashima Nagpal, and James R Carey. Design and usability of a home telerehabilitation system to train hand recovery following stroke. *Journal of Medical Devices*, 3(4):041003, 2009.

103 M Patrice Eiff, Allen T Smith, and Gary E Smith. Early mobilization versus immobilization in the treatment of lateral ankle sprains. *The American Journal of Sports Medicine*, 22(1):83–88, 1994.

104 George ElKoura and Karan Singh. Handrix: Animating the human hand. In *Proceedings of the 2003 ACM SIGGRAPH/Eurographics Symposium on Computer Animation*, SCA '03, pages 110–119, Aire-la-Ville, Switzerland, Switzerland, 2003. Eurographics Association.

105 Bridget Ellis and Anne Bruton. A study to compare the reliability of composite finger flexion with goniometry for measurement of range of motion in the hand. *Clinical Rehabilitation*, 16(5):562–570, 2002.

106 Christina Engstrand, Barbro Krevers, and Joanna Kvist. Interrater reliability in finger joint goniometer measurement in Dupuytren's disease. *Am J Occup Ther*, 66:98–103, 2012.

107 E Ernst. A review of stroke rehabilitation and physiotherapy. *Stroke*, 21(7):1081–5, 1990.

108 Adolf Faller and Gabriele Schünke. *The Human Body: An Introduction to Structure and Function*. Thieme, 2004.

109 Paul M. Fitts. The information capacity of the human motor system in controlling the amplitude of movement. *Journal of Experimental Psychology*, 47(6):381–391, 1954.

110 Martha Flanders. *Encyclopedia of Neuroscience*, chapter Voluntary Movement, pages 4371–4375. Springer, Berlin, Heidelberg, 2009.

111 D J Ford, S El-Hadidi, P G Lunn, and F D Burke. Fractures of the phalanges: results of internal fixation using 1.5 mm and 2 mm A. O. screws. *The Journal of Hand Surgery*, 12:28–33, 1987.

112 F Frenet. Sur les courbes à double courbure. *Journal de Mathématiques Pures et Appliquées*, pages 437–447, 1852.

113 F Frenet. Sur les courbes à double courbure. *Journal de Mathématiques Pures et Appliquées*, pages 437–447, 1852.

114 C Frey. *Clinical Examination of the Foot and Ankle*, pages 15–20. Springer, 1998.

115 Laurent Frossard, Nathan Stevenson, John Sullivan, Maggie Uden, and Mark Pearcy. Categorization of activities of daily living of lower limb amputees during short-term use of a portable kinetic recording system: A preliminary study. *JPO: Journal of Prosthetics and Orthotics*, 23(1):2–11 10.1097/JPO.0b013e318207914c, 2011.

116 Daniele Giansanti, Sandra Morelli, Giovanni Maccioni, and Giovanni Costantini. Toward the design of a wearable system for fall-risk detection in telerehabilitation. *Telemedicine and e-Health*, 15(3):296–299, 2009.

117 H.B. Gibson. A form of behaviour therapy for some states diagnosed as affective disorder. *Behaviour Research and Therapy*, 16(3):191–195, 1978.

118 D. Girbau, A. Lazaro, A. Ramos, and R. Villarino. Remote sensing of vital signs using a doppler radar and diversity to overcome null detection. *Sensors Journal, IEEE*, 12(3):512–518, March 2012.

119 C. G. Goetz, G. T. Stebbins, H. M. Shale, A. E. Lang, D. A. Chernik, T. A. Chmura, J. E. Ahlskog, and E. E. Dorflinger. Utility of an objective dyskinesia rating scale for parkinson's disease: Inter- and intrarater reliability assessment. *Movement Disorders*, 9(4):390–394, 1994.

120 Germán Gómez-Herrero, Zbynek Koldovský, Petr Tichavský, and Karen Egiazarian. A fast algorithm for blind separation of non-Gaussian and time-correlated signals. In *Proccedings of the 15th European Signal Processing Conference. EUSIPCO*, pages 2007–2007. Citeseer, 2007.

121 H. Gonzalez-Jorge, B. Riveiro, E. Vazquez-Fernandez, J. Martínez-Sánchez, and P. Arias. Metrological evaluation of Microsoft Kinect and Asus Xtion sensors. *Measurement*, 46(6):1800–1806, 2013.

122 John Goodfellow and John O'Connor. The mechanics of the knee and prosthesis design. *Journal of Bone & Joint Surgery, British Volume*, 60(3):358–369, 1978.

123 ST Green. Patellofemoral syndrome. *Journal of Bodywork and Movement Therapies*, 9(1):16–26, 2005.

124 Paul Grossman. Respiration, stress, and cardiovascular function. *Psychophysiology*, 20(3):284–300, 1983.

125 P. Grossmann. Depth from focus. *Pattern Recognition Letters*, 5(1):63–69, 1987.

126 Changzhan Gu. 2.4 ghz multifunctional software-defined radar sensor system. http://www.webpages.ttu.edu/chgu/resource.html.

127 Changzhan Gu, Jiang Long, Jiangtao Huangfu, S. Qiao, W.Z. Cui, W. Ma, and Lixin Ran. An instruments-built doppler radar for sensing vital signs. In *Antennas, Propagation and EM Theory, 2008. ISAPE 2008. 8th International Symposium on*, pages 1398–1401, 2008.

128 Changzhan Gu, Guochao Wang, Yiran Li, Takao Inoue, and Changzhi Li. A hybrid radar-camera sensing system with phase compensation for random body movement cancellation in Doppler vital sign detection. *IEEE Transactions on Microwave Theory and Techniques*, 61(12):4677–4688, 2013.

129 Ye Gu, Ha Do, Yongsheng Ou, and Weihua Sheng. Human gesture recognition through a kinect sensor. In *Robotics and Biomimetics (ROBIO), 2012 IEEE International Conference on*, pages 1379–1384, 2012.

130 Chen Guang, Li Jituo, Wang Bei, Zeng Jiping, Lu Guodong, and Zhang Dongliang. Reconstructing 3d human models with a kinect. *Computer Animation and Virtual Worlds*, 27(1):72–85.

131 G. Guerra-Filho and Y. Aloimonos. A language for human action. *Computer*, 40(5):42–51, 2007.

132 Gutemberg Guerra-Filho, Cornelia Fermuller, and Yiannis Aloimonos. Discovering a language for human activity. In *Proceedings of the AAAI 2005 Fall Symposium on Anticipatory Cognitive Embodied Systems*, Washington, DC, 2005.

133 A Gupta, G S Rash, N N Somia, M P Wachowiak, J Jones, and A Desoky. The motion path of the digits. *The Journal of Hand Surgery*, 23(6):1038–42, November 1998.

134 Fredrik Gustafsson, Fredrik Gunnarsson, Niclas Bergman, Urban Forssell, Jonas Jansson, Rickard Karlsson, and P-J Nordlund. Particle filters for positioning, navigation, and tracking. *Signal Processing, IEEE Transactions on*, 50(2):425–437, 2002.

135 Umut Guzelkucuk, Iltekin Duman, Mehmet Ali Taskaynatan, and Kemal Dincer. Comparison of Therapeutic Activities with Therapeutic Exercises in the Rehabilitation of Young Adult Patients With Hand Injuries. *Journal of Hand Surgery (American Volume)*, 32:1429–1435, 2007.

136 A. Hadjidj, M. Souil, A. Bouabdallah, Y. Challal, and H. Owen. Wireless sensor networks for rehabilitation applications: Challenges and opportunities. *Journal of Network and Computer Applications*, 36(1):1–15, 2013.

137 P Hahn, H Krimmer, A Hradetzky, and U Lanz. Quantitative analysis of the linkage between the interphalangeal joints of the index finger. An in vivo study. *Journal of Hand Surgery*, 20(5):696–9, October 1995.

138 Hai Trieu Pham, Pubudu N Pathirana, and Terry Caelli. Functional range of movement of the hand: Declination angles to reachable space. In *36th Annual International*

Conference of the IEEE Engineering in Medicine and Biology Society, pages 6230–6233, 2014.

139 J. Han, L. Shao, D. Xu, and J. Shotton. Enhanced computer vision with Microsoft Kinect sensor: A review. *IEEE Transactions on Cybernetics*, 43(5):1318–1334, 2013.

140 J. Han, L. Shao, D. Xu, and J. Shotton. Enhanced computer vision with Microsoft Kinect sensor: A review. *IEEE Transactions on Cybernetics*, 43(5):1318–1334, Oct. 2013.

141 Shao-Li Han, Meng-Jie Xie, Chih-Cheng Chien, Yu-Che Cheng, and Chia-Wen Tsao. Using mems-based inertial sensor with ankle foot orthosis for telerehabilitation and its clinical evaluation in brain injuries and total knee replacement patients. *Microsystem Technologies*, pages 1–10, 2015.

142 Darlene Hertling and Randolph M Kessler. *Management of Common Musculoskeletal Disorders: Physical Therapy Principles and Methods*. Lippincott Williams & Wilkins, 2006.

143 Todd Hester, Richard Hughes, Delsey M Sherrill, Bethany Knorr, Metin Akay, Joel Stein, and Paolo Bonato. Using wearable sensors to measure motor abilities following stroke. In *Wearable and Implantable Body Sensor Networks, 2006. BSN 2006. International Workshop on*, pages 4–8. IEEE.

144 P Hobson, A Holden, and J Meara. Measuring the impact of parkinson's disease with the parkinson's disease quality of life questionnaire. *Age and Ageing*, 28(4):341–346, 1999.

145 Fay Horak, Laurie King, and Martina Mancini. Role of body-worn movement monitor technology for balance and gait rehabilitation. *Physical Therapy*, 95(3):461–470, 2015.

146 Roger A Horn and Charles R Johnson. *Matrix Analysis*. Cambridge University Press, 1985.

147 Xinping Huang. On transmitter gain/phase imbalance compensation at receiver. *Communications Letters, IEEE*, 4(11):363–365, 2000.

148 Pamela L. Hudak, Peter C. Amadio, Claire Bombardier, Dorcas Beaton, Donald Cole, Aileen Davis, Gillian Hawker, Jeffrey N. Katz, Matti Makela, Robert G. Marx, Laura Punnett, and James Wright. Development of an upper extremity outcome measure: The dash (disabilities of the arm, shoulder, and head). *American Journal of Industrial Medicine*, 29(6):602–608, 1996.

149 Paul E Hughes, Jim C Hsu, and Matthew J Matava. Hip anatomy and biomechanics in the athlete. *Sports Medicine and Arthroscopy Review*, 10(2):103–114, 2002.

150 M C Hume, H Gellman, H McKellop, and R H Brumfield. Functional range of motion of the joints of the hand. *The Journal of Hand Surgery*, 15(2):240–3, March 1990.

151 Aapo Hyvärinen. Fast and robust fixed-point algorithms for independent component analysis. *Neural Networks, IEEE Transactions on*, 10(3):626–634, 1999.

152 Aapo Hyvärinen, Juha Karhunen, and Erkki Oja. *Independent Component Analysis*, volume 46. John Wiley & Sons, 2004.

153 Aapo Hyvärinen and Erkki Oja. Independent component analysis: algorithms and applications. *Neural Networks*, 13(4):411–430, 2000.

154 T. Ito. Walking motion analysis using 3d acceleration sensors. In *Computer Modeling and Simulation, 2008. EMS '08. Second UKSIM European Symposium on*, pages 123–128, 2008.

155 Christopher J James and Oliver J Gibson. Temporally constrained ica: an application to artifact rejection in electromagnetic brain signal analysis. *Biomedical Engineering, IEEE Transactions on*, 50(9):1108–1116, 2003.

156 Joseph Jankovic and Anthony ELang. Classification of movement disorders. *Neurosurgical Treatment of Movement Disorders*. Park Ridge, IL: AANS, pages 3–18, 1998.

157 O.C. Jenkins and M.J. Mataric. Deriving action and behavior primitives from human motion data. In *Intelligent Robots and Systems, 2002. IEEE/RSJ International Conference on*, volume 3, pages 2551–2556, 2002.

158 Odest Chadwicke Jenkins and Maja J Mataric. Automated derivation of behavior vocabularies for autonomous humanoid motion. In *Proceedings of the Second International Joint Conference on Autonomous Agents and Multiagent Systems*, pages 225–232. ACM, 2003.

159 Crispin Jenkinson, Ray Fitzpatrick, Viv Peto, Richard Greenhall, and Nigel Hyman. The pdq-8: Development and validation of a short-form Parkinson's disease questionnaire. *Psychology & Health*, 12(6):805–814, 1997.

160 D. S. Jevsevar, P. O. Riley, W. A. Hodge, and D. E. Krebs. Knee kinematics and kinetics during locomotor activities of daily living in subjects with knee arthroplasty and in healthy control subjects. *Physical Therapy*, 73(4):229–39; discussion 240–2, 1993.

161 Per Jonsson and Peter W. Johnson. Comparison of measurement accuracy between two types of wrist goniometer systems. *Applied Ergonomics*, 32:599–607, 2001.

162 R. Phillip Dellinger Joseph E. Parrillo. *Critical Care Medicine: Principles of Diagnosis and Management in the Adult*. Elsevier Health Sciences, 2007.

163 Rudolph Emil Kalman et al. A new approach to linear filtering and prediction problems. *Journal of Basic Engineering*, 82(1):35–45, 1960.

164 D G Kamper, E G Cruz, and M P Siegel. Stereotypical fingertip trajectories during grasp. *Journal of Neurophysiology*, 90(Morasso 1981):3702–3710, 2003.

165 Juha Karhunen, Erkki Oja, Liuyue Wang, Ricardo Vigario, and Jyrki Joutsensalo. A class of neural networks for independent component analysis. *Neural Networks, IEEE Transactions on*, 8(3):486–504, 1997.

166 Masami Kato, Ayumu Echigo, Hisaaki Ohta, Sumio Ishiai, Mitsuhiro Aoki, Sadako Tsubota, and Eiichi Uchiyama. The accuracy of goniometric measurements of proximal interphalangeal joints in fresh cadavers: Comparison between methods of measurement, types of goniometers, and fingers. *Journal of Hand Therapy*, 20:12–19, 2007.

167 Kenji Kawamura and Kevin C. Chung. Treatment of scaphoid fractures and nonunions. *The Journal of Hand Surgery*, 33(6):988–997, 2008.

168 Motoaki Kawanabe, Klaus-Robert Müller, and Aapo Hyvärinen. Estimating functions for blind separation when sources have variance dependencies. *Journal of Machine Learning Research*, 6(4), 2005.

169 Florence Peterson Kendall, Elizabeth Kendall McCreary, Patricia Geise Provance, Mary McIntyre Rodgers, and William Anthony Romani. Muscles, testing and function. *British Journal of Sports Medicine*, 18(1):25, March 1984.

170 Eamonn J. Keogh and Michael J. Pazzani. Derivative dynamic time warping. In *In First SIAM International Conference on Data Mining (SDM 2001)*, 2001.

171 Samyra HJ Keus, Bastiaan R Bloem, Erik JM Hendriks, Alexandra B Bredero-Cohen, and Marten Munneke. Evidence-based analysis of physical therapy in Parkinson's disease with recommendations for practice and research. *Movement Disorders*, 22(4):451–460, 2007.

172 T. Khan, P. Ramuhalli, and S.C. Dass. Particle-filter-based multisensor fusion for solving low-frequency electromagnetic NDE inverse problems. *Instrumentation and Measurement, IEEE Transactions on*, 60(6):2142–2153, June 2011.

173 A. Kim and M. F. Golnaraghi. A quaternion-based orientation estimation algorithm using an inertial measurement unit. In *Position Location and Navigation Symposium, 2004. PLANS 2004*, pages 268–272, April 2004.

174 P T Ali Kitis, Nihal Buker, and Inci Gokalan Kara. Comparison of two methods of controlled mobilisation of repaired flexor tendons in zone 2. *Scandinavian Journal of Plastic and Reconstructive Surgery and Hand Surgery. Supplementum*, 43:160–165, 2009.

175 N. Kitsunezaki, E. Adachi, T. Masuda, and J. Mizusawa. Kinect applications for the physical rehabilitation. In *Medical Measurements and Applications Proceedings (MeMeA), 2013 IEEE International Symposium on*, pages 294–299, 2013.

176 Anne Klintworth, Zénó Ajtay, Alina Paljunite, Sándor Szabados, and László Hejjel. Heart rate asymmetry follows the inspiration/expiration ratio in healthy volunteers. *Physiological Measurement*, 33(10):1717, 2012.

177 B. Knorr, R. Hughes, D. Sherrill, J. Stein, M. Akay, and P. Bonato. Quantitative measures of functional upper limb movement in persons after stroke. In *Neural Engineering, 2005. Conference Proceedings. 2nd International IEEE EMBS Conference on*, pages 252–255.

178 A. Kolahi, M. Hoviattalab, T. Rezaeian, M. Alizadeh, M. Bostan, and H. Mokhtarzadeh. Design of a marker-based human motion tracking system. *Biomedical Signal Processing and Control*, 2(1):59–67, 2007.

179 Morey J Kolber and William J Hanney. The reliability and concurrent validity of shoulder mobility measurements using a digital inclinometer and goniometer: a technical report. *International Journal of Sports Physical Therapy*, 7:306–313, 2012.

180 Zbynek Koldovsky, Petr Tichavsky, and Erkki Oja. Efficient variant of algorithm fastica for independent component analysis attaining the cram&# 201; r-rao lower bound. *Neural Networks, IEEE Transactions on*, 17(5):1265–1277, 2006.

181 T Kondo, T Uhlig, P Pemberton, and PD Sly. Laser monitoring of chest wall displacement. *European Respiratory Journal*, 10(8):1865–1869, 1997.

182 M. Kristan, S. Kovacic, A. Leonardis, and J. Pers. A two-stage dynamic model for visual tracking. *Systems, Man, and Cybernetics, Part B: Cybernetics, IEEE Transactions on*, 40(6):1505–1520, 2010.

183 J.J. Kuch and T.S. Huang. Human computer interaction via the human hand: a hand model. In *Proceedings of 1994 28th Asilomar Conference on Signals, Systems and Computers*, vol. 2, pages 1252–1256. IEEE Comput. Soc. Press, 1994.

184 Neelesh Kumar, Dinesh Pankaj, Ankit Mahajan, Amod Kumar, and B. S. Sohi. Evaluation of normal gait using electro-goniometer. *Journal of Scientific and Industrial Research India*, 68:696–698, 2009.

185 Deepa Kundur and Dimitrios Hatzinakos. Digital watermarking using multiresolution wavelet decomposition. In *Acoustics, Speech and Signal Processing, 1998. Proceedings of the 1998 IEEE International Conference on*, volume 5, pages 2969–2972. IEEE, 1998.

186 Li Chieh Kuo, Haw Yen Chiu, Cheung Wen Chang, Hsiu Yun Hsu, and Yun Nien Sun. Functional workspace for precision manipulation between thumb and fingers in normal hands. *Journal of Electromyography and Kinesiology*, 19(5):829–839, 2009.

187 Yushiro Kuratomi, Nobuo Okazaki, Teruo Ishihara, Tatsuo Arai, and Shiro Kira. Variability of breath-by-breath tidal volume and its characteristics in normal and diseased subjects. ventilatory monitoring with electrical impedance pneumography. *Japanese Journal of Medicine*, 24(2):141, 1985.

188 Ann Kusoffsky, Ingmarie Apel, and Helga Hirschfeld. Reaching-lifting-placing task during standing after stroke: coordination among ground forces, ankle muscle activity, and hand movement. *Archives of Physical Medicine and Rehabilitation*, 82(5):650–660, 2001.

189 E Lachat, H Macher, MA Mittet, T Landes, and P Grussenmeyer. First experiences with kinect v2 sensor for close range 3d modelling. *International Archives of the Photogrammetry, Remote Sensing and Spatial Information Sciences (ISPRS)*, 2015.

190 AC Lai and Paola Loreti. From discrete to continuous reachability for a robot's finger model. *Communications in Applied and Industrial Mathematics*, 3(2), 2013.

191 Anna Chiara Lai and Paola Loreti. Robot's finger and expansions in non-integer bases. *Networks and Heterogeneous Media*, 7(1):71–111, February 2012.

192 Richard D. Lane, William M. Glazer, Thomas E. Hansen, William H. Berman, and Stephen I. Kramer. Assessment of tardive dyskinesia using the abnormal involuntary movement scale. *The Journal of Nervous and Mental Disease*, 173(6):353–357, 1985.

193 Peter Langhorne, Fiona Coupar, and Alex Pollock. Motor recovery after stroke: a systematic review. *The Lancet Neurology*, 8(8):741–754, 2009.

194 J. William Langston, Hakan Widner, Christopher G. Goetz, David Brooks, Stanley Fahn, Thomas Freeman, and Ray Watts. Core assessment program for intracerebral transplantations (capit). *Movement Disorders*, 7(1):2–13, 1992.

195 W Laupattarakasem, W Sirichativapee, W Kowsuwon, S Sribunditkul, and C Suibnugarn. Axial rotation gravity goniometer. A simple design of instrument and a controlled reliability study. *Clinical Orthopaedics and Related Research*, 251:271–274, 1990.

196 Joseph J LaViola Jr. A comparison of unscented and extended kalman filtering for estimating quaternion motion. In *American Control Conference, 2003. Proceedings of the 2003*, volume 3, pages 2435–2440. IEEE, 2003.

197 Yee Siong Lee, P.N. Pathirana, Evans Robin.J., and C.L. Steinfort. Noncontact detection and analysis of respiratory function using microwave Doppler radar. *Journal of Sensors*, 2015.

198 Yee Siong Lee, P.N. Pathirana, C.L. Steinfort, and T. Caelli. Monitoring and analysis of respiratory patterns using microwave Doppler radar. *Translational Engineering in Health and Medicine, IEEE Journal of*, 2:1–12, 2014.

199 Yee Siong Lee, Pubudu N Pathirana, Terry Caelli, and R Evans. Doppler radar in respiratory monitoring: Detection and analysis. In *ICCAIS 2013: Proceedings of the 2013 2nd International Conference on Control Automation and Information Sciences*, pages 199–203. IAMI, 2013.

200 Yee Siong Lee, Pubudu N Pathirana, Terry Caelli, and Saiyi Li. Further applications of Doppler radar for non-contact respiratory assessment. In *Engineering in Medicine and Biology Society (EMBC), 2013 35th Annual International Conference of the IEEE*, pages 3833–3836. IEEE, 2013.

201 Giovanni Legnani, Bruno Zappa, Federico Casolo, Riccardo Adamini, and Pier Luigi Magnani. A model of an electro-goniometer and its calibration for biomechanical applications. *Medical Engineering & Physics*, 22:711–722, 2000.

202 N. Leibowitz, N. Levy, S. Weingarten, Y. Grinberg, A. Karniel, Y. Sacher, C. Serfaty, and N. Soroker. Automated measurement of proprioception following stroke. *Disability and Rehabilitation*, 30(24):1829–1836, 2008.

203 Samuel T Leitkam, Tamara Bush, and Laura Bix. Determining functional finger capabilities of healthy adults: Comparing experimental data to a biomechanical model. *Journal of Biomechanical Engineering*, December 2013.

204 Edward Lemaire. The physical rehabilitation distance communication initiative. 2000.

205 Russell V. Lenth. On a form of piecewise linear regression. *The American Statistician*, 29(3):116–117, 1975.

206 Yafi Levanon. The advantages and disadvantages of using high technology in hand rehabilitation. *Journal of Hand Therapy*, 26:179–183, 2013.

207 T. Lewiner, J. D. Gomes, H. Lopes, and M. Craizer. Curvature and torsion estimators based on parametric curve fitting. *Computers & Graphics-Uk*, 29(5):641–655, 2005.

208 Changzhi Li and Jenshan Lin. Random body movement cancellation in doppler radar vital sign detection. *IEEE Transactions on Microwave Theory and Techniques*, 56(12):3143–3152, 2008.

209 S. Li, M. Ferraro, T. Caelli, and P. Pathirana. A syntactic two-component encoding model for the trajectories of human actions. *IEEE Journal of Biomedical and Health Informatics*, 18(6):1903–1914, 2014.

210 T.S. Li, Yu-Te Su, Shao-Hsien Liu, Jhen-Jia Hu, and Ching-Chang Chen. Dynamic balance control for biped robot walking using sensor fusion, Kalman filter, and fuzzy logic. *Industrial Electronics, IEEE Transactions on*, 59(11):4394–4408, Nov 2012.

211 Zhang Licong, J. Sturm, D. Cremers, and Lee Dongheui. Real-time human motion tracking using multiple depth cameras. In *Intelligent Robots and Systems (IROS), 2012 IEEE/RSJ International Conference on*, pages 2389–2395.

212 Hwai-Ting Lin, Li-Chieh Kuo, Hsin-Yi Liu, Wen-Lan Wu, and Fong-Chin Su. The three-dimensional analysis of three thumb joints coordination in activities of daily living. *Clinical Biomechanics*, 26(4):371–6, May 2011.

213 IM Lin, LY Tai, and SY Fan. Breathing at a rate of 5.5 breaths per minute with equal inhalation-to-exhalation ratio increases heart rate variability. *International Journal of Psychophysiology*, 91(3):206–211, 2014.

214 Jill M. Thein Lori Thein Brody. Nonoperative treatment for patellofemoral pain. *Journal of Orthopaedic & Sports Physical Therapy*, 28(5):336–344, 1998.

215 Elan D. Louis. Essential tremor. *The Lancet Neurology*, 4(2):100–110, 2005.

216 P. S. Lum, S. Mulroy, R. L. Amdur, P. Requejo, B. I. Prilutsky, and A. W. Dromerick. Gains in upper extremity function after stroke via recovery or compensation: Potential differential effects on amount of real-world limb use. *Topics in Stroke Rehabilitation*, 16(4):237–253, 2009.

217 Laura Luna-Oliva, Rosa María Ortiz-Gutiérrez, Roberto Cano-de la Cuerda, Rosa Martínez Piédrola, Isabel M. Alguacil-Diego, Carlos Sánchez-Camarero, and María del Carmen Martínez Culebras. Kinect Xbox 360 as a therapeutic modality for children with cerebral palsy in a school environment: A preliminary study. *NeuroRehabilitation*, 33(4):513–521, 2013.

218 Zhiqiang Luo, Chee Kian Lim, Weiting Yang, Ke Yen Tee, Kang Li, Chao Gu, Kim Doang Nguen, I-Ming Chen, and Song Huat Yeo. An interactive therapy system for arm and hand rehabilitation. In *Robotics Automation and Mechatronics (RAM), 2010 IEEE Conference on*, pages 9–14. IEEE.

219 Ronald C Lyle. A performance test for assessment of upper limb function in physical rehabilitation treatment and research. *International Journal of Rehabilitation Research*, 4(4):483–492, 1981.

220 Sebastian OH Madgwick, Andrew JL Harrison, and Ravi Vaidyanathan. Estimation of imu and marg orientation using a gradient descent algorithm. In *Rehabilitation Robotics (ICORR), 2011 IEEE International Conference on*, pages 1–7. IEEE, 2011.

221 Andrea Mannini, Stephen S Intille, Mary Rosenberger, Angelo M Sabatini, and William Haskell. Activity recognition using a single accelerometer placed at the wrist or ankle. *Medicine and Science in Sports and Exercise*, 45(11):2193, 2013.

222 RichardA Marder and GeorgeJ Lian. *Tendon Disorders*, book section 5, pages 123–144. Springer, New York, 1997.

223 F Landis Markley and Daniele Mortari. Quaternion attitude estimation using vector observations. *Journal of the Astronautical Sciences*, 48(2):359–380, 2000.

224 F.L Markley. Attitude determination using vector observations and the singular value decompositions. In *AAS/AIAA Astrodynamics Specialist Conference*, pages 487–490, 1987.

225 F.L Markley and D Mortari. Three-axis attitude determination from vector observations. *Journal of the Astronautical Sciences*, 2(48):359–380, 2000.

226 Diane P. Martin, Ruth Engelberg, Julie Agel, Deborah Snapp, and Marc F. Swiontkowski. Development of a musculoskeletal extremity health status instrument: The musculoskeletal function assessment instrument. *Journal of Orthopaedic Research*, 14(2):173–181, 1996.

227 P. Martínez-Martín, A. Gil-Nagel, L. Morlán Gracia, J. Balseiro Gómez, J. Martínez-Sarriés, and F. Bermejo. Unified Parkinson's disease rating scale characteristics and structure. *Movement Disorders*, 9(1):76–83, 1994.

228 RG Marx, C Bombardier, and JG Wright. What do we know about the reliability and validity of physical examination tests used to examine the upper extremity? *Journal of Hand Surgery (American Volume)*, 24:185–193, 1999.

229 W. Massagram, N. Hafner, V. Lubecke, and O. Boric-Lubecke. Tidal volume measurement through non-contact Doppler radar with dc reconstruction. *Sensors Journal, IEEE*, 13(9):3397–3404, 2013.

230 W. Massagram, V.M. Lubecke, and O. Boric-Lubecke. Microwave non-invasive sensing of respiratory tidal volume. In *Engineering in Medicine and Biology Society, 2009. EMBC 2009. Annual International Conference of the IEEE*, pages 4832–4835, 2009.

231 Larry Matthies, Takeo Kanade, and Richard Szeliski. Kalman filter-based algorithms for estimating depth from image sequences. *International Journal of Computer Vision*, 3(3):209–238, 1989.

232 JM McCarthy. The differential geometry of curves in an image space of spherical kinematics. *Mechanism and Machine Theory*, 22(3):205–211, 1987.

233 Philip W McClure, Lori A Michener, Brian J Sennett, and Andrew R Karduna. Direct 3-dimensional measurement of scapular kinematics during dynamic movements in vivo. *Journal of Shoulder and Elbow Surgery*, 10(3):269–277, 2001.

234 Gerald McGinty, James J Irrgang, and Dave Pezzullo. Biomechanical considerations for rehabilitation of the knee. *Clinical Biomechanics*, 15(3):160–166, 2000.

235 Stan Melax, Leonid Keselman, and Sterling Orsten. Dynamics based 3d skeletal hand tracking. In *Proceedings of Graphics Interface 2013*, GI '13, pages 63–70, Toronto, Ont., Canada, 2013. Canadian Information Processing Society.

236 M D Mermelstein and J A Blodgett. Single-mode optical fiber goniometer. *Opt Lett*, 17:85–87, 1992.

237 Cheryl D Metcalf, Rebecca Robinson, Adam J Malpass, Tristan P Bogle, Thomas a Dell, Chris Harris, and Sara H Demain. Markerless motion capture and measurement of hand kinematics: validation and application to home-based upper limb rehabilitation. *Biomedical Engineering, IEEE Transactions on*, 60(8):2184–92, Aug. 2013.

238 Gail D. Deyle Michael D. Bang. Comparison of supervised exercise with and without manual physical therapy for patients with shoulder impingement syndrome. *Journal of Orthopaedic & Sports Physical Therapy*, 30(3):126–137, 2000.

239 F.M. Mirzaei and S.I. Roumeliotis. A kalman filter-based algorithm for imu-camera calibration: Observability analysis and performance evaluation. *Robotics, IEEE Transactions on*, 24(5):1143–1156, Oct 2008.

240 Susanna Mondini and Christian Guilleminault. Abnormal breathing patterns during sleep in diabetes. *Annals of Neurology*, 17(4):391–395, 1985.

241 Francis C Moon. *The Machines of Leonardo Da Vinci and Franz Reuleaux: kinematics of machines from the Renaissance to the 20th Century*, vol. 2. Springer, 2007.

242 A. P. Moore. Classification of movement disorders. *Neuroimaging Clinics of North America*, 20(1):1–6, 2010.

243 Vincent Morelli and Victoria Smith. Groin injuries in athletes. *American Family Physician*, 64(8):1405–1414, 2001.

244 Meg E Morris. Movement disorders in people with parkinson disease: A model for physical therapy. *Physical Therapy*, 80(6):578–597, 2000.

245 Mosby. *Mosby's Medical Dictionary*. Elsevier Inc, 2009.

246 Michael J Mullaney, Malachy P McHugh, Christopher P Johnson, and Timothy F Tyler. Reliability of shoulder range of motion comparing a goniometer to a digital level. *Physiother Theory Pract*, 26:327–333, 2010.

247 Y.P. Munjal, S.K. Sharma, M.D. A. K. Agarwal, and P. Gupta. *Api Textbook of Medicine*. G - Reference, Information and Interdisciplinary Subjects Series. Jaypee Brothers, Medical Publishers, 2012.

248 Kevin Murphy et al. The Bayes net toolbox for Matlab. *Computing Science and Statistics*, 33(2):1024–1034, 2001.

249 H Nagasaki. Asymmetric velocity and acceleration profiles of human arm movements. *Experimental Brain Research*, 74(2):319–326, 1989.

250 K Nakajima, T Tamura, and H Miike. Monitoring of heart and respiratory rates by photoplethysmography using a digital filtering technique. *Medical Engineering & Physics*, 18(5):365–372, 1996.

251 Peter A. Nathan, Kenneth D. Meadows, and Richard C. Keniston. Rehabilitation of carpal tunnel surgery patients using a short surgical incision and an early program of physical therapy. *The Journal of Hand Surgery*, 18(6):1044–1050, 1993.

252 Justine M Naylor, Victoria Ko, Sam Adie, Clive Gaskin, Richard Walker, Ian A Harris, and Rajat Mittal. Validity and reliability of using photography for measuring knee range of motion: a methodological study. *BMC Musculoskelet Disord*,12:77, 2011.

253 Andrew Nealen and TU Darmstadt. An as-short-as-possible introduction to the least squares, weighted least squares and moving least squares methods for scattered data approximation and interpolation. *URL:* http://www.nealen.com/projects, 2004.

254 CHARLES S. NEER. Anterior Acromioplasty for the Chronic Impingement Syndrome in the Shoulder, volume 54 of *A Preliminary Report*. 1972.

255 Charles S Neer. Impingement lesions. *Clinical Orthopaedics and Related Research*, 173:70–77, 1983.

256 G Nelen, M Martens, and A Burssens. Surgical treatment of chronic Achilles tendinitis. *The American Journal of Sports Medicine*, 17(6):754–759, 1989.

257 R. Nerino, L. Contin, W. J. Gon, W. J. Gonçalves da Silva Pinto, G. Massazza, M. Actis, P. Capacchione, A. Chimienti, and G. Pettiti. A BSN based service for post-surgical knee rehabilitation at home, 2013.

258 Andrew S. Neviaser and Jo A. Hannafin. Adhesive capsulitis: A review of current treatment. *The American Journal of Sports Medicine*, 38(11):2346–2356, 2010.

259 Ashish D Nimbarte, Rodrigo Kaz, and Zong-Ming Li. Finger joint motion generated by individual extrinsic muscles: A cadaveric study. *Journal of Orthopaedic Surgery and Research*, 3:27, January 2008.

260 Cynthia C. Norkin and D. Joyce White. *Measurement of Joint Motion A Guide to Goniometry*. F.A. Davis Company, India, 4th edition, 2004.

261 Jorge Nunez, Xavier Otazu, Octavi Fors, Albert Prades, Vicenc Pala, and Roman Arbiol. Multiresolution-based image fusion with additive wavelet decomposition. *Geoscience and Remote Sensing, IEEE Transactions on*, 37(3):1204–1211, 1999.

262 L. Nyberg, L. Lundin-Olsson, B. Sondell, A. Backman, K. Holmlund, S. Eriksson, M. Stenvall, E. Rosendahl, M. Maxhall, and G. Bucht. Using a virtual reality system to study balance and walking in a virtual outdoor environment: A pilot study. *Cyberpsychology & Behavior*, 9(4):388–395, 2006.

263 S Obdrzalek, Gregorij Kurillo, Ferda Ofli, Ruzena Bajcsy, Edmund Seto, Holly Jimison, and Michael Pavel. Accuracy and robustness of kinect pose estimation in the context of coaching of elderly population. In *Engineering in Medicine and Biology Society (EMBC), 2012 Annual International Conference of the IEEE*, pages 1188–1193. IEEE, 2012.

264 Matthew Ockendon and Robin Gilbert. Validation of a Novel Smartphone Accelerometer-Based Knee Goniometer. *Am J Knee Surg*, 25:341–346, 2012.

265 Maurice M Ohayon and Thomas Roth. Prevalence of restless legs syndrome and periodic limb movement disorder in the general population. *Journal of Psychosomatic Research*, 53(1):547–554, 2002.

266 Iasonas Oikonomidis, Nikolaos Kyriazis, and Antonis A. Argyros. Markerless and efficient 26-dof hand pose recovery. In *Proceedings of the 10th Asian Conference on Computer Vision - Volume Part III*, ACCV'10, pages 744–757, Berlin, Heidelberg, 2011. Springer-Verlag.

267 Erienne V. Olesh, Sergiy Yakovenko, and Valeriya Gritsenko. Automated assessment of upper extremity movement impairment due to stroke. *Plos One*, 9(8):e104487–e104487, 2014.

268 Rosa Ortiz-Gutiérrez, Roberto Cano-de-la Cuerda, Fernando Galán-del Río, Isabel María Alguacil-Diego, Domingo Palacios-Ceña, and Juan Carlos Miangolarra-Page. A telerehabilitation program improves postural control in multiple sclerosis patients: a spanish preliminary study. *International Journal of Environmental Research and Public Health*, 10(11):5697–5710, 2013.

269 Joon Ho Oum, Dong-Wook Kim, and Songcheol Hong. Two frequency radar sensor for non-contact vital signal monitor. In *Microwave Symposium Digest, 2008 IEEE MTT-S International*, pages 919–922, june 2008.

270 T. O'Haver. Interactive fourier filter. http://www.mathworks.com/matlabcentral/fileexchange/12377, September 2006.

271 Andrew K Palmer and Frederick W Werner. The triangular fibrocartilage complex of the wrist-anatomy and function. *The Journal of Hand Surgery*, 6(2):153–162, 1981.

272 G. Panahandeh, N. Mohammadiha, A. Leijon, and P. Handel. Continuous hidden Markov model for pedestrian activity classification and gait analysis. *Instrumentation and Measurement, IEEE Transactions on*, 62(5):1073–1083, 2013.

273 Panel, Ralph L. Sacco, Emelia J. Benjamin, Joseph P. Broderick, Mark Dyken, J. Donald Easton, William M. Feinberg, Larry B. Goldstein, Philip B. Gorelick, George Howard, Steven J. Kittner, Teri A. Manolio, Jack P. Whisnant, and Philip A. Wolf. Risk factors. *Stroke*, 28(7):1507–1517, 1997.

274 Andrew E Park, John J Fernandez, Karl Schmedders, and Mark S Cohen. The Fibonacci sequence: Relationship to the human hand. *The Journal of Hand Surgery*, 28(1):157–60, January 2003.

275 Byung-Kwon Park, Shuhei Yamada, and Victor Lubecke. Measurement method for imbalance factors in direct-conversion quadrature radar systems. *Microwave and Wireless Components Letters, IEEE*, 17(5):403–405, 2007.

276 Katerina Pastra, Panagiotis Dimitrakis, Eirini Balta, and Georgios Karakatsiotis. Praxicon and its language-related modules. In *Proceedings of Companion Volume of the 6th Hellenic Conference on Artificial Intelligence (SETN)*, pages 27–32, 2010.

277 P.N. Pathirana, S.C.K. Herath, and A.V. Savkin. Multitarget tracking via space transformations using a single frequency continuous wave radar. *Signal Processing, IEEE Transactions on*, 60(10):5217–5229, 2012.

278 P.N. Pathirana, S.C.K. Herath, and A.V. Savkin. Multitarget tracking via space transformations using a single frequency continuous wave radar. *Signal Processing, IEEE Transactions on*, 60(10):5217–5229, 2012.

279 Judy Pearsall and Patrick Hanks. *The New Oxford Dictionary of English*. Clarendon Press, 1998.

280 L. M. Pedro and G. A. de Paula Caurin. Kinect evaluation for human body movement analysis. In *Biomedical Robotics and Biomechatronics (BioRob), 2012 4th IEEE RAS & EMBS International Conference on*, pages 1856–1861, 2012.

281 W. E. Pentland and L. T. Twomey. Upper limb function in persons with long term paraplegia and implications for independence: Part i. *Paraplegia*, 32(4):211–8, 1994.

282 I.R. Petersen and A.V. Savkin. *Robust Kalman Filtering for Signals and Systems with Large Uncertainties*. Birkhauser, Boston, 1999.

283 Jakub Petkov and Zbyněk Koldovský. Bssgui–a package for interactive control of blind source separation algorithms in matlab. In *Cross-Modal Analysis of Speech, Gestures, Gaze and Facial Expressions*, pages 386–398. Springer, 2009.

284 V. Peto, C. Jenkinson, R. Fitzpatrick, and R. Greenhall. The development and validation of a short measure of functioning and well being for individuals with Parkinson's disease. *Quality of Life Research*, 4(3):241–248, 1995.

285 Alexandra Pfister, Alexandre M West, Shaw Bronner, and Jack Adam Noah. Comparative abilities of Microsoft Kinect and Vicon 3D motion capture for gait analysis. *Journal of Medical Engineering & Technology*, 38:1–7, 2014.

286 Trieu Pham, Pubudu N. Pathirana, Hieu Trinh, and Pearse Fay. A non-contact measurement system for the range of motion of the hand. *Sensors*, 15(8):18315, 2015.

287 Diana Barbara Piazzini, I Aprile, Paola Emilia Ferrara, Carlo Bertolini, P Tonali, Loredana Maggi, Alessia Rabini, Sergio Piantelli, and Luca Padua. A systematic review of conservative treatment of carpal tunnel syndrome. *Clinical Rehabilitation*, 21(4):299–314, 2007.

288 Pietro Picerno, Valerio Viero, Marco Donati, Tamara Triossi, Virginia Tancredi, and Giovanni Melchiorri. Ambulatory assessment of shoulder abduction strength curve using a single wearable inertial sensor. *Journal of Rehabilitation Research & Development*, 52(2), 2015.

289 S. M. Pincus. Approximate entropy as a measure of system-complexity. *Proceedings of the National Academy of Sciences of the United States of America*, 88(6):2297–2301, 1991.

290 L. Piron, A. Turolla, P. Tonin, F. Piccione, L. Lain, and M. Dam. Satisfaction with care in post-stroke patients undergoing a telerehabilitation programme at home. *Journal of Telemedicine and Telecare*, 14(5):257–260, 2008.

291 Alex Pollock, Gill Baer, Valerie M Pomeroy, and Peter Langhorne. Physiotherapy treatment approaches for the recovery of postural control and lower limb function following stroke. *The Cochrane Library*, 2007.

292 Leslie G. Portney and Mary P. Watkins. *Foundations of Clinical Research: Application to Practice*. Pearson/Prentice Hall, USA, 3rd edition, 2009.

293 Friedemann Pulvermüller. Brain mechanisms linking language and action. *Nature Reviews Neuroscience*, 6(7):576–582, 2005.

294 Lawrence R Rabiner. A tutorial on hidden markov models and selected applications in speech recognition. *Proceedings of the IEEE*, 77(2):257–286, 1989.

295 Ali H. Rajput and Alex Rajput. Medical treatment of essential tremor. *Journal of Central Nervous System Disease*, 6(4165-JCNSD-Medical-Treatment-of-Essential-Tremor.pdf):29–39, 2014.

296 NB Reese and WD Bandy. *Joint Range of Motion and Muscle Length Testing*. Saunders, USA, 2nd edition, 2009.

297 Liu Ren, Gregory Shakhnarovich, Jessica K Hodgins, Hanspeter Pfister, and Paul Viola. Learning silhouette features for control of human motion. *ACM Transactions on Graphics (ToG)*, 24(4):1303–1331, 2005.

298 Lars Reng, Thomas B Moeslund, and Erik Granum. Finding motion primitives in human body gestures. In *Gesture in Human-Computer Interaction and Simulation*, pages 133–144. Springer, 2006.

299 Arthur C Rettig, Erich J Weidenbener, and Robert Gloyeske. Alternative management of midthird scaphoid fractures in the athlete*. *The American Journal of Sports Medicine*, 22(5):711–714, 1994.

300 Albert A. Rizzo, Dorothy Strickland, and Stéphane Bouchard. The challenge of using virtual reality in telerehabilitation. *Telemedicine Journal and e-Health*, 10(2):184–195, 2004.

301 Asbjørn Roaas and Gunnar B. J. Andersson. Normal range of motion of the hip, knee and ankle joints in male subjects, 30-40 years of age. *Acta Orthopaedica*, 53(2):205–208, 1982.

302 Cristina Rodriguez, Antonio Miguel, Horacio Lima, and Kristinn Heinrichs. Osteitis pubis syndrome in the professional soccer athlete: a case report. *Journal of Athletic Training*, 36(4):437, 2001.

303 Marco Rogante, Mauro Grigioni, Daniele Cordella, and Claudia Giacomozzi. Ten years of telerehabilitation: A literature overview of technologies and clinical applications. *Neurorehabilitation*, 27(4):287–304, 2010.

304 B. Rohrer, S. Fasoli, H. I. Krebs, R. Hughes, B. Volpe, W. R. Frontera, J. Stein, and N. Hogan. Movement smoothness changes during stroke recovery. *J Journal of Neuroscience*, 22(18):8297–304, 2002.

305 K Rome and F Cowieson. A reliability study of the universal goniometer, fluid goniometer, and electrogoniometer for the measurement of ankle dorsiflexion. *Foot & Ankle International*, 17:28–32, 1996.

306 Douglas C Ross, Ralph T Manktelow, Mark T Wells, and J Brian Boyd. Tendon function after replantation: prognostic factors and strategies to enhance total active motion. *Annals of Plastic Surgery*, 51:141–146, 2003.

307 A. K. Roy, Y. Soni, and S. Dubey. Enhancing effectiveness of motor rehabilitation using kinect motion sensing technology. In *2013 IEEE Global Humanitarian Technology Conference: South Asia Satellite (GHTC-SAS)*, pages 298–304, Aug 2013.

308 T. G. Russell, G. A. Jull, and R. Wootton. The diagnostic reliability of internet-based observational kinematic gait analysis. *Journal of Telemedicine and Telecare*, 9 (Suppl 2):S48–51, 2003.

309 Trevor G Russell, Peter Buttrum, Richard Wootton, and Gwendolen A Jull. Rehabilitation after total knee replacement via low-bandwidth telemedicine: the patient and therapist experience. *Journal of Telemedicine and Telecare*, 10(Suppl 1):85–87, 2004.

310 David Salmond, Neil Gordon, D Salmond, N Gordon, D Salmond, and N Gordon. An introduction to particle filters. *State Space and Unopbserved Component Models Theory and Applications*, pages 1–19, 2005.

311 Stan Salvador and Philip Chan. Toward accurate dynamic time warping in linear time and space. *Intelligent Data Analysis*, 11(5):561–580, October 2007.

312 G. Salvi, L. Montesano, A. Bernardino, and J. Santos-Victor. Language bootstrapping: Learning word meanings from perception - action association. *Systems, Man, and Cybernetics, Part B: Cybernetics, IEEE Transactions on*, 42(3):660–671, 2012.

313 A. Sant'Anna and N. Wickström. Developing a motion language: Gait analysis from accelerometer sensor systems. In *Pervasive Computing Technologies for Healthcare, 2009. PervasiveHealth 2009. 3rd International Conference on*, pages 1–8, 2009.

314 Veronica J Santos and Francisco J Valero-Cuevas. Reported anatomical variability naturally leads to multimodal distributions of Denavit-Hartenberg parameters for the human thumb. *Biomedical Engineering, IEEE Transactions on*, 53(2):155–63, February 2006.

315 M.R.U. Saputra, W. Widyawan, G.D. Putra, and P.I. Santosa. Indoor human tracking application using multiple depth-cameras. In *Advanced Computer Science and Information Systems (ICACSIS), 2012 International Conference on*, pages 307–312, Dec. 2012.

316 L. Savard, A. Borstad, J. Tkachuck, D. Lauderdale, and B. Conroy. Telerehabilitation consultations for clients with neurologic diagnoses: cases from rural Minnesota and American Samoa. *NeuroRehabilitation*, 18(2):93–102, 2003.

317 Abraham Savitzky and Marcel JE Golay. Smoothing and differentiation of data by simplified least squares procedures. *Analytical Chemistry*, 36(8):1627–1639, 1964.

318 A. V. Savkin and I. R. Petersen. Robust state estimation and model validation for discrete-time uncertain systems with a deterministic description of noise and uncertainty. *Automatica*, 34(2):271–274, 1998.

319 R.W. Schafer. What is a Savitzky-Golay filter? [lecture notes]. *Signal Processing Magazine, IEEE*, 28(4):111–117, 2011.

320 Sevara Schepina. Human body muscular system. http://www.aireurbano.com/human-muscular-system-and-its-description/human-body-muscular-system//, 2014. Accessed: 2015-05-22.

321 Mary V. Seeman. Tardive dyskinesia: Two-year recovery. *Comprehensive Psychiatry*, 22(2):189–192, 1981.

322 Samsu Sempena, Nur Ulfa Maulidevi, and Peb Ruswono Aryan. Human action recognition using dynamic time warping. In *Electrical Engineering and Informatics (ICEEI), 2011 International Conference on*, pages 1–5. IEEE, 2011.

323 J-A Serret. Sur quelques formules relatives à la théorie des courbes à double courbure. *Journal de Mathématiques Pures et Appliquées*, pages 193–207, 1851.

324 J-A Serret. Sur quelques formules relatives à la théorie des courbes à double courbure. *Journal de Mathématiques Pures et Appliquées*, pages 193–207, 1851.

325 Peiyun She and Gina A. Livermore. Long-term poverty and disability among working-age adults. *Journal of Disability Policy Studies*, 19(4):244–256, 2009.

326 Marc A Sherry and Thomas M Best. A comparison of 2 rehabilitation programs in the treatment of acute hamstring strains. *Journal of Orthopaedic & Sports Physical Therapy*, 34(3):116–125, 2004.

327 Jamie Shotton, Toby Sharp, Alex Kipman, Andrew Fitzgibbon, Mark Finocchio, Andrew Blake, Mat Cook, and Richard Moore. Real-time human pose recognition in parts from single depth images. *Communications of the ACM*, 56(1):116–124, 2013.

328 A. Shpunt and Z. Zalevsky. Depth-varying light fields for three dimensional sensing, 2008.

329 M. D. Shuster and S. D. OH. Three-axis attitude determination from vector observations. *Journal of Guidance, Control, and Dynamics*, 4(1):70–77, 2013/08/05 1981.

330 Malcolm D Shuster. Approximate algorithms for fast optimal attitude computation. In *AIAA Guidance and Control Conference*, pages 88–95. AIAA New York, NY, 1978.

331 Malcolm D Shuster. A survey of attitude representations. *Navigation*, 8(9), 1993.

332 G. M. Simpson, J. H. Lee, B. Zoubok, and G. Gardos. A rating scale for tardive dyskinesia. *Psychopharmacology*, 64(2):171–179, 1979.

333 Satish Sinha, Partha S Routh, Phil D Anno, and John P Castagna. Spectral decomposition of seismic data with continuous-wavelet transform. *Geophysics*, 70(6):P19–P25, 2005.

334 M.D. Stanley J. Swierzewski, III. Sleep disorders. http://www.healthcommunities.com/sleep-disorders/types.shtml, 2011.

335 L. Stead and S.H. Thomas. *Emergency Medicine*. Board review series. Lippincott Williams & Wilkins, 2000.

336 Oscar Steila. Automatic in-phase quadrature balancing. http://www.qsl.net/ik1xpv/dsp/pdf/aiqben.pdf, 2006.

337 D L Streiner and G R Norman. *Health Measurement Scales: A Practical Guide to Their Development and Use*. Oxford University Press, USA, 5th edition, 2008.

338 Kathleen Strong, Colin Mathers, and Ruth Bonita. Preventing stroke: Saving lives around the world. *The Lancet Neurology*, 6(2):182–187, 2007.

339 Chuan Jun Su. Personal rehabilitation exercise assistant with kinect and dynamic time warping. *International Journal of Information and Education Technology*, 3(4):448–454, 2013.

340 Yu Su, C.R. Allen, D. Geng, D. Burn, U. Brechany, G.D. Bell, and R. Rowland. 3-d motion system ("data-gloves"): application for Parkinson's disease. *Instrumentation and Measurement, IEEE Transactions on*, 52(3):662–674, 2003.

341 Satoshi Suzuki, Takemi Matsui, Hayato Imuta, Maki Uenoyama, Hirofumi Yura, Masayuki Ishihara, and Mitsuyuki Kawakami. A novel autonomic activation measurement method for stress monitoring: non-contact measurement of heart rate variability using a compact microwave radar. *Medical & Biological Engineering & Computing*, 46(7):709–714, 2008.

342 Satoshi Suzuki, Takemi Matsui, Hiroshi Kawahara, Hiroto Ichiki, Jun Shimizu, Yoko Kondo, Shinji Gotoh, Hirofumi Yura, Bonpei Takase, and Masayuki Ishihara. A non-contact vital sign monitoring system for ambulances using dual-frequency microwave radars. *Medical & Biological Engineering & Computing*, 47(1):101–105, 2009.

343 Heidi Sveistrup, Joan McComas, Marianne Thornton, Shawn Marshall, Hillel Finestone, Anna McCormick, Kevin Babulic, and Alain Mayhew. Experimental studies of virtual reality-delivered compared to conventional exercise programs for rehabilitation. *CyberPsychology & Behavior*, 6(3):245–249, 2003.

344 Marc F. Swiontkowski, Ruth Engelberg, Diane P. Martin, and Julie Agel. Short musculoskeletal function assessment questionnaire: Validity, reliability, and responsiveness. *Journal of Bone and Joint Surgery*, 81(9):1245–1260, 1999.

345 Gentaro Taga. A model of the neuro-musculo-skeletal system for human locomotion. *Biological Cybernetics*, 73(2):97–111, 1995.

346 Yaqin Tao and Huosheng Hu.A novel sensing and data fusion system for 3-d arm motion tracking in telerehabilitation. *Instrumentation and Measurement, IEEE Transactions on*, 57(5):1029–1040, 2008.

347 A. Tariq and H. Ghafouri-Shiraz. Vital signs detection using doppler radar and continuous wavelet transform. In *Antennas and Propagation (EUCAP), Proceedings of the 5th European Conference on*, pages 285–288, 2011.

348 Gabriel Taubin. Estimation of planar curves, surfaces, and nonplanar space curves defined by implicit equations with applications to edge and range image segmentation. *IEEE Transactions on Pattern Analysis and Machine Intelligence*, 13(11):1115–1138, 1991.

349 Thomas J. A. Terlouw. Roots of physical medicine, physical therapy, and mechanotherapy in the Netherlands in the 19(th) century: A disputed area within the healthcare domain. *The Journal of Manual & Manipulative Therapy*, 15(2):E23–E41, 2007.

350 Deborah Theodoros and Trevor Russell. Telerehabilitation: current perspectives. *Studies in Health Technology and Informatics*, 131:191–209, 2008.

351 P. Tichavsky, A. Yeredor, and Z. Koldovsky. A fast asymptotically efficient algorithm for blind separation of a linear mixture of block-wise stationary autoregressive processes. In *Acoustics, Speech and Signal Processing, 2009. ICASSP 2009. IEEE International Conference on*, pages 3133–3136, April 2009.

352 Petr Tichavskỳ, Eran Doron, Arie Yeredor, and Jan Nielsen. A computationally affordable implementation of an asymptotically optimal BSS algorithm for ar sources. *Proc. EUSIPCO-2006*, pages 4–8, 2006.

353 Petr Tichavskỳ, Zbynek Koldovskỳ, Eran Doron, Arie Yeredor, and Germán Gómez-Herrero. Blind signal separation by combining two ica algorithms: Hos-based efica and time structure-based wasobi. In *Eur. Signal Process. Conf. (EUSIPCO), Florence*, 2006.

354 Petr Tichavsky, Zbynek Koldovsky, Arie Yeredor, Germán Gómez-Herrero, and Eran Doron. A hybrid technique for blind separation of non-gaussian and time-correlated sources using a multicomponent approach. *Neural Networks, IEEE Transactions on*, 19(3):421–430, 2008.

355 C. A. Trombly and C. Y. Wu. Effect of rehabilitation tasks on organization of movement after stroke. *American Journal of Occupational Therapy*, 53(4):333–344, 1999.

356 Catherine A Trombly and Ching-Yi Wu. Effect of rehabilitation tasks on organization of movement after stroke. *American Journal of Occupational Therapy*, 53(4):333–344, 1999.

357 Yu-Chee Tseng, Chin-Hao Wu, Fang-Jing Wu, Chi-Fu Huang, Chung-Ta King, Chun-Yu Lin, Jang-Ping Sheu, Chun-Yu Chen, Chi-Yuan Lo, and Chien-Wen Yang. A wireless human motion capturing system for home rehabilitation. In *Mobile Data Management: Systems, Services and Middleware, 2009. MDM'09. Tenth International Conference on*, pages 359–360. IEEE.

358 School of Medicine University of Virginia. Asthma attacks. http://www.medicine.virginia.edu/clinical/departments/pediatrics/clinical-services/tutorials/asthma/attacks, 2011.

359 Ann Van de Winckel, Hilde Feys, Suzan van der Knaap, Ruth Messerli, Fabio Baronti, Ruth Lehmann, Bart Van Hemelrijk, Franca Pante, Carlo Perfetti, and Willy De Weerdt. Can quality of movement be measured? Rasch analysis and inter-rater reliability of the motor evaluation scale for upper extremity in stroke patients (mesupes). *Clinical Rehabilitation*, 20(10):871–884, 2006.

360 Frans CT van der Helm. A standardized protocol for motion recordings of the shoulder. In *Proceedings of the First Conference of the International Shoulder Group*, pages 27–28. Shaker Publishing BV, Maastricht (Netherlands), 1997.

361 Harry L Van Trees. *Detection, Estimation, and Modulation Theory*. Wiley. com, 2004.

362 A.R. Varkonyi-Koczy and B. Tusor. Human-computer interaction for smart environment applications using fuzzy hand posture and gesture models. *Instrumentation and Measurement, IEEE Transactions on*, 60(5):1505–1514, May 2011.

363 S.C. Venema and B. Hannaford. A probabilistic representation of human workspace for use in the design of human interface mechanisms. *Mechatronics, IEEE/ASME Transactions on*, 6(3):286–294, 2001.

364 Alexander Vergara, Nicolas Petrochilos, Olga Boric-Lubecke, Anders Host-Madsen, and Victor Lubecke. Blind source separation of human body motion using direct conversion Doppler radar. In *Microwave Symposium Digest, 2008 IEEE MTT-S International*, pages 1321–1324. IEEE, 2008.

365 J. A. N. Verhaar. Tennis elbow. *International Orthopaedics*, 18(5):263–267, 1994.

366 B. Vicenzino and A. Wright. Effects of a novel manipulative physiotherapy technique on tennis elbow: a single case study. *Manual Therapy*, 1(1):30–35, 1995.

367 John D Villasenor, Benjamin Belzer, and Judy Liao. Wavelet filter evaluation for image compression. *Image Processing, IEEE Transactions on*, 4(8):1053–1060, 1995.

368 M. J. Volman, A. Wijnroks, and A. Vermeer. Effect of task context on reaching performance in children with spastic hemiparesis. *Clinical Rehabilitation*, 16(6):684–92, 2002.

369 Eric Wade, Avinash Rao Parnandi, and Maja J Mataric. Automated administration of the wolf motor function test for post-stroke assessment. In *Pervasive Computing Technologies for Healthcare (PervasiveHealth), 2010 4th International Conference on*, pages 1–7. IEEE.

370 Grace Wahba. A least squares estimate of satellite attitude. *SIAM Review*, 7(3):409–409, 1965.

371 H Kenneth Walker, W Dallas Hall, and J Willis Hurst. Clinical methods. 1990.

372 Liang Wang, Tieniu Tan, Huazhong Ning, and Weiming Hu. Silhouette analysis-based gait recognition for human identification. *Pattern Analysis and Machine Intelligence, IEEE Transactions on*, 25(12):1505–1518, 2003.

373 Luke Wang, Russell Vanderhout, and Tim Shi. Computer vision detection of negative obstacles with the Microsoft Kinect, 2012.

374 C. Warlow, J. van Gijn, M. Dennis, J. Wardlaw, J. Bamford, G. Hankey, P. Sandercock, G. Rinkel, P. Langhorne, C. Sudlow, and P. Rothwell. *Introduction*, pages 1–5. Blackwell Publishing Ltd, 2008.

375 D. D. Webster. Critical analysis of the disability in Parkinson's disease. *Modern Treatment*, 5(2):257–82, 1968.

376 David Webster and Ozkan Celik. Experimental evaluation of Microsoft Kinect's accuracy and capture rate for stroke rehabilitation applications. In *Haptics Symposium (HAPTICS), 2014 IEEE*, pages 455–460. IEEE, 2014.

377 Frank Weichert, Daniel Bachmann, Bartholomäus Rudak, and Denis Fisseler. Analysis of the accuracy and robustness of the leap motion controller. *Sensors (Basel)*, 13(5):6380–6393, jan 2013.

378 Zhao Wenbing, D. D. Espy, M. A. Reinthal, and Feng Hai. A feasibility study of using a single kinect sensor for rehabilitation exercises monitoring: A rule based approach. In *Computational Intelligence in Healthcare and e-health (CICARE), 2014 IEEE Symposium on*, pages 1–8, 2014.

379 Juyang Weng. Symbolic models and emergent models: A review. *Autonomous Mental Development, IEEE Transactions on*, 4(1):29–53, 2012.

380 Markus Windolf, Nils Götzen, and Michael Morlock. Systematic accuracy and precision analysis of video motion capturing systems-exemplified on the Vicon-460 system. *Journal of Biomechanics*, 41:2776–2780, 2008.

381 J. M. Winters. Motion analysis and telerehabilitation: healthcare delivery standards and strategies for the new millennium. In *Pediatric Gait, 2000. A New Millennium in Clinical Care and Motion Analysis Technology*, pages 16–22.

382 J. M. Winters. Telerehabilitation research: emerging opportunities. *Annual Review of Biomedical Engineering*, 4:287–320, 2002.

383 S.-H.P. Won, W.W. Melek, and F. Golnaraghi. A Kalman/particle filter-based position and orientation estimation method using a position sensor/inertial measurement unit hybrid system. *Industrial Electronics, IEEE Transactions on*, 57(5):1787–1798, May 2010.

384 George W. Stimson. *Introduction to Airborne Radar*. Scitech Publishing Inc, San Francisco, second edition, 1998.

385 B.-F. Wu and C.-L. Jen. Particle filter based radio localization for mobile robots in the environments with low-density wlan aps. *Industrial Electronics, IEEE Transactions on*, PP(99):1–1, 2014.

386 Chris Wylie, Gordon Romney, David Evans, and Alan Erdahl. Half-tone perspective drawings by computer. In *AFIPS '67 (Fall) Proceedings of the November 14-16, 1967, Fall Joint Computer Conference*, pages 49–58, 1967.

387 W. Xu, C. Gu, C. Li, and M. Sarrafzadeh. Robust doppler radar demodulation via compressed sensing. *Electronics Letters*, 48(22):1428–1430, 2012.

388 Xu Xu and Raymond W. McGorry. The validity of the first and second generation microsoft kinect™ for identifying joint center locations during static postures. *Applied Ergonomics*, 49(0):47–54, 2015.

389 Kwok-Yun Yeung, Tsz-Ho Kwok, and Charlie CL Wang. Improved skeleton tracking by duplex kinects: A practical approach for real-time applications. *Journal of Computing and Information Science in Engineering*, 13(4):041007, 2013.

390 Ho-Sub Yoon, Jung Soh, Younglae J Bae, and Hyun Seung Yang. Hand gesture recognition using combined features of location, angle and velocity. *Pattern Recognition*, 34(7):1491–1501, 2001.

391 Xiaogang Yu, Changzhi Li, and Jenshan Lin. Two-dimensional noncontact vital sign detection using Doppler radar array approach. In *Microwave Symposium Digest (MTT), 2011 IEEE MTT-S International*, pages 1–4. IEEE, 2011.

392 YH Song In-Keun Yu, Chang-Il Kim. A novel short-term load forecasting technique using wavelet transform analysis. *Electric Machines &Power Systems*, 28(6):537–549, 2000.

393 George Yuan, Nicole A Drost, and R Andrew McIvor. Respiratory rate and breathing pattern. *McMaster University Medical Journal*, 10:23–28, 2013.

394 Xiaoping Yun and E. R. Bachmann. Design, implementation, and experimental results of a quaternion-based kalman filter for human body motion tracking. *IEEE Transactions on Robotics*, 22(6):1216–1227, December 2006.

395 J. Zariffa, N. Kapadia, J. L. K. Kramer, P. Taylor, M. Alizadeh-Meghrazi, V. Zivanovic, U. Albisser, R. Willms, A. Townson, A. Curt, M. R. Popovic, and J. D. Steeves. Relationship between clinical assessments of function and measurements from an upper-limb robotic rehabilitation device in cervical spinal cord injury. *IEEE Transactions on Neural Systems and Rehabilitation Engineering*, 20(3):341–350, 2012.

396 Zhen-Xin Zhang, Gustavo C. Roman, Zhen Hong, Cheng-Bing Wu, Qiu-Ming Qu, Jue-Bing Huang, Bing Zhou, Zhi-Ping Geng, Ji-Xing Wu, Hong-Bo Wen, Heng Zhao, and Gwendolyn E. P. Zahner. Parkinson's disease in China: prevalence in Beijing, Xian, and Shanghai. *The Lancet*, 365(9459):595–597, 2005.

397 Zhiqiang Zhang and Jiankang Wu. A novel hierarchical information fusion method for three-dimensional upper limb motion estimation. *Instrumentation and Measurement, IEEE Transactions on*, 60(11):3709–3719, 2011.

398 He Zhao and Zheyao Wang. Motion measurement using inertial sensors, ultrasonic sensors, and magnetometers with extended kalman filter for data fusion. *Sensors Journal, IEEE*, 12(5):943–953, May 2012.

399 Ruoyin Zheng and Jiting Li. Kinematics and workspace analysis of an exoskeleton for thumb and index finger rehabilitation. *2010 IEEE International Conference on Robotics and Biomimetics, ROBIO 2010*, (Cmc):80–84, 2010.

400 Zhang Zhengyou. Microsoft kinect sensor and its effect. *MultiMedia, IEEE*, 19(2):4–10, 2012.

401 Huiyu Zhou, Huosheng Hu, Honghai Liu, and Jinshan Tang. Classification of upper limb motion trajectories using shape features. *Systems, Man, and Cybernetics, Part C: Applications and Reviews, IEEE Transactions on*, 42(6):970–982, 2012.

402 Jilin Zhou, F. Malric, and S. Shirmohammadi. A new hand-measurement method to simplify calibration in cyberglove-based virtual rehabilitation. *Instrumentation and Measurement, IEEE Transactions on*, 59(10):2496–2504, Oct 2010.

403 C. Zhu and Weihua Sheng. Wearable sensor-based hand gesture and daily activity recognition for robot-assisted living. *Systems, Man and Cybernetics, Part A: Systems and Humans, IEEE Transactions on*, 41(3):569–573, 2011.

404 R Zhu and Z Zhou. A real-time articulated human motion tracking using tri-axis inertial/magnetic sensors package. *IEEE Transactions on Neural Systems and Rehabilitation Engineering*, pages 295–302, 2004.

405 Loredana Zollo, Luca Rossini, Marco Bravi, Giovanni Magrone, Silvia Sterzi, and Eugenio Guglielmelli. Quantitative evaluation of upper-limb motor control in robot-aided rehabilitation. *Medical & Biological Engineering & Computing*, 49(10):1131, 2011.

Index

Human Motion Capture and Identification for Assistive Systems Design in Rehabilitation, First Edition.
Pubudu N. Pathirana, Saiyi Li, Yee Siong Lee and Trieu Pham.